Knowledge Diffusion

in the U.S. Aerospace Industry

Managing Knowledge for Competitive Advantage

Part A

Contemporary Studies in Information Management, Policy, and Services

(formerly Information Management, Policy, and Services Series)

Peter Hernon, series editor

Statistics: A Component of the Research Process
Peter Hernon, 1991
Statistics: A Component of the Research Process, Second Edition
Peter Hernon, 1994
Service Quality in Academic Libraries
Peter Hernon and Ellen Altman, 1996
Microcomputer Software for Performing Statistical Analysis: A Handbook
for Supporting Library Decision Making
Peter Hernon and John V. Richardson (editors), 1998
Evaluation and Library Decision Making
Peter Hernon and Charles R. McClure, 1990
Public Access to Government Information, Second Edition
Peter Hernon and Charles R. McClure, 1988
Federal Information Policies in the 1990s: Views and Perspectives
Peter Hernon, Charles McClure, and Harold C. Relyea, 1996
Statistics for Library Decision Making: A Handbook
Peter Hernon, et al., 1989
Understanding Information Retrieval Interactions: Theoretical and Practical Implications
Carol A. Hert, 1997
Reclaiming the American Library Past: Writing the Women In
Suzanne Hildenbrand (editor), 1996
Libraries: Partners in Adult Literacy
Deborah Johnson, Jane Robbins, and Douglas L. Zweizig, 1991
The National Research and Education Network (NREN): Research
and Policy Perspectives
Charles R. McClure, Ann P. Bishop, Philip Doty and Howard Rosenbaum (editors), 1991
Library and Information Science Research: Perspective and Strategies for Improvement
Charles R. McClure and Peter Hernon (editors), 1991
U.S. Government Information Policies: Views and Perspectives
Charles R. McClure, Peter Hernon, and Harold C. Relyea, 1989
U.S. Scientific and Technical Information Policies: Views and Perspectives
Charles R. McClure and Peter Hernon, 1989
Gatekeepers in Ethnolinguistic Communities
Cheryl Metoyer-Duran, 1993
Knowledge Diffusion in the U.S. Aerospace Industry: Managing Knowledge
for Competitive Advantage
Thomas E. Pinelli, Rebecca O. Barclay, John M. Kennedy and Ann P. Bishop, 1997
Silencing Science: National Security Controls and Scientific Communication
Harold C. Relyea, 1994
Records Management and the Library: Issues and Practices
Candy Schwartz and Peter Hernon, 1993
Assessing the Public Library Planning Process
Annabel K. Stephens, 1996
Depository Library Use of Technology: A Practitioner's Perspective
Jan Swanbeck and Peter Hernon, 1993
For Information Specialists
Howard White, Marcia Bates, and Patrick Wilson, 1992
Public Library Youth Services: A Public Policy Approach
Holly G. Willett, 1995

In Preparation:

Knowledge Diffusion

in the U.S. Aerospace Industry

Managing Knowledge for Competitive Advantage

Thomas E. Pinelli
NASA Langley Research Center

Rebecca O. Barclay
Knowledge Management Associates

John M. Kennedy
Indiana University

Ann P. Bishop
University of Illinois at Urbana–Champaign

In Collaboration With:

Claire J. Anderson	*Old Dominion University*
Myron Glassman	*Old Dominion University*
Vicki L. Golich	*California State University, San Marcos*
Keith Hayward	*Staffordshire University*
Laura M. Hecht	*California State University, Bakersfield*
W.D. Kay	*Northeastern University*
Michael L. Keene	*University of Tennessee, Knoxville*
Mindy L. Kotler	*Japanese Information Access Project*
Elizabeth W. Morrison	*New York University*
Daniel J. Murphy	*SUNY Institute of Technology at Utica/Rome*
John R. Webb	*Concurrent Communications*

Ablex Publishing Corporation
Greenwich, Connecticut
London, England

Printed in the United States of America

Library of Congress Cataloging-in-Publication Data

Pinelli, Thomas E.
  Knowledge diffusion in the U.S. aerospace industry : managing
knowledge for competitive advantage / Thomas E. Pinelli.
   p. cm. — (Contemporary studies in information management,
policy, and services)
  Includes bibliographical references and index.
  ISBN 1-56750-225-3 (cloth). — ISBN 1-56750-226-1 (paper)
  1. Aerospace industries—Technological innovations—United States—
Management. 2. Diffusion of innovations—United States.
3. Aerospace industries—Technological innovations—Management—Case
studies. 4. Diffusion of innovations—Case studies.
5. Competition, International. I. Title. II. Series.
HD9711.5.U6P56 1997
338.4'76291'0973—dc21   97-25175
                  CIP

Ablex Publishing Corporation
55 Old Post Road #2
P.O. Box 5297
Greenwich, CT 06830

Published in the U.K. and Europe by:
JAI Press Ltd.
38 Tavistock Street
Covent Garden
London WC2E 7PB
England

CONTENTS

ix Contents

FOREWORD

There is general agreement among such diverse groups as economists, corporate and government policymakers, management theorists and strategists, and behaviorists that knowledge is a critical element in innovation, technological progress, and economic competitiveness. At both the theoretical and practical levels, knowledge was traditionally taken for granted; for some knowledge was only a by-product of research and development (R&D), and for others it was a public good free for the taking. Its characteristics, its behavior, and its role in the process of technological innovation were largely ignored. With competitiveness becoming an increasingly important issue for firms and nations, interest in the role and importance of knowledge as a factor in innovation has grown. Increasing interest in knowledge indicates how much is yet to be learned about its role and behavior in technological innovation and progress before knowledge can be used for competitive advantage. Supporters of basic research, as well as a number of recent books, advocate the creation of more knowledge as the key to successful innovation. However, critics of this approach point out that successful innovation depends more on the application of existing knowledge than on the creation of new knowledge, because technological innovation occurs incrementally rather than radically. When breakthrough innovations do occur, they are usually the result of "trial and error" invention rather than new knowledge that was created as a result of basic research. The project team for the *NASA/DoD Aerospace Knowledge Diffusion Research Project* believe that the diffusion of knowledge is the key to successful innovation. Because much useful and relevant knowledge needed for innovation resides external to (outside) a firm, understanding the process by which firms and individuals identify, absorb, assimilate, and exploit knowledge (e.g., the results of federally funded aerospace R&D) is essential for understanding technological innovation and ultimately improving economic competitiveness.

This book broadens our understanding of knowledge diffusion as an important component of the process of technological innovation. In focusing on aerospace as a technological community, the contributors to this book have studied a variety of social structures and relationships within the community to understand how knowledge diffuses at the individual, organizational, national, and international levels of the community. Previous research has focused on the diffusion of a single innovation, for example, the diffusion of hybrid-seed corn; the diffusion of a specific innovation within an industry, for example, the diffusion of a new process or technique in the steel industry; or the diffusion of knowledge within a firm to understand how and why the firm enjoys a competitive advantage over its rivals. However, few if any studies have examined the diffusion of knowledge within an entire technological community, the role of U.S. public policy in shaping the external environment of the community, and the influence of the community's actors (i.e., the producers and users of knowledge) on technological progress via the diffusion of knowledge. Recognizing how public policy has shaped the environment of U.S. aerospace and understanding how knowledge diffuses over time through communication channels among the members of the aerospace community can help firms and the nation better manage privately and publicly funded knowledge for competitive advantage.

As Alexis de Tocqueville wrote in *Democracy in America* (1835), the Americans "judge that the diffusion of knowledge must necessarily be advantageous, and the consequences of ignorance fatal." I recommend this book if you are interested in the diffusion of knowledge and innovation in a high-technology industry.

Everett M. Rogers
University of New Mexico

PREFACE

This book presents the results of a decade of work conducted under the auspices of the *NASA/DoD Aerospace Knowledge Diffusion Research Project*. The scholars involved in this project have examined multiple aspects of knowledge diffusion—specifically, its production, transfer, and use—in the large commercial aircraft (LCA) sector of the United States aerospace industry. This book, which reports the results of these studies, concentrates on the diffusion of knowledge as a source of innovation and competitiveness for the U.S. aerospace industry. This industry, principally the LCA sector, is unique in that it has been the beneficiary of federally funded research and development (R&D) for nearly a century. Although millions of taxpayer dollars are spent annually for this purpose, little is known about how the knowledge and technology resulting from federally funded aerospace R&D diffuse at the individual, organizational, national, and international levels. The project's results are a logical first step in understanding the diffusion process. We believe that an empirical understanding of the process by which the results of U.S. government performed and sponsored aerospace R&D diffuse has practical and pragmatic implications for policymakers.

The U.S. aerospace industry, in particular the LCA sector, is important to the well-being of the nation. For decades, it has generated the largest trade surplus of any U.S. manufacturing industry. In 1996, with sales totaling $113 billion, the industry took first place, generating a $37.4 billion trade surplus. It employs a disproportionately large number of highly-paid, highly-skilled workers, engineers, and scientists—258 000 in 1996 out of nearly 806 000 aerospace employees—in knowledge intensive production jobs. Finally, as a rich source of knowledge, product and process technologies, and sophisticated manufacturing and production techniques, the industry is a critical component of the U.S. industrial base.

Over the last 30 years, aerospace firms have formed a variety of transnational strategic alliances and partnerships to research, develop, and produce (RD&P) LCA, to spread financial risks, to acquire capital, to gain market access, and to obtain externally produced knowledge and technology. Risk-sharing foreign partners of U.S.

aerospace firms reap similar benefits. A proliferation of joint RD&P arrangements has contributed to the globalization of knowledge and technology and their diffusion. One consequence of globalization is the sharing not only of privately held knowledge and technology, but also of publicly held knowledge and technology that result from U.S. federally funded R&D. Access to increasingly sophisticated and more tightly networked transportation and communications systems has spurred the portability of knowledge. Worldwide, a skyrocketing number of public and private organizations and individuals produce, transfer, and use knowledge in their work. To serve security and commercial interests, nearly every industrialized nation has enacted public policies that directly and indirectly support and influence aerospace R&D and the consequent production, transfer, and use of knowledge and technology.

Once largely ignored or discounted, knowledge is becoming an important component of economic, innovation, and management theories. Developed and developing nations are devoting more resources to knowledge creation, thus increasing the global pool of knowledge. Organizations and governments are beginning to recognize the value of knowledge as a leveragable resource. Knowledge has become an important determinant of competitiveness and, by extension, of a nation's economic well-being. The ability of a firm to absorb, assimilate, and apply internal and external knowledge for commercial purposes is critical to its innovative capability. Nowhere is this more evident than in the aerospace industry, whose complex and ambitious technological developments and products incorporate a wide range of scientific and technical, explicit and tacit, process and product, and systems integration and managerial knowledge. The Boeing Company's 777 typifies the absorption, assimilation, and integration of privately funded, internal knowledge and publicly funded, external knowledge for commercial purposes.

Researchers in a variety of disciplines have begun to investigate such aspects of knowledge as its flow within the firm, its absorption and assimilation through interfirm alliances, its management in process and product design, and its basis as a dynamic theory of the firm itself. Additionally, researchers have examined knowledge from the standpoint of firm size, industry type, products and services offered, and a number of cultural variables. Building on the seminal work of Everett Rogers and others in the diffusion of knowledge and innovation, our research draws on behavioral theory to understand the production, transfer, and use of federally (publicly) funded aerospace R&D at the individual, organizational, national, and international levels in the LCA sector of the industry. These research results form the basis of this book.

Organization of the Book

Because of its length, the book is divided—albeit arbitrarily—into two parts. Part A contains Chapters 1–9. Part B contains Chapters 10–19. Chapter 1 traces the evolution of LCA manufacturing in the U.S. from its beginnings as a collection of small entrepreneurial businesses to its present status as a strategic industry of global proportions. Chapter 2 explores the influence of U.S. public policy on technological innovation in LCA and the role of public policy in shaping U.S. innovation and transportation systems. Chapter 3 examines U.S. public policy and the production of federally funded aeronautical research and technology (R&T). Chapter 4 investigates the influence of U.S. public policy on the dissemination of federally funded aeronautical R&T. The focus of this research on technology rather than science prompted us to study engineers as both producers and users of knowledge. Chapter 5 builds the case for an engineering knowledge community distinct from a science knowledge community and reviews previous research into engineers' information production and use behaviors.

Chapter 6 describes the study populations and selected demographic characteristics of the survey participants. A critical element in understanding knowledge diffusion, the communications practices and the information production and use activities of aerospace engineers and scientists are examined in Chapter 7. The aerospace industry, in particular the LCA sector, is characterized by a high degree of systemic complexity embodied in product design and development. Industries like aerospace must deal with technical and market uncertainty from outside the organization as well as work-related uncertainty about problem solving within the organization. Chapter 8 explores the relationship between technical uncertainty and information use by aerospace engineers and scientists.

The ability to communicate knowledge effectively is critical to the professional success of aerospace engineers and scientists. Feedback from the engineering community about professional competencies consistently ranks communications skills *high* in terms of their importance to engineering practice, but this same feedback ranks the communications skills of entry-level engineers *low*. Chapter 9 explores the career goals, communications practices, and information-seeking behaviors of early-career stage U.S. aerospace engineers and scientists. Chapter 10 investigates the career goals, communications practices, and information-seeking behaviors of U.S. aerospace engineering and science students to gather adequate and generalizable data about the instruction they receive and the skills they develop in communication and information use. Chapter 11 extends the research into communications practices and information-seeking behaviors among aerospace engineering professionals and students across international boundaries. Chapter 12 examines the effects of years of academic work experience on information production and use by U.S. aerospace engineering faculty. Chapter 13 looks at the communication environment within U.S. aerospace and the influence of analyzability, equivocality, and uncertainty on communication within aerospace organizations. Aerospace firms, like other organizations involved in innovation and R&D, are investing in computers and computer networks to increase productivity, facilitate communication, and improve competitiveness. Chapter 14 investigates the use of computer networks in aerospace, and whether their acquisition and use achieves those goals. Chapter 15 examines the role of information intermediaries in the knowledge diffusion process, using U.S. academic and industry libraries and librarians as the survey populations. The technical reports published by the DoD and NASA are a primary means by which the results of federally funded aerospace R&D are documented and distributed to U.S. aerospace. Chapter 16 reports data that provide an empirical basis for understanding the role of the U.S. government technical report in the diffusion of knowledge resulting from federally funded aerospace R&D.

The production of LCA enjoys a special niche in the national innovation systems and the domestic and foreign policies of several countries. We limit our analysis to those countries with significant involvement in the RD&P of LCA. Chapters 17 and 18 examine knowledge diffusion and public policies affecting the RD&P of LCA in Western Europe and Japan, respectively. In both chapters we examine the role of government in the RD&P of LCA, focusing on the use of knowledge and technology to foster competitiveness, and the role of strategic alliances—domestic and transnational—in the absorption, assimilation, and management of knowledge. Chapter 19 explores knowledge management as a strategy for diffusing the results of (U.S.) federally funded aeronautical R&T and for bolstering the competitive status of the LCA sector of the industry.

Who Should Read This Book

Completing this project was an exercise in collaboration. We assembled scholars from a number of different disciplines including business and strategic management; communications; economics; international political economy; library and information science; organizational science and learning theory; political science; public policy; science, technology, and society; and sociology. We sought input from each of these disciplines to ensure balance and objectivity. Collectively, the contributors constitute an epistemic community concerned with the role of knowledge in a global economy that is now driven by high technology. The contents of this book reflect the necessity of a multidisciplinary approach when trying to comprehend a process as complex as knowledge diffusion in aerospace. Individuals interested in knowledge as an intellectual asset and as a source of innovation and competitiveness should find this book worthwhile. It should broaden the view of students, serve as a valuable resource for scholars and researchers, and help practitioners shape public policy and manage knowledge.

Acknowledgements

Several scholars have influenced our research, including John A. Alic, Thomas J. Allen, Lewis G. Branscomb, David C. Mowery, Ikujiro Nonaka, Everett M. Rogers, Richard R. Nelson, Nathan Rosenberg, Walter G. Vincenti, and Sidney G. Winter. Our apologies if we misused or misrepresented their ideas. We acknowledge the technical assistance provided by Nanci Glassman and John E. Lamar, and we are deeply indebted to Robert A. Kilgore, without whose help we could not have surveyed aerospace engineering communities in seven foreign countries. We are also grateful to the following individuals who helped make the surveys in the foreign countries possible: Keisuke Asai, David Elazar, B. Ewald, Michael J. Goodyer, A. Kawabe, H. Kubota, Sergey M. Novikov, M. A. Ramaswamy, J. Sato, John L. Stollery, Axel S.T. Tan, and E.K. Tulapurkara. We also thank the members of the several professional and technical societies who participated in the project. Karen Holloway, American Institute of Aeronautics and Astronautics (AIAA); Dave Cyback, Society of Automotive Engineers (SAE); and Gene Korte, Society of Manufacturing Engineers (SME) deserve special thanks.

Thanks also to the contributors for their patience and perseverance. We appreciate the time and assistance of those who offered valuable comments and constructive criticism: Peder Andersen, Jerry Bernstein, Matthew Dixon, John Fischer, Peter Hernon, Sarah Kadec, Philip Murray, Knute Oxnevad, Simon Reich, Yuko Sato, and Tim Sprehe. Walter R. Blados (retired) and R. Paul Ryan from the DoD and Randy Graves, John Stokes, and Jerry Hansbrough (all retired) from NASA provided financial support for our research as did the Council on Library Resources and the Society for Technical Communication. Many people at Indiana University contributed to the project's success. George Walker, Vice-President for Research, provided substantial support. Thanks to all of the staff members at the Center for Survey Research who supported the project. We are especially indebted to Catherine Kabe, Nancy Bannister, Lois Kelly, Jen Lengacher, Terry White, Heidi Hansen, Tammi Taylor, Jennifer Inghram, and Matt Knorr. We are most grateful to Phyllis Thomas for her continuing contributions over the life of the project. The book's superb indexing was performed by Jean Moody. Finally, a project as ambitious as this one was arduous, contentious, frustrating, time-consuming, and exhausting. But, as Jeanette Picard, wife of the famous balloonist, noted, "Without turbulence, nothing flies."

We thank the individuals who made the production of this book possible. The professional services of Susan Adkins, Cecelia Grzeskowiak, Jason Jacobs, Nancy (Rinker) Kaplan, and Garland Gouger of the NASA Langley Research Center Technical Library and those of William L. Cooper of the Marshall–Wythe School of Law Library, College of William & Mary were invaluable. Denise Beasley, Dee Bullock, Lee Fellers, Carol Fowler, Karen Freidt, C. Lynn Jenkins, Eloise Johnson, Bill Kluge, Fay Satterthwaite, Harriet Machie, Peggy Sipes, Jennifer Upton, and Cheryl Winstead provided administrative, editorial, and graphic support. Their patience, intelligent queries, and attention to detail improved our work. Lastly, the views expressed in this book are those of the respective contributors and are not necessarily those of the DoD or NASA.

Thomas E. Pinelli
Rebecca O. Barclay
John M. Kennedy
Ann P. Bishop

November 1997

Chapter 1

The Evolution of Large Commercial Aircraft in the U.S.—An Overview

Vicki L. Golich
Thomas E. Pinelli
Rebecca O. Barclay

SUMMARY

Chapter 1 presents an overview of the evolution of large commercial aircraft manufacturing in the U.S. from its beginnings as a collection of entrepreneurial small businesses to its present status as a strategic industry of global proportions characterized by a complex oligopolistic international production structure. In telling this intriguing tale of transformation, we first explain its current strategic importance to a nation's balance of trade, employment, and technological synergy as well as its vulnerability to market risks. Second, we consider the evolution of this industry in seven phases: the early days of flight (1903-1930), the Great Depression and industry retrenchment (1931-1938), wartime (1939-1945), post World War II industry retrenchment and adjustment (1946-1965), an era of economic challenge (1966-1980), the rise of international collaboration (1980-1995), and the institutionalization of international collaboration (1996-). We close with some implications for knowledge diffusion that derive from international collaboration in the research, development, and production of large commercial aircraft.

INTRODUCTION

On April 9, 1994, The Boeing Company rolled out its most recent new-generation aircraft, the Boeing 777. The B-777 is remarkable for four reasons. *First*, it embodies the latest technology including composites, fly-by-wire operational systems, and extraordinary flexibility of configuration to meet customers' needs in terms of interior design, range, and capacity. *Second*, Boeing used state-of-the-art manufacturing technologies, ranging from computer-aided parts design to laser alignment to achieve a fit so precise that its variance has been "less than the expansion resulting from sunshine heating one side of the transport" (Cole, 1995, p. 21; Proctor, 1994b, p. 42). *Third*, Boeing's 777 promises "to be the last all-new Western heavy transport to take flight this century" and ranks as "probably the largest current endeavor in the world funded by a single commercial company" (Proctor, 1994b, p. 36). *Fourth*, its research, development and production (RD&P) portends a "profound conceptual shift" (Proctor, 1994b, p. 36) in the management, use, and organization of knowledge, technology, and a skilled workforce to achieve production and profit goals.

The B-777 RD&P processes mark the culmination of nearly a century of technological advances in aeronautics. The airplane—once made by hand of wood and cloth using sewing machines and glue—is now made with composite materials and exotic metals by literally thousands of highly skilled workers using lasers and computers. Analysis of the B-777 RD&P process provides an excellent opportunity to investigate how "data, information, and knowledge are communicated through certain channels over time among members of a social system" (Rogers, 1983, p. 5). The *NASA/DoD Aerospace Knowledge Diffusion Research Project* seeks to understand the processes and consequences of knowledge diffusion resulting from federally funded aerospace research and technology (R&T) using changes in large commercial aircraft manufacturing as a representative case study (Pinelli, Kennedy, and Barclay, 1991). This chapter provides the historical overview critical to any such sector-based analysis.

BACKGROUND

Our selected focus—the large commercial aircraft sector of the U.S. aerospace industry—is comprised of manufacturers of commercial passenger aircraft having 100 or more seats and cargo aircraft weighing more than 15 000 kilograms, typically two-, three- and four-engine jetliners. Globally the industry includes three major and two minor producers in the United States and Western Europe. Two of the major producers are U.S. firms—The Boeing Company and McDonnell Douglas Corporation (whose recent [1997] merger did create the largest aerospace firm in the world); the third major producer is Airbus Industrie, G.I.E., a consortium of six Western European producers. The two minor producers, which manufacture large commercial aircraft having fewer than 120 seats, are the now financially troubled Fokker of the Netherlands, a company acquired by Daimler-Benz of Germany in 1993, and Avro International Aerospace, Inc., a firm which represents a joint venture between British Aerospace, plc of the U.K. and Taiwan Aerospace Corporation (U.S. International Trade Commission, 1993).

Large commercial aircraft manufacturing is an appropriate unit of analysis for the *NASA/DoD Aerospace Knowledge Diffusion Research Project* for three reasons. *First*, it is a strategic industry that plays a critical role in a domestic economy, produces goods or services directly related to national security, and generates "special benefits for the rest of the economy" (Tyson, 1988, p. 112). According to a recent Congressional Research Service study,

> Aircraft production in the United States affects nearly 80 percent of the economy. Directly or indirectly, about 340 sectors of the economy—out of about 429 defined sectors—produce goods and services as a result of the output of aircraft. Of these 340 sectors, 150 supply outputs directly to the aircraft industry. (Cantor, 1992, p. 43)

The large commercial aircraft industry is knowledge-intensive and produces high value-added products. It also has a tremendous and positive impact on our balance of trade; the employment of engineers, scientists, and highly skilled workers; and the preservation of technological synergy across a number of other U.S. high technology industries. *Second*, large commercial aircraft firms operate in a highly competitive, global market characterized by international oligopolies and, for the moment, domestically based oligopsonies. Producer and consumer firms share a symbiotic relationship which, when combined with industry dynamics related to RD&P costs and return on investment, creates significant technical uncertainty and market risk. *Third*, the large commercial aircraft sector is clearly representative of aerospace because it shares a vital knowledge base with other aerospace industry sectors. Moreover, Boeing and McDonnell Douglas have consistently been "the most significant contributors to the trade performance of the U.S. aerospace industry" (U.S. Department of Commerce, 1994, pp. 20-27).

Balance of Trade

Since the late 1950s, aerospace has been the leading industrial contributor to U.S. export earnings: it is the nation's leading exporter of manufactured goods and produces the largest trade surplus of any U.S. manufacturing industry. In 1995, U.S. aircraft sales recorded a trade surplus of $21.3 billion (Napier, 1996). In 1993, roughly 57% of the commercial export volume in terms of dollar value was generated by airline transport sales (Aerospace Industries Association of America, 1994, p. 116). Boeing forecasters now predict that medium-sized twin-aisled aircraft alone—such as the B-777, tri-jets, and the four-engine Airbus A340—represent 40% of an $815 billion market for 12 000 new transports through the year 2010 (Proctor, 1994b, p. 48). If accurate, this prediction suggests that large commercial aircraft firms will continue to play a critical role in ensuring a healthy domestic economy.

Employment

Since 1982, aerospace exports have increased at an average annual rate of $1 billion per year; for every $1 billion of aircraft shipments by U.S. firms in 1991, nearly 35 000 jobs were created (Cantor, 1992, p. 43). The industry provides the second largest number of manufacturing jobs in the United States, behind the automotive industry (U.S. Department of Commerce, 1994). In 1995, aerospace firms in the U.S. employed 778 000 persons. Of those employed in aerospace, 43% were skilled production workers, 22% were engineers and scientists, and 7% were technicians; over 34% were involved in the manufacture of civil aircraft, engines, and parts (Napier, 1995, p. 13).

National Security–Large Commercial Aircraft Sector Synergy

U.S. military and commercial aircraft manufacturers share the same knowledge and a similar production base—a complex infrastructure of more than 15 000 firms that supply sophisticated components, materials, and equipment that includes communications equipment, electronics, and scientific instruments. Commercial aircraft account for 80% of the total aircraft production weight during peacetime. Commercial design and production teams have developed military hardware, and market requirements have triggered technological innovations relevant to military needs and vice versa. The availability of a personnel and production base reduces the cost of providing a military industrial base and wartime mobilization surge capacity (Bacher, 1984; Lopez and Yager, 1987, p. 42; National Academy of Engineering, 1985, pp. 1-2, 25; Neuman, 1984, p. 175; U.S. Executive Office of the President, 1985).

Moreover, manufacturers produce an essential good for the vital air transport service sector. Any business that depends on air transport benefits from increased efficiencies afforded by state-of-the-art equipment. Thus, the "linkage externality" is positive—both the private returns to aircraft manufacturers and the social and private returns to downstream users are increasing. Finally, technological innovations within the large commercial aircraft industry positively affect those core technologies that lead to the development of many products, have a significant impact on production processes, and influence many sectors of the economy (Van Tulder and Junne, 1988). The constant pursuit of technological advances helps maintain a strong knowledge base. Their early application leads to decreased costs and more rapid diffusion to other industries, all of which are thought to be critical to maintaining a competitive position in today's economy (National Academy of Engineering, 1988; Tyson, 1988). Competitiveness across a large number of industry sectors is considered fundamental to national security. Because large commercial aircraft manufacturing affects such a huge percentage of the U.S. economy, it is clearly strategic from a strictly economic perspective. Though historical overlap with military aviation is no longer very significant, the shared knowledge and production base renders the large commercial aircraft sector strategic from a traditional security perspective as well.

Market Risk

Finally, the large commercial aircraft sector merits attention because it is extremely risky financially and the stakes are extraordinarily high. Aerospace firms must make major capital investment decisions despite uncertain future payoffs. They may spend several billion dollars to conceptualize, develop, and build a new airframe; the typical 10 to 15 year return on investment cycle defies accurate prediction; hence, the new aircraft may not fit market needs, as happened after the 1978 economic deregulation when manufacturers were caught in the middle of producing wide body aircraft which airlines no longer wanted. Before making a final design decision, manufacturers typically produce dozens of "paper airplanes" to accommodate as many potential customers as possible, a time-consuming process. For example, the design process for the Boeing 727 took longer than two years and generated "nine separate complete designs for the aircraft" (Mowery and Rosenberg, 1989a, p. 171). The design-definition phase for the Boeing 767 took almost six years. An aircraft design is typically long-lived. For example,

Boeing produced the 727 for 20 years and Douglas produced the DC-8 for 15 years but the design of these aircraft underwent continuous modification to accommodate the changing requirements of the commercial airlines (Mowery and Rosenberg, 1989a).

Despite the high risk associated with long lead times and major capital outlay, timing has been critical to the market success of a new aircraft. Until recently, firms that introduced the first of an aircraft type gained significant competitive advantage. Before the Korean War, the Douglas DC-3 virtually monopolized world commercial air travel. With the introduction of the Boeing 707 jetliner, Boeing supplanted Douglas as the dominant producer of large commercial aircraft, though the comparable Douglas DC-8 was introduced less than a year later. In today's market, performance and maintenance far outweigh any other factor. Nevertheless, Boeing was concerned enough about timing that one of its articulated goals for the B-777 was to avoid the delivery delays and initial service problems that typically accompany the introduction of new generation aircraft.

HISTORICAL OVERVIEW

The marketplace for large commercial aircraft has changed dramatically since the industry's inception, and radically in the last 30 years. The following overview of this transformation emphasizes key events and technological innovations in the U.S., which advanced commercial aviation by improving aircraft performance. Our overview is not intended to be comprehensive, we have left that task to others. (See, for example, Bright, 1978; Davies, 1964; Miller and Sawers, 1970; Mingos, 1930; Simonson, 1968; Solberg, 1979; Vander Meulen, 1991.)

The Early Days of Flight (1903–1930)

Orville Wright's first heavier-than-air flight on December 17, 1903, marked a dramatic technological advance over flight in hot air balloons and helium-filled dirigibles. The Wright Brothers accomplished this feat using an internal combustion engine that powered a propeller-driven adaptation of the biplane gliders that they had built and learned to fly previously. Before powering off the ground at Kitty Hawk, North Carolina, the Wright brothers had proved flight in a wind tunnel in Dayton, Ohio. This success in powered flight derived from their broad study of the science of flight using the scientific method (explicit knowledge) as well as from their natural intuition and engineering experience (tacit knowledge). This remains a key element in the RD&P of all aircraft (see Chapter 3).

The Wright brothers' work inspired a budding U.S. aircraft industry that was characterized by technologically sophisticated, entrepreneurial engineer-owners. Despite operating highly competitive firms, they often shared ideas about how to improve their products. Although not necessarily educated as engineers, their activities mirrored those of modern professional engineers. The first engineer owners consulted the available scientific literature, built models, conducted instrumented tests, and experimented with full-scale prototype aircraft (Biddle, 1991). These entrepreneurs constituted the emerging American aeronautical community.

Dreams of commercial air transport were temporarily supplanted by World War I and a subsequent need to supply military aircraft. In 1918 alone, the U.S. aviation industry built more than 14 000 airplanes for the U.S. military (Stekler, 1985). When the war ended, more than $100 million worth of military airplane orders were cancelled and the industry nearly collapsed; in 1922, U.S. manufacturers built only 300 aircraft (Cunningham, 1951). U.S. aircraft manufacturers benefitted from learning to build airplanes according to precise design specifications, performance characteristics, and scheduled delivery. Nevertheless, only U.S. government intervention provided incentives to invest in new designs. The 1925 Kelly Air Mail Act and a series of subsequent amendments invigorated the industry via a five-year aircraft procurement authorization for the Army and Navy and airmail subsidies, and by charging the federal government with the responsibility to provide the infrastructure—for example, airports and airways—that would support an expanding commercial aviation sector (see Chapter 2).

The development of a U.S. commercial aircraft industry required substantial numbers of professionally trained engineers and a codified body of knowledge, now known as the science of aeronautics. To meet the demand for engineers, a number of engineering schools had expanded their curricula to include aeronautics by the late 1920s, including the California Institute of Technology (CalTech), Stanford University, the University of Michigan, the University of Washington, New York University, and the Georgia School of Technology (later to become the Georgia Institute of Technology). With the exception of CalTech, the schools graduated working engineers, most of whom went to work in the burgeoning aircraft industry. Only CalTech, under the direction of German scientist Theodore von Karman, an internationally recognized expert in theoretical and fundamental aerodynamics, became a preeminent site for the study of scientific aeronautics. Von Karman, Jerome Hunsaker, William F. Durand, and Max Munk, who are widely acknowledged as pioneers of theoretical aerodynamics in the U.S., are credited with building the foundations of aeronautical studies in U.S. colleges and universities during this period (Hanle, 1982).

The Great Depression and Industry Retrenchment (1931–1938)

During the Great Depression of 1932, the demand for commercial aircraft decreased. Many aircraft companies, especially the smaller ones, went out of business, were acquired by, or merged with other manufacturers. Commercial aircraft production plummeted from a high of 5516 units in 1929 to 803 units in 1932; aeronautical exports declined dramatically during the same period (Simonson, 1968). The vertically integrated companies that survived constituted an oligopoly. They received nearly all the orders for military aircraft as well as government contracts for carrying airmail. This oligopoly structure spurred two congressional investigations: the Morrow Board and the Crane Committee, which eventually led to the separation of the manufacturing and service sectors (Mowery and Rosenberg, 1982, p. 105).

By the mid 1930s, aviation was clearly perceived to be strategic and worthy of federal protection to ensure its domestic viability. The U.S. government continued to award air mail and military procurement contracts. Between 1935 and 1940, the Army was authorized to buy 2300 aircraft and the Navy to buy 1200. Aware of a rising tide of nationalism worldwide and of American aircraft manufacturer involvement in the international arms trade, Congress passed the Neutrality Act of 1935. President Roosevelt immediately declared aircraft to be munitions of war. Still, U.S. exports surged upward largely because so many other large producer countries withdrew from "foreign markets, leaving them open to U.S. manufacturers" (Freudenthal, 1940, p. 106).

During this period, passenger transport in the United States finally exceeded mail carriage as the primary activity of commercial airlines. Two events profoundly affected commercial air travel. In

1931, the crash of a Fokker airplane that killed legendary Notre Dame football coach, Knute Rockne, grabbed public attention. The crash was blamed on structural failure—the wood-framed wings of the aircraft had rotted and fallen apart in mid-air. Disclosing the cause of the crash effectively ended the use of the wooden-wing airplanes for commercial air passenger service in the U.S. (Solberg, 1979). The second event followed from the first. Both Boeing and Douglas Aircraft Company introduced all-metal, monocoque airframe transports—the DC-2, the DC-3 and the Boeing 247—a design that represented a major breakthrough in airframe technology (Mowery and Rosenberg, 1982). This single innovation made it possible to incorporate many more new technologies utilizing the best of what was known, proven, and available in fundamental aerodynamics, propulsion, and aircraft production (Rae, 1968). The new aircraft provided greater comfort for passengers and improved operating efficiency for the airlines.

Airlines depended almost exclusively on the Douglas DC-3 for passenger transport, although Boeing, Lockheed, and Curtiss-Wright also produced airframes. Curtiss-Wright and Pratt & Whitney continued to dominate engine production. The DC-3 allowed airlines to generate a profit without having to depend on mail subsidies. It was the first airplane to support itself economically as well as aerodynamically. Within two years, Douglas had sold 803 DC-3 aircraft and they were carrying 95% of the nation's commercial air traffic (Rae, 1968, p. 71).

Wartime (1939–1945)

World War II began in September 1939. Predictably, aviation production focused on military needs in all countries. The demand by foreign governments (e.g., England and France) for U.S.-produced aircraft increased dramatically. In November 1939, the dollar value of aircraft back orders alone amounted to $533 million; more than half of that amount was for orders placed by foreign governments (Lilley, 1947). The large number of airplanes needed focused attention on all aspects of production engineering, including standardization, assembly learning curves, and scales of economies. The aircraft were also increasingly complex to produce. For example, the 18-foot nose section of the B-29 took over 50 000 rivets to assemble and required over 8000 parts procured from more than 1500 vendors (Lilley, 1947, p. 119). In spite of this complexity, the industry was forced to rely on less experienced workers to meet demand. Under the circumstances, no single U.S. company could possibly fill the requisite orders.

To solve this production problem, aircraft manufacturers borrowed from the knowledge, experience, and capability that U.S. automobile manufacturers had gained from mass production processes. Aircraft production shifted from the "job shop" method to an assembly line method. The proprietary control of aircraft designs was replaced by a significant amount of cross-licensing and a fairly substantial transfer of "in-kind" technology to facilitate maximum production. Subcontracting became an integral part of aircraft production due to the economies of scale and other efficiencies that accompanied specialization in component manufacturing. By the end of the war, the aeronautical community in the U.S. had expanded in both size and scope. Hundreds of increasingly skilled and specialized workers shared their explicit and tacit knowledge to build aircraft.

Although war once again disrupted commercial air transportation in the U.S., the technological improvements incorporated into airframes and engines to meet the needs of the military were immediately and easily transferable to the needs of the expanding commercial air transport system. The pre-war grass airfields had been transformed into asphalt or concrete runways, facilitating the landing of more and larger aircraft. New aircraft, such as the DC-4 and the Lockheed Constellation, possessed higher payload capacity, range, and speed, and benefitted from numerous refinements in radio communications, navigation, and instrumentation that had been developed to facilitate long flights over water and navigation at night and in inclement weather (Taneja, 1976).

Wartime production efforts set the stage for the introduction of the jet engine to power large commercial aircraft. Refining the jet engine was among the most hotly pursued efforts in technological innovation. The U.S. contracted with General Electric to perfect the jet engine that had been developed first in Germany and the United Kingdom and introduced in the German Heinkel He 178 in 1939 and the British Gloster Meteor in 1944 (Lederer, 1985, p. 41).

It is not surprising that the most important catalyst for technological change at this time came from the federal government during this period (see Chapter 2 for greater detail). U.S. government requirements for military aircraft and federal financial support made possible a phenomenal leap in technological innovation for the airplane and its physical environment. The engineering "know-how" that came to characterize aircraft production from this point forward was the result of significant borrowing of product and process technology from other industry sectors. The "know-how" was transferred to a variety of firms that produced an assortment of products for the war effort. In addition, the technological evolution and advances that characterized this period had a profound synergistic effect; changes in the operating environment (e.g., concrete runways and improved navigational equipment) facilitated or even necessitated refinements in airframe or engine construction and vice versa.

Post World War II Industry Retrenchment and Adjustment (1946–1965)

The ability to mass produce technologically sophisticated airplanes changed the character of the U.S. aircraft industry. The war gave rise to a three-tiered industry structure, illustrated in Figure 1.1, that is still in place today, consisting of a small number of "prime contractors" at the top, a second tier of subcontractors, and a third tier comprised of a myriad of vendors and suppliers (Bernstein, 1995). Nevertheless, immediately following the war, the future of the industry was uncertain. In particular, it was unclear how the industry and its huge production capacity, as well as its thousands of employees, could be used in a peacetime economy. These concerns were aggravated by a surplus of literally hundreds of thousands of aircraft.

The U.S. aircraft industry, as well as the entire global political economy, was affected by a dramatic restructuring of the international political system following World War II. The U.S. emerged from the war as a superpower that dominated the commercial marketplace and assumed global-level security responsibilities as the leader of the free world. U.S. industry leaders and politicians searched for a "national" policy that would maintain and utilize the existing production capability for national defense, facilitate the transition of the industry to a peacetime economy, and maintain a healthy rate of technical progress. Aircraft industry leaders drew on their wartime experiences that required the rapid incorporation of science and technology and explicit and tacit knowledge into the development of a new aircraft. For a return on investment to be realized, industry leaders were quick to point out, for example, that an improved airfoil developed in a research program must actually be used in producing the wing of an aircraft in order for the airfoil to have any utility. Research results had to be validated through testing and applied to refine or develop an aircraft that was produced in quantity before the fruits of that research could be realized.

A dramatic postwar expansion was predicted for U.S. commercial aviation in terms of air miles traveled, numbers of people employed, and passengers carried. By 1948, domestic air transport had expanded extensively, thanks in part to the government policy executed by the Civil Aeronautics Board (CAB) (see Chapter 2). Equally important, American commercial aviation expanded because great overland distances separated a large, mobile, relatively

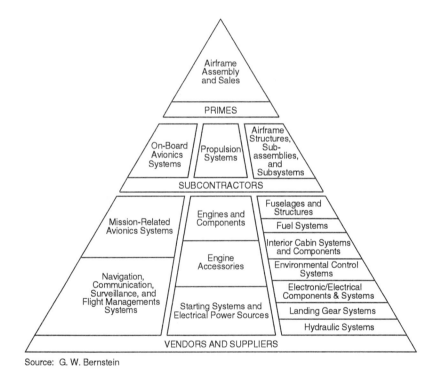

Source: G. W. Bernstein

Figure 1.1. The Large Commercial Aircraft Production Structure in the United States.

wealthy, and relatively egalitarian population (Constant, 1980). The new comforts of flying resulted in a more "air-minded" general public that chose air transportation when travel distances exceeded 200 miles (Davies, 1964, pp. 238-240, 271-272). From 1929 to 1954, the passenger-kilometers generated by the world's scheduled airlines increased at an average annual rate of 25% (Hochmuth, 1974, p. 154).

The military aircraft needed to support America's role as a superpower in the Cold War era had mission and performance requirements far different from those of the aircraft used in World War II. The increased size, weight, and complexity of military jet airplanes such as the B-47 and B-52 had rendered many existing aircraft production facilities obsolete. The introduction of jet aircraft, which, according to Constant (1980), was "the greatest aeronautical revolution since the Wright Brothers" (p. 1), gave rise to noise problems. Runways that had been long enough for propeller-driven aircraft were inadequate for jets. Although jet fighter planes had been used successfully in Korea, the jet engine faced an uncertain future in commercial applications. The fact that jet engines consumed great amounts of fuel was a sticking point in terms of developmental and operational costs. Even though jet engines were quite acceptable for fighter aircraft, in which speed was more important than range, they would require considerable refinement before they could be used in large commercial transports.

The demands of the military for strategic, jet-powered bombers and tankers had pushed the development and early application of jet engine technology. The military financed the cost of the first jet engine (Mowery and Rosenberg, 1982) and provided continued support for the development of specific engines whose cores were adapted for commercial use. Military aircraft such as the KC-135 served as test beds for the technological development of early commercial jet aircraft (March, 1989). The prototype B-707, begun in 1952 and flown in 1954, went into production in two versions: the KC-135 for the Air Force, and the 707-20, the first U.S. jet transport, put into service in October 1958 by Pan American Airways (Rae, 1968).

The technological advances in airframes and engines that increased the potential rewards to be derived from commercial aviation also increased costs and risks. Aircraft and engines were larger and more complicated; longer lead times were required between gestation, production and revenue earning. Attempting to incorporate increasingly sophisticated technology like the jet engine into aircraft design and production made the payoffs uncertain (Golich, 1989, 1992; Hayward, 1983, 1986; Majumdar, 1987, Miller and Sawers, 1968; Mowery and Rosenberg, 1982). Research, development, and production costs rose precipitously, making it increasingly difficult for a single manufacturer to finance a commercial aircraft development program.

The new industry dynamics "put a premium on accurate technical and commercial judgments," while "the price of market failure rose; it became more difficult, and certainly more expensive for a firm missing out on one generation of aircraft to challenge for success in the next" (Hayward, 1983, p. 1). Aircraft manufacturers sought larger markets to sell more airplanes, reduce production costs through the achievement of economies of scale, and eventually lower prices, facilitating the sale of even more aircraft.

An Era of Economic Challenges (1966–1980)

The market for U.S.-built large commercial aircraft became global during this period. By the mid-1960s, U.S.-made commercial aircraft had captured approximately 79% of world markets (Carroll, 1975, p. 153; Mowery and Rosenberg, 1982, p. 113). Until 1958, when jet transports replaced four-engine propellers, Douglas, Lockheed, Convair, and Martin dominated the commercial aircraft market. Once Boeing launched the B-707, followed within a year by the Douglas DC-8, other American manufacturers lost so much market share that they virtually vanished from the commercial sector. Most concentrated on military production or converted to subcontracting for one of the remaining primes. Some merged with other survivors in the second and third tiers; Lockheed eventually turned to developing the wide-body L-1011, a new-generation aircraft.

By 1973, Boeing had replaced Douglas as the dominant U.S. large commercial aircraft producer, controlling more than 63% of the U.S. air carrier market (Mowery and Rosenberg, 1982, p. 111). Facing bankruptcy in 1966, Douglas Aircraft Company merged with defense-oriented McDonnell Aircraft. To prevent a monopoly in the aircraft manufacturing sector, the U.S. government intervened with unusual proactive force. The federal government approved the McDonnell Douglas merger in 1967 and provided a $75 million guaranteed loan to ensure the viability of the new organization. Four years later, in 1971, when Lockheed was reeling from its own bankruptcy threat, the federal government provided a guaranteed loan of $250 million (Golich, 1989; Mowery and Rosenberg, 1982, p. 113).

Several other key factors also altered the market and operating structures for large commercial aircraft during this period. At the domestic level, declining defense spending, the spiraling cost of raising capital, and sharp increases in the price of jet fuel hurt U.S. commercial aircraft manufacturers and airlines. Throughout the 1970s, requirements imposed by the Environmental Protection Agency (EPA) forced aircraft manufacturers and airlines to find ways to reduce the smoke and other emissions from jet engines as well as to decrease the noise they created. Complying with these demands was not only costly and time-consuming, but it also placed additional design requirements upon the engine and airframe manufacturers.

At the production level, given the technical and marketplace uncertainty and the high levels of risk and cost associated with developing a new generation aircraft, manufacturers were decreasingly able to initiate such projects—even with government support through related research and development funding and procurement policies—unless they were assured a significant market share. Only U.S. airlines were large enough in fleet size and served a large enough market to be able structurally to support a launch purchase —the promise to buy enough aircraft to cover a significant percentage of the development costs. Prior to economic deregulation in the U.S., airlines were virtually guaranteed fare increases to cover the increased costs of acquiring a new fleet. Airlines were eager to assume this role because, in the absence of fare price competition, service quality—as reflected in newer, faster, and more comfortable aircraft—was considered a competitive advantage. As a strategy to ensure recoupment of development costs, manufacturers allowed airlines to dictate certain performance characteristics to be built into a new aircraft. Thus, the commercial air carriers began to play a significant role in the early design phases of airframe development (Bluestone, Jordan, and Sullivan, 1981; Golich, 1989; Newhouse, 1982). Although only a few U.S. airlines are now able to serve as guaranteed launch customers, the role of air carriers in determining the performance specifications of new aircraft has increased.

Subcontracting or outsourcing for aircraft components and systems also became more important as a strategy to reduce the cost

and risks associated with introducing a new aircraft. Large commercial aircraft incorporate an incredible number of complex and
sophisticated components and systems. Although primes maintain
in-house design and development expertise, the levels of complexity
and variation in design problems compel most primes to subcontract for a large percentage of the components and systems that will
be used in the final assembly of the aircraft. As primes work with
their subcontractors to determine performance specifications and
characteristics, the latter will typically employ materials, products,
systems, and technologies that they have developed for use in other
industry sectors as they try to reduce costs and risks. Doing so
facilitates the diffusion of science, technology, explicit knowledge,
and tacit knowledge within and across industry sectors.

Initially, the use of subcontracting was limited to domestic
sources, but in recent years subcontracting for aircraft components
has extended to firms outside the U.S. In the early 1930s subcontracting "constituted less than 10% of the industry's operations"
(Rae, 1968, p. 83). Twenty years later, Lockheed subcontracted for
30% to 40% of the Electra's production, and by the mid-1960s,
Boeing employed six major domestic and foreign subcontractors for
70% of its B-747 (Hochmuth, 1974; Mowery and Rosenberg, 1982,
p. 116). In developing the B-747, Boeing asked its subcontractors
to share the risks by providing a substantial portion of the development costs.

For the most part, extensive subcontracting defined the outer
limits of any move toward globalizing production arrangements in
large commercial aircraft by U.S. manufacturers. Several times,
Boeing and McDonnell Douglas initiated discussions regarding more
tightly structured joint venture projects with Britain, France, Italy,
Japan, and The Netherlands. Few were consummated; collapse of
the others was precipitated either by the U.S. firms abruptly termin-
ating discussions or by the European firms deciding to join forces
with Airbus Industrie. In either case, these efforts failed primarily
because "both McDonnell Douglas and Boeing had showed a very
obvious determination to control all aspects of development, production and sale." This "hard-nosed approach to commercial decision-making and central control was ill-suited to the existence of
wider social and political interests underlying European industrial
activity," which included the desire to enhance their own comprehensive design and marketing capabilities (Hayward, 1986, pp. 101,
197). European corporate officials came to believe that U.S. aircraft

firms were "simply playing games with the Europeans in order to sow dissent amongst the Airbus partners and, more importantly, to preempt further European expansion of Airbus Industrie" (Hayward, 1986, p. 98; Newhouse, 1982, pp. 197-198).

The Rise of International Collaboration (1980–1995)

By the mid-1980s, the structural changes in the U.S. aircraft industry begun in the 1960s had intensified and were reinforced by other systemic transformations. International interdependencies, "situations characterized by reciprocal effects among actors in different countries" (Keohane and Nye, 1977, p. 8), proliferated early in the decade. Industry leaders and policymakers concluded that "the national ability to disengage oneself from the network of economic interdependence is much less widespread than it was thought to be in the past" (Rosecrance, 1981, p. 699). Moreover, power capabilities were more equally, if still asymmetrically, distributed among states. Although the U.S. remained "the most powerful actor in the world political economy" and had "greater leeway for autonomous action than other countries" (Keohane, 1984, p. 26), a number of advanced industrial countries had recovered from the war and were becoming highly competitive in several market sectors.

Greater economic parity among the advanced industrial states was accompanied by the growth of aircraft manufacturing capabilities in Europe and the newly industrializing countries of Brazil, Israel, and Indonesia. Several newly independent sovereign states began to spend more of their national budgets on developing a technology base. Although still not strong enough to influence major decision outcomes, the new states began to assert themselves in the global economic arena by demanding participation in the industries that they believed would facilitate economic modernization. Indeed a proliferation of offset agreements, initiated by McDonnell Douglas and Yugoslavia in 1972 which evolved into guarantees of "aircraft purchase in exchange for production participation," characterize this change (Schaufele, 1988).

By the 1980s, RD&P of larger and more complicated airframes and engines took 4 to 5 years and 10 to 15 years to realize a return on investment. The cash-flow needs of aircraft manufacturers began to skyrocket just as the 1970s cuts in defense spending and cancellation of key "high profile" military aircraft projects began to affect corporate coffers. The number of aircraft that had to be sold to earn a "return on investment" multiplied dramatically.

The value of aircraft sales triggered intense competition in a global market decreasingly demarcated by distinct national markets linked by trade. The customer base for large commercial aircraft, now global in scope, continued to shrink—25 major air carriers now accounted for more than 70% of the world travel market, with approximately 67% of the large commercial aircraft market located outside the U.S. (Clarkson, 1992; Taneja, 1994). As Lopez and Yager (1988), research analysts for the Aerospace Industries Association of America, Inc., note, "the marketplace has changed and there is no going back. Success...requires more attention to world markets and to means of better serving those markets" (p. 2).

As foreign markets grew increasingly important for U.S. aircraft producers, foreign airlines gained influence in aircraft design decisions, gradually replacing some U.S. airlines as launch customers. Since foreign airlines were still largely government-owned or supported, aircraft acquisition negotiations nearly always included at least one government. Further, since state policymakers often encouraged airlines to contract with domestic manufacturers, U.S. manufacturers were impelled to pursue some form of international joint production, potentially affecting cross-border transfers of science and technology as well as explicit and tacit knowledge. McDonnell Douglas entered its first foreign procurement agreement with the DC-9 program, signing first with Canada and subsequently with Italy. The Boeing Commercial Aircraft Group initiated foreign subcontracting with the production of the B-707, with Rolls Royce of the United Kingdom furnishing engines, and companies from France and Canada supplying other components.

In the early 1980s, competition for global market shares was intensified by the existence of surplus capacity in all commercial aircraft market segments (Kelly, Oneal, DeGeorge, and Vogel, 1991, pp. 84-85; Lopez and Yager, 1987, p. 5), a phenomenon aggravated, perhaps ironically, by technological progress in the industry that increased the potential life span of a large commercial aircraft from 25 to 50 years and lengthened the replacement cycle. Frustrated manufacturers, to get rid of "white tails"—complete aircraft with no buyer—initiated long-term "walk away leases"; as the leases ran out, these aircraft were added to the pool of relatively cheap air transports.

In addition to the incentives already compelling international collaboration, policymakers now believed that denying a potential foreign partner's access to a production market might propel potential partners to become competitors in the future (National Academy of Engineering, 1985, pp. 63-64), or that if a company did not form an alliance, the competition would. With the impetus toward industry internationalization, learning to manage international projects, including the diffusion and application of knowledge, became extremely important.

Persistent U.S. discomfort with international production originated not only from perceived vulnerabilities associated with mutual dependencies in general, but also from concerns regarding the transfer of knowledge and technology, particularly that funded by U.S. taxpayers for the purposes of maintaining U.S. preeminence in a specific industry or market sector. This perception was reflected in the collapse of the Mitsubishi-General Dynamics Fighter Support Experiment (FS-X) deal that left virtually all participants in limbo regarding guidelines for when collaboration is permissible and when it is not (Prestowiz, 1988). According to a National Academy of Engineering (1985) study, "the entire apparatus of government... tends to reflect the priorities and perceptions of an earlier time.... What is needed...is a change in tone and attitude..." (p. 146).

Trying to balance national security needs with those of commerce and economics created inconsistencies and indecisiveness that may have been detrimental to trade because they called into question the reliability of U.S. manufacturers as partners. For example, Donald Fuqua (1989), President of the Aerospace Industries Association (AIA), noted that the debate over participation in the FS-X sent two clear messages to potential foreign customers and partners and to our own industry: the U.S. has no coherent policy on international trade and cooperation with respect to high-technology industries such as aerospace, and the U.S. is an undependable trading and production partner (p. 3).

Still, U.S. aircraft manufacturers preferred a hierarchically organized international production structure because it allowed them, as senior partners, to exercise control over critical production decisions such as which project(s) to pursue, which designs to adopt, which marketing and distribution methods to use, and how to adjust production schedules to achieve necessary efficiency and flexibility in the assembly of a quality product. It also increased their control over the use and diffusion of "knowledge and technology" assets. As they negotiated international production agreements, they had "no intention of relinquishing control of the program, and insisted on retaining at least a 51% interest" ("Boeing Plans...," 1985, p. 212).

In the end, U.S. corporate executives decided that collaboration with foreign partners was critical to maintaining a competitive position in this strategic industry sector. They did so to secure financial support, avoid potential European tariff barriers, nullify or dilute competition from European firms, and avoid antitrust restrictions. U.S. manufacturers were also motivated by the desire to improve market access; increase risk sharing; lower research, development, and production costs; and gain financial support for sales. U.S. collaboration was also encouraged by the fact that so many aircraft purchases were now being linked to industrial offsets (Hayward, 1986, pp. 31, 94-95; see also Bluestone, Jordan, and Sullivan, 1981, pp. 159-160; Harr, 1972; Schaufele, 1988). "The increasing incidence of joint ventures in the U.S. commercial aircraft industry is a response to the fundamental forces in the evolution of both the world economy and aircraft technology" (Mowery, 1987, p. 150).

The Institutionalization of International Collaboration (1996–)

Today no large commercial aircraft is launched without careful attention to choosing production partners from around the world. The global arrangements include direct investment, co-production, licensing arrangements, and collaborative efforts in the research, development, production, and marketing of the aircraft. U.S. firms build Airbus aircraft components, and European, Canadian, and Asian firms help make U.S. aircraft. McDonnell Douglas currently has international production arrangements with Brazil, India, Italy, Japan, Korea, the People's Republic of China, Poland, Spain, Taiwan, and the former Yugoslavia. The Boeing Company's foreign suppliers include corporations from Australia, Belgium, Brazil, Canada, France, Germany, Greece, Ireland, Israel, Italy, Japan, Korea, The Netherlands, Singapore, Spain, the United Kingdom, and the former Yugoslavia.

The Boeing Company has cautiously expanded its conventional subcontracting arrangements with foreign firms. With the B-767, Boeing inaugurated its first program with non-U.S. risk-sharing participants—Japan's Commercial Airplane Company, a consortium of Mitsubishi Heavy Industries, Kawasaki Heavy Industries, and Fuji Heavy Industries, furnished about 15% of the airframe, while Italy's Aeritalia contributed another 15%. Japanese representatives held membership status on the executive council and program committee and were referred to as "Program Participants" —a transitionary arrangement between a conventional subcontract and a full partner role.

With the research, development, and production of the B-777, Boeing moved further in the direction of institutionalizing its international production structure. Boeing articulated three specific objectives for the on-time delivery of the B-777: (a) to reduce overall production costs, (b) to increase reliability and improve maintainability, thereby reducing maintenance costs, and (c) to avoid the "incredibly painful" delivery delays and initial service problems that had accompanied the introduction of its earlier B-747-400 transport, and virtually every new generation aircraft by every manufacturer since the industry's inception (Proctor, 1994b, p. 39).

To achieve this goal, Boeing first increased its purchases of parts made by outside suppliers with the aim of reducing the in-house content of its transports from 52% to 48% ("Boeing to increase...," 1995, p. 33). Figure 1.2 depicts this shift in contracting strategy. As a corollary strategy, Boeing expanded the number of its international suppliers to nearly 60, creating tighter linkages among some program participants than had been seen previously in the industry. Although program management and control resides with Boeing, the degree of communication and integration in the "working together" teams is unprecedented. For example, the data system designed to integrate engine and airframe together with airline maintenance needs is so interconnected that the airline "can even access the enginemaker's blueprints" (Proctor, 1994b, p. 54; Cole, 1995). The Japanese, who contributed 20% to the program, were able to review its status, progress, and outlook and to influence some design and development decisions early in the process. In return, they accepted significant risk participation, assuming responsibility for both the non-recurring and recurring costs of the hardware items they produced (Benke, 1987).

Another strategy involved massive investment in testing facilities and procedures intended to facilitate the delivery of "service-ready" aircraft from day one. Boeing built a new $370 million Integrated

Aircraft Systems Laboratory (IASL) aimed at eliminating expensive production floor changes and cutting warranty costs (Proctor, 1994b, p. 56). Although Boeing has pursued a policy of testing that exceeds the Federal Aviation Administration (FAA) certification requirements by approximately 60%, one enduring characteristic of commercial aircraft production has been the remarkable learning curve and learn-by-doing advances gained in the first few years of production. Although it seems desirable to point to large increases in productivity and decreases in problems over time, company executives believe the extra testing will save money in the long run by

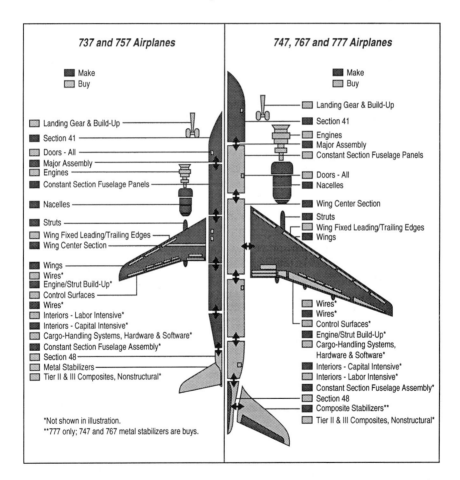

Figure 1.2. A Comparative View of Components Made or Bought for Boeing Large Commercial Aircraft.

finding problems *before* aircraft are built and have to be modified (Proctor, 1994b, p. 43).

The final strategy, and perhaps the boldest, was the introduction of cross-functional design-build teams that included representatives from virtually every group who will come in contact with the plane, namely airlines and their users—pilots, flight attendants, mechanics, baggage and cargo handlers, and passengers. Teams benefitted from Boeing's expanded use of computer-aided design technology: the Dassault/IBM CATIA (Computer-Graphics-Aided Three-Dimensional Interactive Application) linked nearly three terabytes of data from suppliers worldwide. CATIA technology enabled teams to share their ideas and insights regarding the evolving aircraft in real time "early and often," thus avoiding significant amounts of costly "error and rework" incidents (Cole, 1995; Proctor, 1994b, p. 37-40). Every one of the B-777's 132 500 unique engineered parts was designed using CATIA technology; a complete mock-up of the aircraft was created on-line so that no physical prototype was necessary. Given that international suppliers are producing components and that several of the airlines participating in the cross-functional teams are not U.S. domestic carriers (e.g., British Airways, All Nippon Airways, Japan Airlines, and Cathay Pacific Airways), their involvement will, in all likelihood, facilitate significant cross-border diffusion of aeronautical science and technology and explicit and tacit knowledge.

Although it is premature to evaluate how close The Boeing Company has come to achieving all three of its goals, Boeing was able to deliver the first 100% "paperless" aircraft, the world's largest twinjet, and the first new generation aircraft in thirteen years, and the delivery was on time. Moreover, all aircraft performance guarantees were met or exceeded (Cole, 1995; Proctor, 1994a, p. 20; "United Launches...," 1995, p. 50). For example, the B-777 has exceeded fuel efficiency expectations by 5% of what Boeing guaranteed to airline customers ("Boeing Says...," 1994, p. 34). Unprecedented joint FAA and European-type certification was awarded in April 1995, almost exactly five years and an estimated $4 billion after the formal B-777 program was announced by Boeing. This included an extended-range twin-engine operations (ETOPS) authority up to the maximum 180 minutes from an alternate airport, a feat never before achieved at the start of service by any aircraft manufacturer. Previously ETOPS certification followed a minimum of 1.5-2.0 years of safe flight operations (Cole, 1995; Proctor, 1995a, p. 124). Boeing used "service ready" rather than prototype aircraft for test flights throughout the certification process. As a result, they achieved a 25% better flight test per aircraft rate than the best rate achieved by previous Boeing large commercial aircraft test programs (Proctor, 1995b, p. 20). Although test flight totals for the B-777 were roughly double those of the B-757 and the B-767, the early success of all the tests were credited to the extensive IASL noted previously.

An exhaustive list of technologies incorporated into the B-777 is impossible to offer here. The small sample that follows still indicates the extraordinary successes that Boeing achieved in the RD&P of this large commercial aircraft. The avionics system used offers a savings in both weight and volume of 510 pounds, thereby saving an estimated 25% in acquisition costs, lowering maintenance costs, and increasing reliability (Nordwall, 1995, p. 43). A tripling of the use of composite materials added to increased structural strength of the airframe while saving weight and contributing to the greater than anticipated fuel savings (Proctor, 1994b). A blade tail cone replaces the conical shapes used on the 757 and the 767, thus lowering drag and placing less stress on the boundary layer, while the airfoil carries a greater fraction of lift on the aft part than previous Boeing wings (Dornhem, 1994, pp. 46-47). The combination of a fly-by-wire system with traditional flight deck controls has created an aircraft that pilots report is remarkably agile for a heavy transport (Hughes, 1995, p. 42).

Boeing has begun the process of designing and building derivative B-777 aircraft, already posting significant sales, particularly in the Pacific Rim Asian countries. The B-777-300X stretch version will have the capacity to carry 368-550 passengers and a range of 5700 nautical miles; thus, it will serve as an ideal replacement aircraft for B-747-100s and B-747-200s with a 33% lower fuel burn, 40% lower maintenance costs, and considerably greater cargo-carrying capacity of 33% more volume in the lower hold (Shifrin and Sparaco, 1995, p. 31).

IMPLICATIONS FOR KNOWLEDGE DIFFUSION

Collaboration, in the form of global joint production and multinational consortia that involve U.S. airframe and engine manufacturers, raises a number of issues concerning the manufacture of large commercial aircraft and the diffusion of knowledge. Virtually all participants in these "arrangements" recognize that collaboration invariably leads to the transmigration of knowledge and, in some cases, entire technologies. Growing collaboration has resulted in considerable public policy debate about the diffusion of knowledge because knowledge is increasingly recognized as a key to competitive advantage in a global marketplace. Therefore, effective management of knowledge is essential if the U.S. aircraft industry is to maintain its preeminent position in a global economy.

Previously we defined and discussed two types of knowledge: *explicit* and *tacit*. In this section, we now group knowledge into four categories—(a) product-embodied, (b) process, (c) systems integration, and (d) management knowledge. *Product-embodied* knowledge resides within a physical artifact. This category of knowledge tends to lose significance as a competitive factor once it has been incorporated in an airframe or engine and bought by a competitor. Product-embodied knowledge can be protected up to a point as intellectual property by patent, copyright, or licensing. Securing product-embodied knowledge by whatever means is a strategy often used by competitors to obtain technology (Frias, 1995, p. B28). The specific technologies incorporated into the Boeing 777 (e.g., composites and avionics) exemplify *product-embodied* knowledge.

Process knowledge refers to the "how" or "know-how" of RD&P and is considered crucial for achieving and sustaining a national competitive advantage. It combines explicit knowledge, which is derived from science and the application of the scientific method, with tacit knowledge, which is derived from practice and experience, as well as from learning-by-doing. Process knowledge is cumulative to an organization; it is not easily communicated to those outside the organization. It is critical to product development, market share, and the overall competitiveness of a firm. The RD&P of the new B-777 blade tail cone and other specific B-777 technologies represents *process* knowledge.

Systems integration knowledge refers to the ability to incorporate efficiently and effectively exotic metals, "smart materials," and extraordinarily complex and sophisticated products, components, and technologies to manufacture final products which are, by an order of magnitude, even more complicated and sophisticated, for example, large commercial aircraft. The following key technologies must be fully integrated in the assembly of large commercial aircraft: design techniques (i.e., advanced computational analysis), aerodynamics (i.e., active boundary layer management), flight controls (i.e., relaxed stability and active controls), advanced structures (i.e., advanced metallic alloys and metal bonding techniques), propulsion integration (i.e., integrated engines and nacelles), and avionics (i.e., computer-integrated flight management). Leadership in one or several of these technologies does not guarantee success in the marketplace. The ability of U.S. manufacturers, dating back to the early experiences of building military aircraft in World War II, to integrate these technologies consistently over time has been responsible in large part for their competitive success in the large commercial aircraft market. The assembly coordination for a Boeing 777 typifies *systems integration* knowledge.

Management knowledge can be discipline-specific (i.e., organized according to or around subject-matter knowledge) but has more to do with how a firm or company organizes itself (i.e., division of labor and responsibility) to develop and manufacture a product, undertake a project, or to complete a program. In terms of competitiveness, management knowledge is often considered less important in a collaborative or joint production than is product, process, or systems integration knowledge. Nevertheless, companies with a proven management track record are frequently sought out as partners in collaborative or joint production arrangements. From a policy perspective, product-embodied, process, and systems integration knowledge are more likely than management knowledge to be protected (i.e., controlled) by states and firms. Boeing's ability to manage complicated teaming and testing production processes from an international group of subcontractors and vendors reflects *management* knowledge.

The diffusion of knowledge can be promoted or minimized in collaborative ventures, joint productions, and multinational consortia. Promoting or minimizing the diffusion of knowledge in these agreements predictably results in two schools of thought. Those who favor promoting the diffusion of knowledge in these agreements contend that controls should be avoided except in clearly defined and unusual circumstances (Hood, 1995; Julius, 1990, pp. 8-9, 98) on the grounds of three arguments. *First,* the diffusion of knowledge will not necessarily undermine the competitive position of a U.S.

company or firm that participates in these arrangements. Indeed, many corporations continue to develop commercially viable products at the leading edge of technological advancement while licensing current designs, ensuring a continuous flow of production technology, and generating royalties and fees to fund further research and development efforts. Sharing knowledge and technology can be an incentive to continue to innovate. This position is supported by recipient complaints that the knowledge being diffused, while valuable, is not the cutting edge technology associated with design work (Hartley, 1983, pp. 132-133, 144; "Japanese Technology...," 1989, pp. 3-5; "Zeroing In," 1989, pp. 15-16) but may be used to create technological dependence (O'Brien, 1983). States and firms share a universal concern about becoming too dependent upon foreign sources of knowledge and technology and about the corresponding export of capital in terms of royalty payments.

Second, because it is impossible to contain diffusion effectively, advocates suggest that the diffusion of knowledge could be used to help develop and solidify working relations with potential and current foreign production partners. Sharing knowledge could be a confidence-building measure that generates a critical ingredient for stable international joint ventures—trust (Perlmutter and Heenan, 1986, pp. 136-152; Gray, 1989). *Third*, reciprocity could help ensure access to critical knowledge (Mowery and Rosenberg, 1989b). Although the U.S. maintains a strong knowledge base in large commercial aircraft, this position is not overwhelming. Airbus Industrie, for example, also has an excellent knowledge base. Its technologies are equally sophisticated. Indeed,

> civil aerospace is an area of advanced technology where, on technical merit alone, there has never been a "gap" between Europe and the United States...in many respects, Airbus Industrie believes that it has...wrested the technological initiative from the Americans. (Hayward, 1986, p. 200)

According to a McDonnell Douglas official, "technology is not too much of a problem in the commercial transport business. Often we find that other countries have as good or better technology than we have" (Williams, 1991, interview).

Those who advocate minimizing the diffusion of knowledge in these agreements present the following arguments. *First*, nations other than the U.S. tend to view competitiveness in one of two ways when building a market position (Cravens, Downey, and Lauritano, 1992). They can foster close cooperation between government and industry. Japan has successfully used this strategy of competitiveness to enter and develop several industries, one being the aerospace sector (Samuels and Whipple, 1989). Nations can develop multicountry and company alliances, thereby gaining access to subsidy support from larger overall pool of government money. Europe has successfully used this strategy to gain a strong position in the large commercial aircraft market via Airbus Industrie (Tedesco, 1993). In both cases, product-embodied, process, and systems integration knowledge are aggressively sought as a means of increasing nations' technology bases not so much for their use in aerospace as for their "value added" application or "spin-on" to other industries and markets (Frias, 1995).

Second, in spite of existing controls (i.e., bilateral accords, licensing agreements, and patents) on knowledge and technology transfer, protection cannot be guaranteed. Product-embodied, process, and systems integration knowledge, acquired through multi-national joint programs, can quickly be "spun-on" to other sectors or lost to a third-party firm or state. Under certain conditions, a knowledge "advantage" can mean the difference between a firm's staying in or going out of business. Loss of knowledge can be turned into a large (competitive) lead over time by another firm or nation (Krasner, 1990; Tucker, 1991, pp. 83-120). Certainly history demonstrates that knowledge can be exploited by firms and states that did not innovate the advances, as was the case with the jet engine. History also suggests that even the most fiercely protected product secrets—from nuclear technology to Coca Cola syrup—cannot be protected forever. Policymakers will have to balance potential knowledge and technology gains and losses associated with collaborative agreements before committing to any agreements.

State policymakers confront a serious dilemma with respect to the diffusion of knowledge as they consider the merits of international collaborative production agreements. On the one hand, virtually every firm and state recognizes that some form of international production, complete with its attendant diffusion of knowledge, is probably necessary to stay competitive in high technology industries, such as large commercial aircraft manufacturing. Once firms and nations involved in collaborative agreements have experienced initial success, retrenchment to a more "purely" domestic production structure may be difficult because of the political and economic incentives impelling cooperation. Once a successful collaboration structure is in place, the participants become another powerful interest group capable of influencing commercial and political judgment. Fuqua (1990) noted, "cooperation between manufacturers from different nations might, in fact, be the key to our survival and prevent losing further ground to foreign competitors" (p. 3). On the other hand, sustaining some level of control over the diffusion of critical knowledge or core technologies in international production is crucial to the state for reasons of economic competitiveness and national security. In the end, the U.S. must carefully determine the potential consequences associated with knowledge diffusion when international agreements to build large commercial aircraft are negotiated.

CHAPTER REFERENCES

Aerospace Industries Association of America. (1994). *Aerospace Facts & Figures: 1994-1995*. Washington, DC: Aerospace Industries Association of America.

Bacher, T.J. (1984). "The Economics of the Commercial Aircraft Industry." Paper presented at the conference on *The Role of South East Asia in World Airlines and Aerospace Development* sponsored by the *Financial Times Limited* in Singapore, 24-25 September.

Benke, W. (1987). Personal correspondence (December 4) with Manager of International Business for The Boeing Commercial Airplane Group.

Bernstein, G.W. (1995). Personal correspondence (November 8) with Vice President, BACK Management Services.

Biddle, W. (1991). *Barons of the Sky*. New York, NY: Simon & Schuster.

Bluestone, B.; P. Jordan; and M. Sullivan. (1981). *Aircraft Industry Dynamics: An Analysis of Competition, Capital, and Labor*. Boston, MA: Auburn House.

"Boeing Plans Substantial Foreign Involvement in 7-7 Development." (1985) *Aviation Week & Space Technology* (June 3) 122(22): 211-212.

"Boeing Says 777 Fuel Use 5% Better Than Promised." (1994) *Aviation Week & Space Technology* (October 10) 141(15): 34.

"Boeing to Increase Suppliers Workshare." (1995) *Aviation Week & Space Technology* (July 24) 142(28): 33.

Bright, C.D. (1978). *The Jet Makers: The Aerospace Industry from 1945 to 1972*. Lawrence, KS: The Regents Press of Kansas.

Cantor, D.J. (1992). "Aircraft Production and the U.S. Economy." In *Airbus Industrie: An Economic and Trade Perspective*. J.W. Fischer, Coordinator. CRS Report to Congress 92-166 E. Washington, DC: The Library of Congress, Congressional Research Service.

Carroll, S.L. (1975). "The Market for Commercial Aircraft." Chapter 8 in *Regulating the Product*. R.E. Caves and M.J. Roberts, eds. Cambridge, MA: Ballinger, 145-170.

Clarkson, L.W. (1992). "Challenges and Opportunities: Civil Aviation in the Asia Pacific Region." Paper presented at the Kyushu International Aerospace Exhibition, Kyushu, Japan, on January 23.

Cole, R. (1995). "The Boeing 777: From Virtual to Airborne." *UniForum Monthly*. (December) 15(12): 20-23.

Constant, E.W. II. (1980). *The Origins of the Turbojet Revolution*. Baltimore, MD: Johns Hopkins University Press.

Cravens, D.W.; H.K. Downey; and P. Lauritano. (1992). "Global Competition in the Commercial Aircraft Industry: Positioning for Advantage by the Triad Nations." *Columbia Journal of World Business* (Winter) 26(4): 46-58.

Cunningham, W.G. (1951). *The Aircraft Industry: A Study in Industrial Location*. Los Angeles, CA: L.L. Morrison.

Davies, R.E.G. (1964). *A History of the World's Airlines*. London, UK: Oxford University Press.

Dornhem, M.A. (1994). "777 Design Has New Elements." *Aviation Week & Space Technology* (April 11) 140(16): 211-212.

Freudenthal, E.E. (1940). *The Aviation Business*. New York, NY: Vanguard Press.

Frias, M.A. (1995). "Meeting the Growing Foreign Competition." *Aerospace America* (November) 33(11): B28.

Fuqua, D. (1989). "The FSX: Looking Backward, Look Forward." *Aerospace Industries Association Newsletter* (October) 2(4): 3.

Golich, V.L. (1992). "From Competition to Collaboration: The Challenge of Commercial-Class Aircraft Manufacturing." *International Organization* (Autumn) 46(4): 900-934.

Golich, V.L. (1989). *The Political Economy of International Air Safety: Design for Disaster?* London, UK: Macmillan.

Gray, B. (1989). *Collaborating: Finding Common Ground for Multiparty Problems*. San Francisco, CA: Jossey-Bass Publishers.

Hanle, P.A. (1982). *Bringing Aerodynamics to America*. Cambridge, MA: MIT Press.

Harr, K.G. (1972). "Statement." *Industry Week* (17 January), n.p.

Hartley, K. (1983). *NATO Arms Cooperation: A Study in Economics and Politics*. London, UK: George Allen and Unwin, Ltd.

Hayward, K. (1986). *International Collaboration in Civil Aerospace*. New York, NY: St. Martin's Press.

Hayward, K. (1983). *Government and British Civil Aerospace*. Manchester, UK: Manchester University Press.

Hochmuth, M.S. (1974). "Aerospace." Chapter 8 in *Big Business and the State*. R. Vernon, ed. Cambridge, MA: Harvard University Press, 145-169.

Hood, R. (1995). Interview (December 7) with the Acting Director, Subsonic Transportation Division, Aeronautics and Space Technology, Code R, National Aeronautics and Space Administration, Washington, DC.

Hughes, D. (1995). "Fly-by-Wire 777 Keeps Traditional Cockpit." *Aviation Week & Space Technology* (May 1) 142(18): 42-48.

"Japanese Technology Survey: Back to the Drawing Board." *The Economist* (December 2): 3-5.

Julius, D. (1990). *Global Companies and Public Policy: The Growing Challenge of Foreign Direct Investment*. New York, NY: Council on Foreign Relations Press.

Kelly, K.; M. Oneal; G. DeGeorge; and T. Vogel. (1991). "All the Trouble Isn't in the Sky." *Business Week* (March 11) 3203: 84-85.

Keohane, R. (1984). *After Hegemony: Cooperation and Discord in the World Political Economy*. Princeton, NJ: Princeton University Press.

Keohane, R. and Nye, J. (1977). *Power and Interdependence: World Politics in Transition*. Boston, MA: Little, Brown, and Co.

Krasner, S.D. (1990). Personal correspondence dated August 6th.

Lederer, J. (1987). Personal interview.

Lilley, T. (1947). *Problems of Accelerating Aircraft Production During World War II*. Cambridge, MA: Division of Research, Business School, Harvard University.

Lopez, V.C. and L. Yager. (1988). *The U.S. Aerospace Industry and the Trend Toward Internationalization*. Washington, DC: Aerospace Industries Association of America, Inc.

Lopez, V.C. and L. Yager. (1987). "An Aerospace Profile: The Industry's Role in the Economy—The Importance of R&D." *AIA Facts and Perspectives* (April): 1-4.

Majumdar, B.A. (1987). "Upstart or Flying Start? The Rise of Airbus Industrie." *The World Economy* 10(4): 497-518.

March, A. (1989). "The US Commercial Aircraft Industry and its Foreign Competitors." Working Paper Prepared for the Commission Working Group on the Aircraft Industry, MIT Commission on Industrial Productivity.

Miller, R. and D. Sawers. (1968). *The Technical Development of Modern Aviation*. London, UK: Routledge and Kegan Paul.

Mingos, H. (1930). *Birth of an Industry*. New York, NY: W.B. Conkey.

Mowery, D.C. (1987). *Alliance Politics and Economics: Multinational Joint Ventures in Commercial Aircraft*. Cambridge, MA: Ballinger.

Mowery, D.C. and N. Rosenberg. (1989a). "The U.S. Commercial Aircraft Industry." Chapter 5 in *Technology and the Pursuit of Economic Growth*. New York, NY: Cambridge University Press, 169-202.

Mowery, D.C. and N. Rosenberg. (1989b). "New Developments in U.S. Technology Policy: Implications for Competitiveness and International Trade Policy." *California Management Review* (Fall) 32: 107-124.

Mowery, D.C. and N. Rosenberg. (1982). "The Commercial Aircraft Industry." Chapter 3 in *Government and Technical Progress: A Cross-Industry Analysis*. R.R. Nelson, ed. New York, NY: Pergamon Press, 101-161.

Napier, D. (1996). "1995 Year-End Review and Forecast." Washington, DC: Aerospace Industries Association of America.

National Academy of Engineering. (1988). *The Technological Dimensions of International Competitiveness: A Report to the Council of the National Academy of Engineering*. Washington, DC: National Academy Press. (In OCLC: 17949151.)

National Academy of Engineering. (1985). *The Competitive Status of the U.S. Civil Aviation Manufacturing Industry: A Study of the Influences of Technology in Determining International Industrial Competitive Advantage*. Washington, DC: National Academy Press. (Available NTIS: PB88-100334.)

Neuman, S.G. (1984). "International Stratification and Third World Military Industries." *International Organization* (Winter) 38: 167-197.

Newhouse, J. (1982). *The Sporty Game: The High-Risk Competitive Business of Making and Selling Commercial Airliners*. New York, NY: Alfred A. Knopf.

Nordwall, B.D. (1995). "Honeywell's Via 2000 Draws on 777 AIMS." *Aviation Week & Space Technology* (August 7) 143(6): 43-45.

O'Brien, R.C. (ed.). (1983). *Information, Economics and Power: The North-South Dimension*. Boulder, CO: Westview Press.

Perlmutter, H.V. and D.A. Heenan. (1986). "Cooperate to Compete Globally." *Harvard Business Review* (March-April) 86(2): 136-137, 142, 146, 150, 152.

Pinelli, T.E.; J.M. Kennedy; and R.O. Barclay. (1991). "The NASA/DoD Aerospace Knowledge Diffusion Research Project." *Government Information Quarterly* 8(2): 219-233.

Prestowitz, C.V. (1988). *Trading Places: How We Are Giving Our Future to Japan and How to Reclaim It*. New York, NY: Basic Books.

Proctor, P. (1995a). "Early Modeling Helps Speed 777 Flight Tests." *Aviation Week & Space Technology* (June 12) 142(24): 124.

Proctor, P. (1995b). "Certification Tests Continue on GE and Rolls-Powered 777s." *Aviation Week & Space Technology* (May 1) 142(18): 20.

Proctor, P. (1994a). "Test Program Gets Quick Start." *Aviation Week & Space Technology* (June 20) 140(25): 20-21.

Proctor, P. (1994b). "Boeing Rolls Out 777 to Tentative Market." *Aviation Week & Space Technology* (April 11) 140(15): 36-58.

"PW4084-Powered 777 Wins ETOPS Certification." (1995). *Aviation Week & Space Technology* (June 5) 142(23): 31.

Rae, J.B. (1968). *Climb to Greatness: The American Aircraft Industry 1920-1960*. Cambridge, MA: MIT Press.

Rogers, E.M. (1983). *Diffusion of Innovations*. 3rd ed. New York, NY: Free Press.

Rosecrance, R. (1981). "International Theory Revisited." *International Organization* (Autumn) 35(4): 691-713.

Samuels, R.J. and B. C. Whipple. (1989). "Defense Production and Industrial Development: The Case of Japanese Aircraft." Chapter 7 in *Politics and Productivity: The Real Story of Why Japan Works*. C. Johnson, L.D. Tyson, and J. Zysman, eds. Cambridge, MA: Ballinger Publishing, 275-318.

Schaufele, R.D. (1988). Interview (July 7) with the Vice President, Commercial Advanced Products, Douglas Aircraft Company.

Shifrin, C.A. and P. Sparaco. (1995). "Boeing to Build 777-300x for Asia/Pacific Carriers." *Aviation Week & Space Technology* (June 19) 142(25): 31-32.

Simonson, G.R. (ed.). (1968). *The History of the American Aircraft Industry: An Anthology*. Cambridge, MA: MIT Press.

Solberg, C. (1979). *Conquest of the Skies: A History of Commercial Aviation in America*. Boston, MA: Little, Brown and Co.

Stekler, H.O. (1965). *The Structure and Performance of the Aerospace Industry*. Berkeley, CA: University of California Press.

Taneja, N.K. (1976). *The Commercial Airline Industry: Managerial Practices and Regulatory Policies*. Lexington, MA: D.C. Heath.

Tedesco, T. (1991). "International Aviation: Not a Level Playing Field." *Transportation Quarterly* (July) 47(3): 397-412.

Tucker, J.B. (1991). "Partners and Rivals: A Model of International Collaboration in Advanced Technology." *International Organization* (Winter) 45(1): 83-120.

Tyson, L.D. (1988). "Competitiveness: An Analysis of the Problem and a Perspective on Future Policy." Chapter 3 in *Global Competitiveness: Getting the U.S. Back on Track*. M.K. Starr, ed. New York, NY: W.W. Norton, 95-120.

"United Launches 777 Commercial Service." (1995). *Aviation Week & Space Technology* (June 12) 142(24): 50.

U.S. Department of Commerce, International Trade Administration. (1994). *U.S. Industrial Outlook 1994*. 35th ed. Washington, DC: U.S. Department of Commerce. Washington, DC: U.S. Government Printing Office.

U.S. Executive Office of the President, Office of Science and Technology Policy. (1985). *National Aeronautical R and D Goals: Technology for America's Future*. Washington, DC: Office of Science and Technology Policy. (Available NTIS; 87N-12405.)

U.S. International Trade Commission. (1993). *Global Competitiveness of U.S. Advanced-Technology Manufacturing Industries: Large Civil Aircraft.* Washington, DC: U.S. International Trade Commission.

Vander Meulen, J.A. (1991). *The Politics of Aircraft: Building an American Military Industry.* Lawrence, KS: University Press of Kansas.

van Tulder, R. and G. Junne. (1988). *European Multinationals in Core Technologies.* New York, NY: John Wiley.

Williams, D.L. (1991). Interview (June 4) with the Director of Program Development and Marketing, McDonnell Douglas.

"Zeroing In." *The Economist.* (December 2): 15-16.

Chapter 2

The Influence of U.S. Public Policy on Large Commercial Aircraft—Innovation, Transportation, and Knowledge Diffusion

Vicki L. Golich
Thomas E. Pinelli

SUMMARY

Chapter 2 explores how U.S. public policy influences the production and operation of large commercial aircraft (LCA). As background, we describe the expanding role of the U.S. government in the marketplace. Next, we examine how public policy has shaped the innovation and transportation systems within which LCA are produced and operated. We then present a sampling of U.S. policies—some targeted at aviation, others directed toward seemingly unrelated national objectives—that have precipitated changes—some intentional, some unintentional—in LCA. The chapter closes with a set of observations about U.S. public policy and knowledge diffusion that derive from our analysis.

INTRODUCTION

On February 14, 1996, the National Aeronautic Association announced that the Boeing Commercial Airplane Group had won yet another Collier Trophy, this time for its B-777. The twinjet joins the B-757, B-767, and the B-747 as the only commercial aircraft to receive this coveted award. Established in 1911, the trophy is conferred each year in recognition of the greatest achievement in U.S. aeronautics and astronautics in the preceding year. Boeing was honored specifically for "designing, manufacturing, and introducing into service the world's most advanced commercial airplane" (National Aeronautic Association, 1996, n.p.). The Aerospace Industries Association (AIA) nominated the Boeing team, declaring the B-777 "a landmark in commercial aircraft development by virtue of its significantly advanced performance, efficiency, safety, and environmental acceptability" (National Aeronautic Association, 1996, n.p.). The award is testimony to the remarkable success of Boeing's revolutionary production, organization, utilization, and management of knowledge discussed throughout the book. Upon hearing of the latest tribute, Boeing chairman and chief executive officer, Frank Shrontz, comments:

> The 777 is like no other airplane in commercial aviation history. It was created through a team effort that we call "Working To-gether," involving customers, Boeing, program partners, suppliers and regulatory authorities. This approach has not only resulted in an amazing airplane, but has set the standard for managing new airplane programs. (National Aeronautic Association, 1996, p. 1)

The award is also testimony to the unique relationship between the U.S. government and LCA manufacturing. The National Advisory Committee for Aeronautics (NACA)—subsequently incorporated into the National Aeronautics and Space Administration (NASA)—has funded and conducted research on airframe and propulsion technologies since 1917. Table 2.1 lists six U.S. commercial aircraft and the NACA/NASA research and technology (R&T) incorporated in each. Transonic supercritical airfoils, advanced metallic alloys and composite materials, digital flight controls, glass cockpit instrument displays, flight management systems, laminar flow control concepts, fatigue and fracture methodology, advanced composite structures design technology, and structural integrity and analysis methods are some of the NASA aeronautics technologies incorporated in the B-777. In a May 10, 1996, speech at the NASA Langley Research Center (LaRC), Robert Spitzer (1996), Boeing's Vice President for Engineering, Commercial Airplane Group, acknowledged the value of this government-industry partnership. He notes:

> The LaRC is the aeronautical birthplace of working together.... Your relationships with the other government research centers and academia and industry have been quietly showing the way. The relevance and correctness of your working together is evidenced by aviation's incredible growth. Your working together success is evident in every U.S.-made aircraft.

Table 2.1. U.S. Commercial Aircraft Incorporating NACA/NASA Research and Technology (R&T)

Aircraft	NACA/NASA R&T Incorporated
Ford Tri-Motor	The Ford 4-AT/5-T was a popular transport airplane in the late 1920s and early 1930s. The corrugated aluminum skin construction and the use of three motors were among the Tri-Motor's most noticeable features. It used an adaptation of a wing contour and ring cowlings, developed by the NACA, around the engines to improve the airflow characteristics.
Douglas DC-3	The DC-3, built between 1935 and 1946, carried 21 passengers and brought new standards of comfort, reliability, safety, and lower operating costs to the airline industry. It used the low-drag engine cowling design, developed by the NACA.
Lockheed Constellation	The Constellation was the first large transport aircraft with a pressurized cabin and coast-to-coast range. The "Connie" first flew in January 1943 and could carry 91 passengers. The Connie marked the end of piston engine airliners. This airplane used the NACA's aerodynamic drag reduction experimental research results and its low-drag engine cowling design.
Boeing 747	The B-747 was the first wide-body jet airliner, remains the largest ever built for commercial use, and first flew in February 1969. It seats 420 passengers and has a flight range of 13 390 km (8320 miles). NASA research in high by-pass jet engines, low-drag nacelles, swept-wing, airfoils, noise reduction, transonic aerodynamics, and structural research was used in its design.
McDonnell-Douglas MD-11	The MD-11 has a maximum range of 13 355 km (8300 miles) and typically seats 323 passengers. Its first flight for airline service was in 1992. NASA's research in winglet design, supercritical airfoils, digital electronic controls, numerous engine design improvements, high-lift systems, and transonic aerodynamics and structural concepts contributed to its design.
Boeing 777	The B-777 is the first aircraft designed completely by computer technology. United Airlines was the first to fly a B-777 in May 1995. It seats 305 passengers and has a maximum range of 13 667 km (8493 miles). The B-777 benefitted from several NASA research efforts including digital flight controls, the glass cockpit, quiet engine nacelles, aerodynamic design codes, flight management systems, graphite–epoxy structures, and transonic supercritical airfoils.

In testimony before the House Subcommittee on Technology, Environment, and Aviation, AIA civil aviation vice president Bob Robeson agreed. He noted that "NASA is an important partner with industry in precompetitive research and development" (R&D) and urged Congress to continue to fund high-risk, high-return technologies that industry could not afford to undertake on its own. He observed that the availability of affordable, advanced technologies in aerodynamics, materials, computing, and manufacturing allows the industry to compete internationally (Ravitz, 1994, p. 4).

Like every other successful LCA, the B-777 moves "several hundred people cost-effectively and safely from one part of the world to another" (Sabbagh, 1996, p. 18) and represents a compromise of technological, economic, and political feasibility. Like all U.S. LCA, the production and operation of B-777 are influenced by an array of public policies. On the production side, NACA/NASA aeronautical R&T, as well as considerable Department of Defense (DoD) aeronautical R&D and procurement, have benefitted the U.S. aircraft industry (National Science and Technology Council, 1995).

Operationally, airlines conduct business within a national transportation system largely defined by the U.S. government. Legislation passed in the 1920s and 1930s authorized the government to subsidize airlines for carrying the mail, to regulate air carrier fares and route structures to eliminate excessive competition, and to establish aircraft safety and personnel performance standards. Today, the government manages the air traffic control (ATC) system, enforces airport security, and establishes standards that it uses to certify aircraft as safe for flight and air carrier employees—pilots, flight attendants, and mechanics—as competent to perform their aviation-related jobs. Additionally, airlines, like every other U.S. industry, must comply with a myriad of regulations affecting the "general welfare" of U.S. citizens, including those related to smoking, food safety, marketing and sales, and environmental and labor protection.

Globally, the principle of sovereignty dictates that the U.S. government act as the gatekeeper of foreign commerce. Bilateral Air Transport Agreements (BATAs) negotiated by the U.S. Department of Transportation (DoT) and the U.S. Department of State (DoS) govern international flights and specify routes, frequencies, capacity, airport gateways, and cabotage; the latter restricts domestic travel by U.S. citizens to U.S.-owned airlines, thus protecting the

U.S. market—the largest absolute domestic market, accounting for 40% of all world air transport passengers—for American carriers. Although frequently influenced by foreign policy concerns quite distinct from aviation or commerce (Gidwitz, 1980; Golich, 1989), these agreements affect demand for the type and number of aircraft because speed, range, and payload capacity parameters are embedded in the negotiated air transport services.

Chapter 2 explores how U.S. public policy has influenced technological innovation in LCA. As background, we note that the U.S. government has always intervened in the domestic economy. Next, we briefly sketch the role played by public policy in shaping the U.S. innovation and transportation systems within which LCA are produced and operated. We then describe several U.S. public policies—some targeted at aviation, others directed toward seemingly unrelated national objectives—that have precipitated changes in LCA. Sometimes, the effects were intentional, sometimes unintentional. We close with a set of observations about U.S. public policy and knowledge diffusion—as distinct from knowledge creation—that derive from our analysis.

BACKGROUND

United States policymakers ultimately derive their authority to intervene in the market from the U.S. Constitution, which specifies tasks for each branch of government—executive, legislative, and judicial. Its justifications for government action appear to be straightforward: defending the nation, maintaining the infrastructure for the domestic economy, and providing for the general welfare of U.S. citizens. Despite this seemingly simple starting point, the role government plays in shaping economic activity today is much broader in scope and more complicated than the nation's founders envisioned in 1789. The following overview sketches the growing reach of government into the economy with a particular focus on how U.S. public policy came to influence technological innovation in LCA research, development, and production (RD&P).

Following their experience with the Articles of Confederation, U.S. policymakers reluctantly acquiesced in 1789 to the establishment of a relatively strong central government. In addition to providing for national defense, the founders recognized that an efficient market required an economic infrastructure complete with uniform laws and regulations to govern key market components, including currency valuation and exchange rates; standards, weights, and measures; interstate and foreign commerce; and intellectual property rights protection. As they drafted specific tasks, with most of the requisite authority reserved for Congress by Section 8, the framers were influenced by the economic liberalism propounded by Adam Smith. In *The Wealth of Nations*, Smith (1776) argued that government should provide security from internal and external attack and build an infrastructure to facilitate trade, in particular, communications and transportation systems (see Kindleberger, 1978, p. 2).

From this narrowly defined beginning, the federal government gradually expanded its reach into the economy with a series of regulatory policies consciously designed to encourage industrialization, unite the country, and ensure the domestic competition and corporate fair play deemed critical to the evolving capitalist system. A common set of regulations, including tariffs, governing trade with foreign countries was instituted during the first half of the nineteenth century to protect U.S. infant industries from competition with British and European manufacturers offering more advanced products (Tickner, 1987, pp. 112-113). Government grants-in-aid of vacant land, parceled out between 1850 and 1871, encouraged railroads to build a transcontinental transportation system. The promise of free land offered by the Homestead Act of 1862 (12 Stat. 392) persuaded many Americans to settle and farm throughout the western United States. By the end of the century, firms in such key industries as oil, steel, and the railroads had gained control of very large market shares, which allowed them to extract monopoly rents. To redress the anticompetitive effects of this economic structure, Congress established the Interstate Commerce Commission (ICC) in 1887 to administer a set of general and flexible guidelines that would regulate those "industries in which single companies could grow so large that they spread across several states" (McCraw, 1984, pp. 60-61; see also Cooper, 1996, p. A5), and passed the Sherman Act of 1890 (26 Stat. 209) (Hart, 1992, p. 231).

Legislation to provide for the general welfare of U.S. citizens initially sought to protect labor from unsafe or inhumane working conditions. Some industrialized states had enacted such legislation by the late 1800s, but most U.S. labor laws were not adopted until after the Great Depression of 1932. The Sherman Act had been used most effectively against labor unions, not corporate monopolies. The National Labor Relations Act of 1935 (49 Stat. 449) finally protected worker rights to collective bargaining, although it was subsequently weakened by the Labor Relations Act of 1947 (61 Stat. 136, Sect. 101). Most other legislation targeted at providing for the general welfare was not enacted until the 1960s and 1970s.

Policies adopted to strengthen national defense capacities dramatically affected the commercial sector. For example, to meet the demand created by Civil War government military procurement policies, manufacturers initiated product standardization that led to mass production. The Morrill Act of 1862 (12 Stat. 503), 1883 (22 Stat. 484), and 1890 (26 Stat. 417) established "land-grant" colleges to advance "agriculture and the mechanical arts" ensuring sufficient domestic agricultural production and preparing engineers to design and produce military *matériel*, transportation (i.e., bridges, roads, and canals), and communication systems (see Kerr, 1987). Military procurement and education remain among the more popular policies used to achieve multiple national objectives.

By the dawn of the twentieth century, the federal government's role in the economy was well-established. Precipitated by security crises, government decisionmakers adopted policies designed to enhance the country's military preparedness. Perceived economic crises triggered policies to protect U.S. industries against foreign competition while encouraging domestic competition. Eventually, labor and environmental protection laws regulated corporate production processes. The next section examines the influence of U.S. public policy on LCA manufacturing and air transportation service, two inextricably linked and strategic sectors of the U.S. economy, which operate in two critical national systems—innovation and transportation.

U.S. PUBLIC POLICY–INFLUENCE ON LCA
INNOVATION AND AIR TRANSPORTATION

In 1903, when the Wright brothers demonstrated that heavier-than-air flight was possible, government intervention in the U.S. economy was still fairly limited. Hence, the Wright brothers' initial requests for federal support to advance aviation were rejected. Once the military and economic values of the airplane were recognized, however, the government aggressively intervened in its development. Today, governmental interest in the production and operation of LCA is based, in part, on the following factors: (a) LCA manufacturing capability is a vital part of the nation's industrial base and is critical to national security; (b) LCA sales generate significant revenues and produce a positive balance of trade, making the sector a keystone in a healthy domestic economy; (c) tourism, which depends on LCA, consistently contributes about 6% of the U.S. gross domestic product; (d) it facilitates exceptionally efficient trade and commerce by moving millions of people and billions of dollars worth of goods to domestic and foreign markets; U.S. scheduled airlines carried 473 million passengers and cargo exceeding 13 million freight ton miles in 1992; and (e) from 1982–1992, airline employment grew by 64% to 540 000 people (National Commission to Ensure a Strong Competitive Airline Industry, 1993, p. 5).

U.S. public policies concerned with innovation and transportation encompass a larger set of institutions, activities, and objectives than LCA RD&P and air transportation. The U.S. has a lengthy history of enacting policies intended to spur innovation in agriculture, medicine, computers and microelectronics, biotechnology, and synthetic fuels. Transportation policies address all modes of travel, including highways, railroads, and waterways. Our interest in innovation and transportation is limited to the influence of U.S. public policy on commercial aviation as a system that includes both the manufacture and operation of LCA.

The U.S. National Innovation System

Three characteristics distinguish the U.S. innovation system. *First*, its scale is enormous; for a substantial portion of the postwar era, U.S. R&D investment exceeded that of all other Organization for Economic Cooperation and Development (OECD) nations combined. *Second*, the peculiar combination of antitrust policy with military R&D and procurement policies hastened the formation of such giant firms as DuPont and General Electric (GE); these early investors in industrial technology R&D were compelled to bring R&D "in house" because interfirm collaboration on single R&D projects was discouraged or prohibited. *Third*, despite its enormity and the key roles played by industry, academia, and the federal government as funders, performers, and purchasers of R&D, the U.S. national innovation system was neither consciously designed to accomplish an agreed on set of objectives, nor do these actors work together smoothly and coherently (Nelson and Rosenberg, 1993).

The federal government's involvement in the U.S. innovation system began with the Constitutional mandate to protect intellectual property rights with patents and to establish standard weights and measures; it expanded with the establishment of the "land-grant" colleges; it grew exponentially with the advent of World War II and the massive expenditure of funds to develop and purchase the requisite tools and *matériel* to wage war. In *Science—The Endless Frontier*, Vannevar Bush (1945) provided the intellectual rationale for postwar federal involvement in the national innovation system. He argued that federally funded science and technology conducted in partnership with academia and industry should be a catalyst for accomplishing such public policy goals as defending the nation, improving the health of its people, and fueling the economy through the creation of new industries.

The Aviation–Innovation Connection

The federal government first intervened to support aviation when its potential value to national defense was recognized just prior to World War I. Initially, heavier-than-air planes simply replaced hot air balloons as reconnaissance vehicles. Soon, guns were strapped to wings for use as weapons, and the link between aviation and national security was permanently cemented. The military first purchased aircraft in 1914, and, a year later, Congress created the NACA to "supervise and direct the scientific study of the problems of flight with a view to their practical solutions and to give advice to the military air services and other aviation services of government" (Pinelli, 1991, p. 5). Military and commercial flight needs were essentially identical in these early days—increased speed, range, and payload performance. Thus, the NACA's work on problems of aeronautics and aerodynamics paid dividends for both industry sectors. Retractable gear, variable pitch propellers, metal structures, more powerful engines, better instruments and radios, and improved aids to navigation all contributed to more efficient and safer commercial aircraft. (For more on the aviation–innovation connection, see National Academy of Engineering, 1985; U.S. Executive Office of the President, 1987; 1985; 1982a; 1982b.)

During the interwar years, aircraft became increasingly complex and expensive to design and build. Countries with successful aircraft firms and airlines were recognized for their educational systems, their economic resources, and their capacity to organize engineers, scientists, and other skilled workers into successful technology innovation and production social systems. This capacity carried with it an element of prestige, which, although it defied measurement, was considered critical to national status in world affairs (Golich, 1989; March, 1989; Vander Meulen, 1991). Due to the perceived strategic nature of aviation and the growing faith in the promise of science and technology after World War II, U.S. federal government involvement in the industry grew significantly (Bluestone, Jordan, and Sullivan, 1981; Golich, 1989, pp. 47-50).

By the 1970s, economic analyses demonstrated a strong positive correlation between R&D spending and long-term national economic welfare. The connection "between investments in R&D and increases in productivity, success in international economic competition, and new employment opportunities was widely acknowledged" (Logsdon, 1986, p. 13; March, 1989, p. 19; U.S. Congress, Office of Technology Assessment, 1986, p. 3). Technological innovation linkage externalities in aviation flow laterally into diverse industry sectors and downstream into air transportation. Increasingly vital to global as well as domestic trade, air transportation efficiency depends, in large part, on improved LCA performance. Hence, every stakeholder—government, LCA manufacturers, airlines, the traveling public, and ultimately all consumers—benefits from continued technological innovation. However, it is virtually impossible to predict "if, when, and where the results of any piece of fundamental research might find practical application" (Pavitt and Walker, 1976, p. 17). In addition, firms funding the research are seldom able to appropriate any resulting benefits because they are so quickly and so broadly distributed. In the end, high cost, long lead times, and uncertain payoffs dissuade risk-averse firms from pursuing costly technological innovation R&D (Keyworth, 1986, p. 27; Logsdon, 1986, pp. 13, 21; Mowery, 1983, p. 37; Nelson and Langlois, 1983, p. 815; U.S. Congress, Office of Technology Assessment, 1991, p. 344). Given "the magnitude of risk and the enormous staying power required until break even which distinguish commercial aviation from virtually any other industry... countries that elect to participate must expect to provide some form(s) of assistance to their commercial producers" (March, 1989, p. 19; see also Flanigan, 1996).

Primary Government Players in LCA Innovation

The primary federal entities involved in the innovation system that affects commercial aviation are NASA, DoD, and the Federal Aviation Administration (FAA). The aeronautics portion of NASA has the principal responsibility for conducting aeronautical research in the U.S. From 1920 to 1935, the "NACA financed the bulk of research in airframes and engines and made the first wind tunnels and testing facilities available to the industry" (Derian, 1990, p. 64). NASA maintains four aeronautical research centers and a variety of wind tunnels, test facilities, simulators, and flight research aircraft too expensive for any individual LCA manufacturer to build and operate. Federal laboratories, such as those operated by NASA, provide unique features seldom available in other extramural programs supporting academic and industrial R&D projects. They have

> relatively long term and stable funding of research programs; availability of unique facilities; full-time research opportunities without other distractions for staff scientists and engineers; closer links to the missions of their agencies; the ability to sustain programs for longer periods than those specified in the terms of a typical grant; and a capacity for rapid response to emergencies and sudden opportunities. (National Academy of Sciences, 1995, p. 17)

NASA engineers and scientists perform aeronautical R&T in either its Focused Programs or its R&T base. Focused Programs target specific aeronautical objectives such as high-speed technology; NASA researchers partner with LCA manufacturer engineers to conduct relevant research. R&T base work involves basic research in aerodynamics, structures, materials, and human factors affecting flight that is *not* specifically linked to a particular objective; this work is performed mostly in-house at NASA aeronautical research centers, though some limited assignments are contracted to universities and industry. The findings generated through R&T base work are distributed to the U.S. aircraft community through NASA technical reports; NASA-sponsored workshops, conferences, and symposia; professional and technical journals; and society-sponsored conferences. Dissemination of Focused Program research results is limited to the "circle of friends" who participated in the work (Hood, 1995). (For more on the role of NACA/NASA aeronautical R&T, see Anderton, 1980; Becker, 1980; Booz-Allen Applied Research, Inc., 1971a, 1971b; Fraser and Maggin, 1982;

Hansen, 1987; Hudson, 1972; Loftin, 1985; Mowery, 1985; Roland, 1985; Rosenberg, Thompson, and Belsley, 1978; Wiegel, 1995.)

DoD research has yielded indirect, but very important, innovative spillovers to the U.S. commercial aircraft sector, most notably in the areas of airframe development, propulsion, avionics, and flight control systems (March, 1989). The development of the first jet engine in the U.S. was financed entirely by DoD, reflecting "both the perceived military urgency of the project and the lack of interest in the development of such an engine expressed by commercial aircraft firms prior to 1940" (Mowery and Rosenberg, 1982). Boeing's first jet-powered LCA, the B-707, was a derivative of the Air Force's KC-135; Boeing spent roughly $180 million and the government approximately $2 billion on its RD&P (March, 1989; Mowery and Rosenberg, 1982; Rae, 1968). Turbofan engine research for Lockheed's military transport—the C-5A—led to the development of high-bypass-ratio engines, which were adapted for use on Boeing's 747, 757, 767, and 777 (Mowery, 1985). The B-747 began as an unsuccessful competitive bid for this airplane contract; Boeing persisted with its R&D and eventually built the B-747, still the largest jumbo jet in commercial service. Boeing's monopoly of this niche market has helped sustain its profitability during the inevitable, but persistently unpredictable, low points in the cycle of LCA production.

The Federal Aviation Act of 1958 as amended (P.L. 85-726) and the Aviation Safety Research Act of 1988 (P.L. 100-591) define the FAA's R&D role as primarily mission-oriented. Its key purpose is problem solving, not the development of fundamental knowledge or core technologies. FAA R&D also focuses on its provision, maintenance, and improvement of those systems, procedures, and facilities needed for safe and efficient navigation, ATC, airspace and landing area utilization, and for protection against terrorism (U.S. Congress, Office of Technology Assessment, 1994; U.S. Congress, House of Representatives, 1988a).

NASA and the FAA coordinate aviation research programs in "ATC-cockpit integration, human factors, severe weather, airworthiness, environmental issues, and program support" (U.S. Congress, Office of Technology Assessment, 1994, p. 48; U.S. Department of Transportation, Federal Aviation Administration, 1994, p. A-3). The FAA has field offices at NASA's Ames and Langley Research Centers to monitor joint programs, reduce duplication of effort, and help conserve scarce research resources. Typically, the FAA identifies its needs, and NASA determines the feasibility of providing the necessary support. Field office projects have dealt with simulation capabilities, human factors, windshear, microwave landing systems, and global positioning satellite navigation systems (U.S. Congress, Office of Technology Assessment, 1994; U.S. Department of Transportation, Federal Aviation Administration, 1994).

U.S. National Transportation System

The federal government's responsibility to create and maintain a transportation system is embedded in the Constitutional requirements to provide for the national defense and to regulate interstate and foreign commerce. In these early days, government's responsibility was limited to harbors, roads, and canals; eventually, it expanded to include not only new modes of transportation (e.g., trains, cars, and planes) but also the increasingly sophisticated infrastructure each required. The task of coordinating the vast network of roads, railroads, waterways and ocean shipping lanes, airways, and pipelines that comprise the national transportation system is staggering, in part because of the enormous geographic space involved. The job is rendered more difficult because the oversight of each transportation component was split among 35 different departments, agencies, and programs until 1966, when the DoT was established. Much like the U.S. innovation system, the national transportation system was not consciously designed, nor do the primary federal institutions responsible for it work together smoothly and coherently.

The air transportation system is a vital national resource, consisting of air carriers, airports, air routes, ATC, and the LCA that carry passengers and cargo. To provide the infrastructure, the U.S. government controls the use of navigable air space; installs, operates, and maintains air navigation facilities; develops and operates ATC and navigation systems; and develops and implements regulations to control aircraft noise and other environmental effects of civil aviation. (For more on the aviation portion of the national transportation system see Bailey, 1986; Brenner, 1988; Dempsey and Goetz, 1992; Gidwitz, 1980; Kahn, 1988a, 1988b; Kasper, 1988; Kent, 1980; Komons, 1978; National Commission to Ensure a Strong Competitive Airline Industry, 1993; Rochester, 1976; Schwartz,

1973; Taneja, 1976; Thayer, 1965; Thornton, 1970; U.S. Department of Transportation, Federal Aviation Administration, 1995.)

The Aviation–Transportation Connection

Aviation's first contribution to the U.S. economy was quick and reliable mail delivery. A U.S. airmail route operated by government personnel was initiated on May 15, 1918; soon mail transport was expanded to include other goods as well as people. Policymakers, impressed that aviation served the dual goals of providing national security and an economic infrastructure, passed legislation to encourage the industry through military procurement and airmail subsidies, including the Air Mail Acts of 1925 (43 Stat. 960) and 1930 (46 Stat. 259), and the Air Commerce Act of 1926 (44 Stat. 568). In addition, the Department of Commerce (DoC) was directed to respond to public concern over safety resulting from a disconcerting rate of fatal crashes and to standardize an irregular and high fare structure perceived to be an additional obstacle to air travel. The DoC created the Bureau of Air Commerce to "set up federal licensing standards for planes and pilots," and to "establish and maintain flyways" just as the government "maintained waterways and highways" (Solberg, 1979, p. 64; see also Davies, 1964; Golich, 1989; McCraw, 1984).

Nearly a decade later, the Civil Aeronautics Safety Regulation Act of 1938 (52 Stat. 973, Sect. 601) reinforced the federal government's obligation to regulate air transport, administer ATC, certify aircraft, and license airline personnel according to strict safety standards, and provide an airways and airports infrastructure. The Civil Aeronautics Board (CAB) supplanted the DoC regulatory structure. By virtually guaranteeing route and fare structures, the CAB protected U.S. air carriers from excessive foreign and domestic competition. Airline managers knew that any increased cost in delivering air transport services would be covered by an increase in fares. Both aircraft manufacturers and airlines benefitted: manufacturers knew airlines could afford new generation aircraft, and airlines—competing primarily on the basis of service quality—constantly demanded new generation aircraft complete with the latest that technological innovation could offer in the way of increased performance (Bluestone, Jordan, and Sullivan, 1981; Golich, 1989; McCraw, 1984; National Academy of Engineering, 1985; Newhouse, 1982; Schwartz, 1973).

Nearly two decades later another disturbing string of air accidents led to the 1958 Federal Aviation Act, which limited the CAB
to airline economic regulation and created the Federal Aviation
Agency to promote aviation's economic development, foster air
safety, and investigate air accidents. Eventually, conflict of interest
concerns inherent in the Agency's dual responsibility to promote
aviation and ensure safety led to efforts to separate these functions.
In 1967, the Federal Aviation Agency was transferred to DoT and renamed the Federal Aviation Administration (FAA). The FAA Administrator was given authority to regulate aviation safety and the
FAA's promotional duties found in Section 305 of its charter were
specifically not among the enumerated portions delegated to the
FAA; however, the DoT Secretary immediately delegated the responsibility to foster air commerce back to the FAA (Golich, 1989, pp.
26-27; U.S. Congress, House Committee on Government Operations, 1966; U.S. Congress, House of Representatives, 1980, pp. 89-
90; U.S. Congress, Office of Technology Assessment, 1991). In
1975, the National Transportation Safety Board (NTSB) was created
as an independent agency to investigate all transportation accidents—including aviation—and to determine their cause(s).

A few years later, the trend toward liberalization reached the air
transport sector. Studies by academic and government economists
compared the fares of economically unregulated *intra*state flights to
those of *inter*state flights that were similar in stage length but were
subject to federal economic regulation (Bailey, 1986; Derthick and
Quirck, 1985; Kahn, 1988a, 1988b; Levine, 1981, 1987a, 1987b;
Wilkins, 1984). They determined that economically deregulated airlines could drastically reduce service costs, yielding significant
benefits for a broad range of industries as a result of increased
demand for air transport: Rising airline profits would finance new
aircraft purchases; upstream industries that supplied airlines and
aircraft manufacturers would benefit as they met the new demand;
and, all the downstream users of air transport services would benefit from more efficient service. The Airline Deregulation Act of 1978
(P.L. 95-504) liberalized airline fare and route structures in a
conscious attempt to make U.S. air carriers more competitive, both
domestically and internationally (Kasper, 1988; National Commission to Ensure a Strong Competitive Airline Industry, 1993). Although initial airline responses to economic deregulation had a
negative effect on U.S. LCA manufacturers, the point here is that
the federal government actively implemented a policy intended to

promote both the service and manufacturing sectors of commercial aviation. (For more on the costs and benefits of economic deregulation on U.S. airlines, see Brenner, 1988; Brenner, Leet, and Schott, 1985; Brown, 1987; Dempsey and Goetz, 1992; Kahn, 1988a, 1988b; Kelly, Oneal, DeGeorge, and Vogel, 1991; Lopez and Yager, 1987; National Academy of Engineering, 1985; Peterson and Glab, 1994.)

Primary Government Players in Air Transportation

The primary government players in air transportation are the FAA and its home, the DoT. As noted earlier, DoS also plays a key role with respect to negotiating BATAs. History notwithstanding, today's FAA remains responsible for both promoting safety and fostering air commerce. It regulates virtually the entire spectrum of aviation activities, from airports and airways to aircraft and the people who work in and around them. The FAA operates the nation's ATC system, a complex network of equipment and people that work 24 hours a day, every day of the year, in hundreds of locations across the U.S. and its territories. To fulfill its broad mandate, the FAA issues and enforces rules, regulations, and minimum safety standards related to manufacturing, operating, and maintaining aircraft. (For more extensive information on the history of the FAA, see Horwitch, 1982; Kent, 1980; Komons, 1978; Preston, 1987; Rochester, 1976; Schwartz, 1973.)

Though often viewed by industry personnel as a costly nuisance, safety standards and other governmental regulations have, at times, protected the industry (Lowndes, 1982) and served as a catalyst for significant technological innovation. In addition to enhancing air travel safety, standards raise the cost of flying in the U.S., thus protecting domestic airlines against foreign and "fly-by-night, shoestring" competitors (Bluestone, Jordan, and Sullivan, 1981, p. 166). Moreover, FAA directives and recommendations have frequently compelled manufacturers to incorporate new technologically sophisticated components into their LCA, including collision avoidance equipment, flight data recorders, and navigational instrumentation.

The Innovation–Transportation Connection

Technological innovation—a critical component of aviation—affects greater aircraft production efficiency as well as enhanced flight performance and safety. Incorporating advanced technologies in military aircraft enhances their mission capabilities; doing so in commercial aircraft enables airlines to provide more efficient transportation services, transferring benefits to a wide range of downstream industries. Although technological innovation could be financed solely by the immediate beneficiaries—the private sector—with R&D costs passed on to the consumer, the history of government involvement in aviation has created an RD&P structure that depends on public support for its continued success.

As a component of the nation's evolving innovation and transportation systems, aviation-related U.S. public policies are often reactive, *ad hoc*, and even incoherent. Nevertheless, they have combined to support two of the country's most successful and strategic industry sectors. In the case of LCA manufacturing, public policies have created a "sheltered economy," where firms build complex custom products for a limited number of large customers in a context of significant government involvement (Derian, 1990). In the case of air transportation, government has provided the infrastructure that enables private firms to offer a critical travel service. The following section focuses on the broader context of U.S. public policy within which the nation's innovation and transportation systems reside. Five broad areas of public policy—national security, economic infrastructure and well-being, science and technology, general welfare, and foreign affairs—directly and indirectly influence the production of LCA and the system in which they operate. They also influence—directly and indirectly—the diffusion of knowledge. We provide a brief analysis of each issue-area except for science and technology which is examined in Chapters 3 and 4.

U.S. PUBLIC POLICY–EXTRINSIC TO INNOVATION AND TRANSPORTATION

It would be a sterile exercise to try to describe and analyze the nation's innovation or transportation system separate from its political and economic context, or to depict the policies concerned with innovation and transportation as quite apart from those concerned with, for example, national security or the economy. Public policy is neither developed nor implemented in isolation. Policy decisions influence and are influenced by a multitude of other issue-areas, previously implemented policies, ideology, and world events. Decisions are made in a fluid, dynamic environment in which policies directed at achieving a particular set of goals in one arena may have unintended, but nevertheless significant, consequences for activities and goal achievement in another. Whatever policies are implemented, private actors will adjust their behavior to survive in the evolving political-economic environment. The relationship between the public and private sectors is characterized by a continuous dynamic reciprocity of action and reaction. Changes in the U.S. innovation and transportation systems frequently result from policies targeted at other separate and seemingly unrelated issue-areas.

Policymakers respond to shifting and often conflicting demands from both domestic and foreign constituencies (Golich, 1992). For example, decisions to fund military projects are subject to pressures from the domestic public sector—Congress, the Pentagon collectively, and the armed services individually—and private sector—defense contractors, formal trade associations like the Aerospace Industries Association, the Air Transport Association, and unions—as well as the international public sector—inter- and intra-state conflicts, trade negotiations, and social concerns—and private sector—firms involved in strategic production alliances with U.S. corporations. As a result, even a fairly narrow slice of policy—the R&D budget—does not represent any comprehensive examination of total federal R&D spending. Rather, it presents in one place the outcomes of negotiations among a plethora of individual executive branch departments and agencies, the Office of Management and Budget, and the White House regarding their separate budget plans. In Congress alone, 14 committees authorize R&D activities in major R&D agencies. In the United States

> technology policy has normally been a derivative category, shaped by decisions made on other grounds. Year-by-year, case-by-case budgetary decisions by DoD, NASA, and other mission agencies, by the Office of Management and Budget, and by Congress drive much of U.S. technology policy. (Alic, Branscomb, Brooks, Carter, and Epstein, 1992 p. 45)

Policy options are constrained not only by issue-area linkages and interest group pressures, but also by ideology and the consequences of prior policy implementation. In the U.S., an "ideology that celebrates the virtues of *laissez-faire*, and says government involvement in the market is imprudent" and the "lack of a predictive theory that can guide effective government actions" (Kline and

Kash, 1992, p. 12) proscribe overt government support for innovation or transportation, such as the direct capitalization of a specific commercial project in its development stage. Still, the U.S. has long had self-conscious technology and innovation policies—"government actions aimed at assisting delivery of competitive goods or services by sociotechnical systems" (Kline and Kash, 1992, p. 12)—directed at defense, medicine, agriculture, and aviation.

National Security

The U.S. Constitution clearly directs the federal government to provide for the nation's defense. To do so, policymakers must coordinate three components critical to military preparedness: (a) a technology base to research and design superior defense equipment, (b) an industrial base to produce the equipment, and (c) an economic infrastructure that can mobilize each in response to a crisis. Public policies designed to ensure national security may not directly target the U.S. innovation or transportation system or the LCA manufacturing or service sectors. However, the consequences of defense-related policies often affect both systems and sectors.

Throughout the post-World War II era, most federal technology innovation money funded military R&D as an obvious element of national security. The fact that defense technologies frequently "spun-off" into the commercial sector supported the prevailing assumption that commercial sector advances were dependent on those in the defense sector. By the 1960s and 1970s, a number of studies challenged this norm and argued that, in fact, the defense sector was increasingly dependent on commercial technology innovation (Alic, Branscomb, Brooks, Carter, and Epstein 1992; Branscomb, 1993; Gummett and Reppy, 1988; Mowery, 1985; Reppy, 1985; U.S. Congress, Office of Technology Assessment, 1994, 1991, 1989a, 1989b; 1988). Although most of the technology that is engineered into defense systems is still developed in DoD laboratories and by DoD contractors, the ability to build the systems is often contingent on developments in the commercial sector (U.S. Congress, Office of Technology Assessment, 1989b). Defense business alone is no longer big enough to keep a large number of DoD contractors afloat; moreover, most corporations involved in high-technology industries now compete in a global, not domestic, market. Hence, policymakers confront a three-part dilemma: (a) most technologies that meet specific high performance military requirements are considered to be so exotic and specialized as to have little or no use in the commercial sector; (b) many of the technologies that do have military value are actually developed in the commercial sector where they are refined to satisfy a demanding and sophisticated global market characterized by intense competition; and (c) strategic economic alliances that cross national boundaries make it difficult "to safeguard the civil technology base as a necessary foundation for defense purposes" (Gummett and Reppy, 1988).

In search of a solution, the DoD launched the "Manufacturing Technology" (ManTech) program in 1968. Designed to bridge the gap between fundamental R&D and full-scale production applications of new, more efficient technologies, ManTech is grounded on two premises: the cost of some process technology development exceeds the normal risk that firms are willing to pursue, and process technologies often precede key advances in product technology and performance. Hence, DoD funds contractors to upgrade their manufacturing technology to produce high-quality weapons systems, shorten production times, and reduce *matériel* acquisition costs. Directed by the Defense Advanced Research Projects Agency (DARPA), now the Advanced Research Projects Agency (ARPA), ManTech catalyzed precision structural castings, large scale composite materials applications, numerically controlled machines, computer integrated manufacturing, and advances in turbine blade production processes that led to high performance engines used in the latest generation LCA (National Academy of Engineering, 1981; National Research Council, 1987; 1986; Paulisick, 1972).

DoD also introduced the Independent Research and Development (IR&D) and Bid and Proposal (B&P) programs which allow firms to include defense-related R&D costs in the final price of an item. These funds are independently controlled by individual companies. IR&D covers the full spectrum of R&D activities, including expansion of basic knowledge, exploitation of scientific discoveries, improvement of existing technologies, and creation of new ones. With B&P, companies develop and support specific technical bids and proposals—solicited and unsolicited—to potential customers, including the government. Even with a government ceiling on recoverable IR&D and B&P costs, most defense R&D expenses are reimbursed.

The size and relative security of government contracts enable corporations to pursue commercial projects that otherwise might have been too risky. Military procurement increases profit margins and provides significant "learn-by-doing and learn-by-using" opportunities. Special rules restrict "dual use"—military and civilian— technology RD&P to ensure that public funds only support public goods (in this case national security); still, money is fungible. Profits can be used to finance commercial as well as military aircraft RD&P, and, although increasingly rare, defense technologies can sometimes be adapted to commercial use. Equally as important, the teams of engineers, scientists, and other skilled workers gain invaluable tacit knowledge from the processes of experimentation and assembly associated with the RD&P of military equipment (Golich, 1989; Hayward, 1986; Mowery and Rosenberg, 1982; Rosenberg, 1982; Tyson, 1992; U.S. Congress, Office of Technology Assessment, 1991; von Hippel, 1976). The connection between government procurement and corporate pursuit of civil projects is reflected in the Clinton Administration's decision to purchase 80 McDonnell Douglas C-17s in addition to the existing procurement of 40; the infusion of an additional $17 billion allowed McDonnell Douglas to stay afloat financially, and hire over 2000 additional workers (Chambers, 1996; Peltz and Vartabedian, 1996, pp. D1-D7).

Finally, the military benefits from "a strong, profitable, technically up-to-date commercial airline industry" and LCA manufacturing industry (Thornton, 1970, p. 80). Commercial airlines have "a fleet of planes, supporting equipment, and a staff of trained personnel which can be used in flying personnel and *matériel* to a combat theater" (Stratzheim, 1969, p. 15). The Civil Reserve Air Fleet authorizes the federal government to use commercial airlines for military purposes, which it did during World War II, the Korean and the Vietnam conflicts, and most recently the Gulf War. In exchange, the government "provides added spare parts backup and supports communication facilities it would not otherwise have established" (Thornton, 1970, pp. 80-81; see also Thayer, 1965, p. 123). Similarly, a competitive LCA manufacturing sector reduces the cost of providing an essential military industrial base and wartime mobilization surge capacity. Teams of engineers and scientists are kept in a high state of readiness by the continuing requirements of the commercial market.

The synergy between national security and LCA manufacturing and operation has costs as well as benefits for these strategic industry sectors. Dependency on government support for national defense makes LCA producers vulnerable to decreases in military budgets. Such a pattern has been repeated in the U.S. several times since World War II. Between 1968 and 1971, industry sales to the U.S. DoD dropped 24%, and jet transport sales dropped 68%. The collapse of the Soviet Union accelerated further cuts in defense spending, with new orders for aerospace products and services falling an estimated 23% in 1993 to less than half the new business received in 1989 (Napier, 1995). Cuts in public support—particularly when aggravated by a paucity of private funds resulting from the slumping airline industry that registered its first profitable year in 1995, following 16 years of incredible losses—can decrease corporate capacity to launch new generation LCA RD&P. Lacking sufficient domestic funds, U.S. firms have accessed capital through transnational joint venture arrangements; some policymakers and analysts have voiced concern about increasing vulnerabilities to foreign suppliers of strategic goods that might result from a proliferation of such arrangements. (For more on the synergy between national security and LCA, see Bacher, 1984; Bluestone, Jordan, and Sullivan, 1981; Chambers, 1996; Defense Science Board, 1988; Gansler, 1989; Golich, 1989; 1992; Gummett and Reppy, 1988; Hayward, 1986; Irwin, 1993; Moran, 1990; Moran and Mowery, 1994; Mowery, 1985; Mowery and Rosenberg, 1982; National Academy of Engineering, 1985; National Research Council 1987; 1986; 1981; Newhouse, 1982; Paulisick, 1972; Tyson, 1992; U.S. Congress, Office of Technology Assessment, 1994; 1991; 1989a, 1989b; 1988; U.S. Department of Defense, 1988).

Economic Infrastructure and Well-Being

Providing the infrastructure that facilitates commerce and promoting economic well-being rank a distant second to national security as a rationale for U.S. government intervention in the market. Nevertheless, policymakers have treated the domestic transportation system as a public good, the provision of which has generated significant benefits for the domestic economy as a whole. Commercial aviation is an integral component of the U.S. national transportation system; therefore, a significant number of public policies have been directed at multiple facets of LCA production and operation. In addition to the government roles described in the transportation system section earlier, from 1946 to 1970 the federal government provided grants-in-aid to state and local governments for airport development. Since 1970, federally imposed user taxes on air fares, air freight, and aviation fuel have underwritten the Airport and Airway Trust Fund that finances airport construction and improvement projects as well as capital improvements to the airway system (National Academy of Sciences, 1994).

As in the case of national security, several U.S. economic public policies do not specifically target commercial aviation but still influence the industry substantially. For example, building on the traditions of supporting domestic economic activities established early on, the U.S. has implemented several policies designed to bolster exports and to protect domestic sales for domestic producers (Goodrich, 1966; Heaton, 1989; Rosenberg, Thompson, and Belsley, 1978; Smith, 1985; Tickner, 1987). As early as 1934, the U.S. created the Export-Import Bank to promote exports using guaranteed low interest loans. In 1994, the ExIm Bank was authorized to finance exports of nonlethal dual use products used primarily for civilian purposes. As a result, U.S. companies can now use ExIm Bank financing to compete for a growing number of contracts that were previously the domain of foreign competitors, including ATC, communications, surveillance, and utility air transport aircraft and helicopters.

In the years leading up to World War II, "Buy American" legislation (47 Stat. 1489) overtly protected domestic markets. Since then, more subtle forms of market protection have been employed. In addition to BATAs mentioned previously, other international agreements concluded under the auspices of the International Civil Aviation Organization (ICAO) and the International Air Transport Association (IATA) have affected the demand for LCA. Among the most contentious for foreign airlines competing with U.S. air carriers are those related to antitrust and insurance regulations (Gidwitz, 1980; Golich, 1990; Kasper, 1988; "Liberal....," 1985; Proctor, 1988; Shifrin, 1987; Taneja, 1970). Because LCA production is characterized by strong learning curve effects that sharply reduce RD&P costs over time, any policy that enhances LCA sales, including market protection—either directly or indirectly—should lead to lower prices, more sales, and increased profit. The profit, of course, can be used to help finance the RD&P of technologically differentiated, new generation aircraft (see Mowery, 1985).

Since the end of World War II, policymakers have assumed that federally funded scientific research leads to broad economic benefits for the entire country.

> The strength of a company is generally measured in terms of its market share and profits. Similarly, the strength of an industrial base can be measured by a country's trade balance and Gross National Product. These characteristics depend on product quality, cost delivery time, and special features, which in turn are largely determined by manufacturing technology. Advancing the state of the art in manufacturing technology, therefore, can help a company or country become more competitive. (National Research Council, 1986, p. 4)

In addition to direct R&D spending, several other policies seek to sustain technological innovation, including tax credits; joint research ventures among industry, academia, and government; relaxation of antitrust; and a strengthening of patent and intellectual property protections, and international trade laws. Though not necessarily aimed at LCA production or operation, these innovation or technology policies have provided indirect support for commercial aviation because they bolster sectors that develop and use transferable technologies. For example, technologies enabling the installation of more precise weather radar, navigational and landing equipment (e.g., global positioning satellite and microwave landing systems), and ATC communications systems, derive from space exploration and ocean shipping R&D. (For more on linkage between scientific R&D and economic benefits, see Alic, 1986; Council on Competitiveness, 1994; 1988; Evenson, Waggoner and Ruttan, 1979; Flanigan, 1996; Kleinman, 1995; Mansfield, 1986; 1980a, 1980b, 1980c; 1968; 1965; Mansfield, Rapoport, Romeo, Villani, Wagner, and Husic, 1977; Mowery, 1985; National Academy of Engineering, 1985; National Academy of Sciences, 1995; Nelson, 1984; Terleckyj, 1985; 1974; U.S. Congress, House of Representatives, 1988b; U.S. Congress, Office of Technology Assessment, 1986).

Initially, the Reagan Administration opposed a federal role in the private sector, but later launched several initiatives aimed at increasing the development of technologies perceived to have high potential value in commercial sectors (Heaton, 1989; Mowery and Rosenberg, 1993; U.S. Executive Office of the President, 1982b). The Stevenson-Wydler Technology Innovation Act in 1980 (P.L. 96-480) made technology transfer to industry a mission of all federal laboratories. The Tax Reform Acts of 1981 (Economic Recovery Tax Act, Section 221; IRC Section 44F) and 1986 (Section 231[c], IRC Section 41) established corporate tax credits for increases in R&D; the 1986 Act added a "basic research credit" to encourage industry-sponsored basic research at universities. Combined, these policies represented "a-several-billion-dollar-per-year incentive" (Heaton, 1989, pp. 90-91; see also Clinton and Gore, 1993; Cordes, 1988; Helm, 1995; Knezo, 1996; Mowery, 1994; Rothwell, 1982; Schacht, 1996). The 1982 Bayh-Dole Patent Act (P.L. 97-256) and the 1984 National Cooperative Research Act (P.L. 98-462) reduced antitrust penalties for precommercial collaboration by private firms (Mowery and Rosenberg, 1993). Soon thereafter, the Federal Technology Transfer Act of 1986 (P.L. 99-502) and the National Competitiveness Technology Transfer Act of 1989 (P.L. 101-189) authorized federal laboratories to enter into cooperative R&D agreements (CRADAS) with companies and consortia of companies to pursue projects of mutual interest (National Academy of Sciences, 1995). Subsequently, a series of reports issued by a Defense Science Board Task Force on Semiconductor Dependency—convened to assess the "impact on U.S. national security if any leading edge of technologies are no longer in this country" (Hart, 1992, p. 275)—expressed concern over growing U.S. dependence on foreign sources of advanced integrated circuits. As a result, the DoD supported Sematech (Semiconductor Manufacturing Technology) and its mission to return U.S. firms to the technological frontier in semiconductor manufacturing.

During the Bush Administration, the Omnibus Trade and Competitiveness Act of 1988 (15 U.S.C. 3704) transformed the DoC's National Bureau of Standards—founded in 1901 to "create advantages for national industry through the establishment of standards that could serve as barriers against imports" (Alic, Branscomb, Brooks, Carter, and Epstein, 1992, p. 49)—into the National Institute of Standards and Technology (NIST). NIST's primary mission is to promote U.S. economic growth by working with industry to develop and apply technology, measurements, and standards. It carries out this mission through four major programs: an Advanced Technology Program (ATP) that provides cost-shared awards to industry to develop high-risk technologies that can enable significant commercial advances; a grassroots Manufacturing Extension Partnership that helps small- and medium-sized companies to access critical information about production technologies; a strong laboratory effort planned and implemented in cooperation with industry that focuses on infrastructural technologies (e.g., electronics and electrical engineering, manufacturing engineering, chemical science and technology physics, materials science and engineering, and computer systems); and a quality improvement program associated with the Malcolm Baldrige National Quality Award that recognizes achievements in the areas of manufacturing and service. President Clinton expanded these efforts, establishing a new defense conversion program to get military technology into the private sector and increasing the ATP budget from $68 million in 1992 to $431 million in 1995 (Helm, 1995).

Other economic policies that affect LCA production and operation include financial capital cost, supply of skilled labor, and trade restrictions (U.S. Department of Commerce, 1983). Although the primary responsibility for the advance and use of commercial technologies rests with the private sector, federal support for precommercial innovation contributes in important ways to U.S. corporate competitiveness (Carnegie Commission on Science, Technology, and Government, 1991). (For more on the influence of economic public policies on technological innovation in LCA production and operation, see Alic, 1986; Bacher, 1984; Dertouzos, Lester, and Solow, 1989; Evenson, Waggoner and Ruttan, 1979; Flanigan, 1996; Golich, 1992; 1989; Hayward, 1986; Heaton, 1989; Mansfield, 1986, 1980a, 1980b, 1980c; 1968; 1965; Mowery, 1985; Mowery and Rosenberg, 1982; National Academy of Engineering, 1985; National Research Council 1987; 1986; 1981; Nelson, 1984; Schwartz, 1992; Terleckyj, 1985; 1974; Tyson, 1992; U.S. Congress, House of Representatives, 1988c; U.S. Congress, Office of Technology Assessment, 1994; 1991; 1989a, 1989b; 1988; 1986).

General Welfare

General welfare covers a seemingly infinite array of protections for U.S. citizens. In keeping with our intent to provide representative policies, we focus here on environmental protection policies affecting aviation. In 1970 President Nixon reacted to mounting public concerns about pollution by creating the Environmental Protection Agency as an independent agency pursuant to Reorganization Plan No. 3 of 1970 (5 U.S.C. app.) with the mandate to regulate both noise and emissions pollution. Corresponding legislation required aircraft manufacturers to use more fuel efficient engines and to reduce emissions from the toxic chemicals used in the production and operation of LCA.

In 1952, a Presidential Commission chaired by James H. Doo-little identified aircraft noise as the airport planning problem most likely to increase in the future (President's Airport Commission, 1952). In response, the NACA established a Special Subcommittee on Aircraft Noise and pursued a three-pronged program aimed at "(a) understanding the mechanisms for generating jet noise; (b) developing devices for attenuating it; and (c) studying the impact that noise had on vehicle structures" (Langford, 1989, p. 58). In 1966, the U.S. Office of Science and Technology (a precursor to the U.S. Office of Science and Technology Policy) published *Alleviation of Jet Aircraft Noise Near Airports*; it concluded that aircraft noise could only be reduced with the help of the federal government:

> The FAA and/or NASA, using qualified contractors as necessary, (should) establish and fund...an urgent program for conducting the physical, psycho-acoustical, sociological, and other research results needed to provide the basis for quantitative noise evaluation techniques which can be used...for hardware and operational specifications. (U.S. Executive Office of the President, 1966, p. 8)

NASA pursued three programs that contributed significantly to the industry's ability to reduce noise: the Acoustically Treated Nacelle Program, The Quiet Engine Program, and the Steep Approach Program (see Langford, 1989, pp. 61-63 for a detailed discussion of each of these efforts). In 1968, Congress authorized the FAA to regulate noise emissions as part of its aircraft certification process. Aided by the results of the NASA research programs, in 1969 the FAA promulgated Part 36 of the Federal Airworthiness Regulation (FAR Part 36), which established noise standards based on weight. The Noise Control Act of 1972 (P.L. 92-574) and amendments restrict noise pollution even further (Langford, 1989).

Concern about emissions pollution led to the Clean Air Act of 1963 (P.L. 88-206); as many as 18 major pieces of federal legislation and a plethora of presidential executive orders followed in its wake (Lopez, 1994). Intended to ensure that the air we breathe is clean, the resulting federal regulations have dramatically affected commercial aviation. As many as 30 000 different chemicals—many toxic or potentially toxic to the environment—are used and emitted in the production and operation of LCA at an estimated 3000 facilities in the U.S. alone and in thousands more facilities worldwide. LCA utilize a variety of sealants, lubricants, coolants, and fuels, each selected for its special contribution to product efficiency, durability, and safety. In addition, an increasing number of structural parts are made of composites—chemically treated materials—that are stronger and lighter than the metals they replace. As with safety standards, environmental standards are often considered a costly nuisance by the industry but can also serve to protect or create markets for new products (Carey and Regan, 1995; National Academy of Sciences, 1995). For example, to meet volatile organic compound (VOC) emissions requirements, United Technology Corporation's Sikorsky Aircraft division developed new paint and painting processes. The company now uses half the paint for the same amount of coverage, which reduces aircraft coating weight by 35% and VOC emissions by 34% ("AIA Member...," 1994).

Noise and emissions standards rendered industry workhorses such as Boeing's 707s and 727s obsolete unless the aircraft were retrofitted with new generation "hush kits" and emissions reduction equipment. The standards precipitated a new industry sector aimed at muffling the take-off and landing noise of these airplanes, as well as the RD&P of new generation aircraft which could satisfy both regulations and air carrier requirements. (For more on the relationship between policies designed to protect and enhance the general welfare of U.S. citizens and LCA production and operation, see "AIA Member...," 1994; Carey and Regan, 1995; Langford, 1989; Lopez, 1994; National Academy of Sciences, 1995; President's Airport Commission, 1952; U.S. Executive Office of the President, 1966).

Foreign Policy

Foreign policies that affect LCA RD&P typically focus on such macro-level goals as national security and global economic competitiveness. Several of the policies already discussed under national security, economic infrastructure, economic well-being, and general welfare transcend national boundaries and become foreign policies with direct and indirect consequences for LCA RD&P. Occasionally, governments have targeted both airlines and aircraft manufacturers in their efforts to achieve broader goals. Airlines have been used to execute foreign policy because they are inherently mobile, flexible, and span "a wide geographic sweep" (Gidwitz, 1980). In 1977, the U.S. granted an operating permit and landing rights for a joint Syrian-Jordanian air service to New York over the objections of aviation advisers because the U.S. was concerned with the progress of the ongoing Middle East peace negotiations (Gidwitz, 1980, p. 24).

Such policies may also affect the proclivity of a country to buy aircraft from "friendly" states.

More specifically targeting manufacturers, countries use procurement policies to achieve national goals. Airlines may be compelled to purchase foreign-built aircraft to gain favor with a country, even if the planes do not fit airline needs. In 1978, Japan's foreign policy bureaucracy coerced Toa Domestic Airlines (TDA) to buy European Airbus aircraft to calm European Community concerns over its large trade deficit with Japan. The airline preferred the McDonnell Douglas DC-10, and the U.S. was also experiencing a large trade deficit with Japan. "Several European governments applied greater pressure and the Japanese government responded accordingly" (Gidwitz, 1980, p. 26). In 1994, the U.S. DoS reversed its decision to cut off defense sales to Saudi Arabia for failure to pay for previous shipments when U.S. firms complained that this disadvantaged them vis-à-vis their foreign competitors; shortly thereafter, Boeing announced one of its first billion dollar foreign sales of the B-777 to Saudi Arabia. More recently, Beijing postponed purchases of up to $3.5 billion in Boeing planes, partly to retaliate against Washington's tough stance on trade (Engardio and Roberts, 1996, p. 125).

Environmental regulations and safety standards that transcend boundaries via international agreements and treaties can influence production and sales costs. The ICAO—created in 1944 to develop an international aviation transportation network—has assumed the task of harmonizing the extant set of disparate national noise and emission standards; to date, only limited progress has been made. With respect to safety standards, aircraft producers complain that, although their planes are global in design, construction, and operation, their safety is governed by an obsolete system. Currently, a separate airworthiness certificate is required for each country in which an aircraft is sold, increasing costs with no noticeable improvement in safety. European Union members have successfully harmonized several requirements through their Joint Airworthiness Requirements (JARs)—the equivalent of the U.S. FAA's Federal Aviation Requirements. For the last six years, U.S. and European authorities have worked with less success to harmonize international certification standards.

Multilateral and bilateral trade agreements influence the marketing and sales strategies of LCA firms. LCA sales negotiations are complicated and include such a large number of factors—ranging from interior configurations to engine preferences to maintenance support and exterior paint choices—that it is impossible to identify a standard price for any particular plane. However, certain international rules established under the auspices of the General Agreement on Tariffs and Trade—now the World Trade Organization (WTO)—regarding acceptable uses of subsidy to promote industry sectors apply to LCA. Three target LCA—The Standstill (1974), The Arrangement (1976), and The Commonline (1981); others apply to all goods and services governed by WTO rules (Beard, 1982; Golich, 1992; 1989; Tyson, 1992). U.S. aircraft manufacturers and their primary foreign competitor—Airbus Industrie—are locked in bitter debate over the use of subsidies in support of the RD&P of their respective LCA. U.S. manufacturers complain that Airbus Industrie is subsidized at so many levels of production that it can offer "flexible and imaginative financing programs" not available to privately owned U.S. firms. Airbus Industrie officials counter that U.S. producers enjoy similar subsidies, if indirectly via the military (Mann, 1981a, 1981b). Because LCA sales generate significant revenues that can affect a firm's ability to finance innovation in the RD&P of new generation LCA, countries and companies are likely to scrutinize trade practices for the foreseeable future.

LCA production is increasingly globalized in structure because no single company or country can afford to capitalize the entire cost of LCA RD&P, particularly in light of the high risk of failure. The U.S. Congress, Office of Technology Assessment (1989b) noted years ago, that "high technology is a worldwide enterprise. The United States no longer has a monopoly on it" (p. 10). As a consequence, the national identities of firms are less clear or meaningful (Reich, 1991; 1990). Through global partnering foreign firms become vested in the success of U.S. firms, and vice versa. Hence, protecting U.S. interests is not a simple matter; the task is complicated by the fact that other nations seek to protect their companies and home markets in this confusing world as well (Bergsten, Horst, and Moran, 1978). As U.S. officials seek to formulate policies to deal with this situation, they will need to consider such factors as

international patterns of trade, manufacturing, and corporate ownership; the costs and opportunities of maintaining domestic capabilities; existing relations with other nations; and the effects of policy choices on foreign relations. (U.S. Congress, Office of Technology Assessment, 1989b, p. 11)

They must be sensitive to the role of foreign companies and multinationals as technology drivers and of the potential for international markets to generate huge profits that can finance R&D of new products and underlying technologies. The mutual dependencies among states are significant and growing.

U.S. Public Policy and the Consequences for Aviation

The combined effect of U.S. public policies has been to create both "demand-pull" and "supply-push" incentives for LCA manufacturers. Procurement, trade, and regulatory policies have, individually and collectively, created demand for technologically enhanced LCA. Defense-related U.S. government procurement has provided secure markets and reliable cash flow, which have helped finance technological innovation benefiting both military and commercial aviation. "Buy American" policies protected domestic aircraft producers; economic regulation guaranteed fare increases, and route structures eliminated price competition and limited entry, so airlines competed largely on the basis of service quality. Because the increased cost of "new and improved" aircraft was virtually assured, aircraft manufacturers were willing to invest in constant and continual upgrading of their planes. Shifting safety and environmental standards created demand for ever more sophisticated and meticulously designed and produced airframes, components, and engines.

In the case of LCA, where RD&P costs are very high, technological innovation is virtually impossible for individual firms to appropriate, and return-on-investment cycles—assuming they are successful—can take as long as 10-15 years, creating demand may not be enough to impel firms to innovate. National governments with active LCA producers—whether at the prime or major component level (see Figure 1.1, p. 12)—all invest public money in an effort to underwrite this risk because LCA manufacturing is a strategic industry sector. It plays an important role in the domestic economy generating a positive balance of payments and pioneering technologies with broad application; it produces goods and services directly related to national security; and it generates "special benefits for the rest of the economy (Tyson, 1988). U.S. public policies encourage firms to participate in LCA RD&P by providing a critical R&T base, including research and testing facilities, a pool of engineers and scientists, and accessible knowledge about technological possibilities. The largest amount of funding for this R&T base flows through NASA, although some is also channeled through the DoD. The cost of constructing, operating, and maintaining NASA's aeronautical research facilities alone is beyond the capacity of firms, individually or collectively, to furnish. Moreover, the cumulative knowledge gained by LCA manufacturers performing R&D under DoD and NASA contracts is incalculable. According to one recent study, in the future, "funding for the nation's science and technology [supply-push incentives] base may contribute more to stimulating new sectors of economic growth than will federal procurement and the 'demand-pull' on an emerging technology" (National Academy of Sciences, 1995, p. 21).

Despite their overall success in supporting LCA RD&P, most U.S. public policies have focused on knowledge creation, with little attention given to how and by whom the knowledge may be used. In addition, the NACA and NASA notwithstanding, most policies targeted air transport, not LCA manufacturing, by providing the requisite infrastructure. This continues to be the case. Although one could argue the merits of the current "open skies" efforts to liberalize global air transportation, it is clearly a policy intended to sustain and enhance the status of U.S. airlines (Kasper, 1988).

As detailed throughout this chapter, other policies that have affected LCA RD&P were selected to achieve such seemingly unrelated goals as national security, a healthy economic infrastructure and competitive industries, the general welfare of U.S. citizens, and foreign affairs. Moreover, this plethora of policies was designed and implemented by a wide array of policy players at the federal level; no single federal entity has ultimate responsibility for coordinating or even overseeing the effects of these policies on commercial aviation. The various players frequently have conflicting objectives; for example, the DoC, DoT, and the DoD may disagree over the value of transnational production of LCA or of foreign ownership of U.S. airlines. Although LCA production is, and has been for years, considered a strategic industry sector by most policymakers and analysts, no comprehensive, coordinated, or coherent policy or set of policies has been designed or implemented with the explicit purpose of sustaining or enhancing LCA manufacturing performance.

OBSERVATIONS ABOUT U.S. PUBLIC POLICY
AND KNOWLEDGE DIFFUSION

Knowledge creation and knowledge diffusion constitute distinct though equally complex and multifaceted processes. The former concentrates on producing new knowledge, whether basic or applied, whereas the latter focuses on the transfer and use of that knowledge. Our analysis clearly reveals that U.S. public policy places much greater emphasis on knowledge creation than on knowledge diffusion. Although many countries explicitly manage knowledge diffusion to achieve carefully defined national goals, in the United States the process remains more anarchic and chaotic. In keeping with our *laissez faire* ideology, "we prefer to create the knowledge and let the market decide if and how to use it" (Soderquist, 1996).

Nevertheless, when concerns about global competitiveness gained national attention in the 1980s, policymakers revisited the question of how best to capitalize on the knowledge created by public policies. Legislation proliferated a series of programs and projects intended, on the one hand, to enhance or refine existing technologies and, on the other, to commercialize knowledge resulting from federally funded science and technology (see Chapter 3 for greater detail). Despite the flurry of federal activities and significant resources that seek to improve the production, transfer, and use of knowledge, a key debate continues over how to ensure that knowledge resulting from federally funded science and technology is best utilized. Six issues from this debate are particularly relevant to LCA RD&P and speak to the question of managing knowledge as an intellectual or public asset that is at least as important as creating knowledge in the first place.

First, knowledge diffusion in the United States involves "dozens of government agencies; hundreds of federal laboratories and universities; thousands of industrial corporations; and hundreds of thousands of engineers, technicians, managers, and scientists" (Alic, Branscomb, Brooks, Carter, and Epstein, 1992, p. 28), making it difficult to track and link knowledge creators and users in any meaningful way. In the U.S., shifting political alliances prevent the government from successfully participating in long-term, focused projects (Tornatsky and Fleischer, 1990; Tornatsky and Luria, 1992). The failure of the supersonic transport (SST) development efforts exemplifies U.S. attitudes toward government involvement in projects with explicit commercialization of a new technology as an ultimate goal (Horwitch, 1982, p. 3; see also Alic, Branscomb, Brooks, Carter, and Epstein, 1992; Heaton, 1991; National Academy of Engineering, 1985).

Second, as defined in Chapter 1, both explicit and tacit knowledge—embedded in product-embodied, process, systems integration, and management—are critical to LCA RD&P. Most explicit knowledge—certainly that created within universities and NASA aeronautical research centers—is easily accessed through open literature (journals, papers, and technical reports), conference and workshop presentations, and final contract and grant reports. This free exchange of knowledge is a key value of academia and of democracies. Policies that seek to restrict the flow of explicit knowledge through such tactics as security classification are seldom successful. Whereas explicit knowledge is important to R&D, technological innovation, and the RD&P of the next generation LCA, tacit knowledge—experience-based judgement, heuristics, and intuition at both the individual and organizational levels—is much more significant (Teece, 1981; von Hippel, 1994).

To gain the requisite tacit knowledge, countries or corporations employ a wide array of mechanisms, including direct investment, turnkey arrangements, licensing and co-production, RD&P joint ventures, recruiting and hiring retired engineers and scientists, commercial visits, as well as outright industrial espionage. Policies that seek to limit access to legal means of acquiring critical tacit knowledge may work at cross purposes to other foreign, economic, or military policies aimed at cementing friendships, gaining market access, and rewarding allies. (For additional source material on public policies directed at knowledge acquisition and transfer, see Alic, Branscomb, Brooks, Carter, and Epstein, 1992; Branscomb, 1993; 1992; Ergas, 1987; Gellman and Price, 1978; Gummett and Reppy, 1988; Hall and Johnson, 1970; Heaton, 1991; Irwin, 1993; March, 1989; Moran, 1990; Moran and Mowery, 1994; Mowery and Rosenberg, 1989; 1985; National Academy of Engineering, 1987; 1985; U.S. Congress, Office of Technology Assessment, 1989a, 1989b; U.S. Department of Commerce, International Trade Administration, 1983; U.S. Executive Office of the President, Office of Science and Technology Policy, 1982b.)

Third, knowledge diffusion can follow both a linear or a circular, more stochastic path as it moves from the creator to the user. Knowledge can easily transcend national borders, particularly when it is treated as a public good and made freely available. Although certain countries and companies have well-deserved reputations for particular knowledge expertise (e.g., in chemical or aeronautical engineering), history teaches us that others who are committed to doing so and who have the requisite resources, can supplant the original leaders. Hence, it is wise never to become complacent: policymakers are well-advised to encourage efficient use of knowledge and to monitor knowledge creation and its use from around the world.

In the U.S. when federal policies *have* addressed knowledge transfer, they have tended to focus on specific domestic "missions"—military R&D, space exploration, and biomedical developments. Those policies that do address the issue of transferring knowledge to domestic players in academia and industry, do little more than ensure that the information is available (i.e., accessible) in some documented format. Rather than proactively linking producers and users of knowledge, except perhaps in agriculture, U.S. public policy has tried—generally without success—to limit knowledge transfer to foreign economic or military competitors. Perhaps it is ironic then, that U.S. policy has seldom sought to identify, absorb, and diffuse knowledge that resides external to our nation, despite the fact that we clearly recognize others are trying to acquire and use knowledge we created at public expense. It is particularly ironic because we consistently borrowed from aeronautical R&T discoveries in Europe via immigration and targeted information gathering in the early years of aviation. Specifically, jet engines, swept wing, and laminar flow technologies were initiated in Europe but were developed and used in market-successful products in the United States (see e.g., Davies, 1964; Hayward, 1986; Loftin, 1985; Mowery, 1994; Rae, 1968; Solberg, 1979).

Fourth, the value and usability of knowledge changes over time; our policies need to be relevant to these advances. For example, the synergy between military and commercial aeronautics was once very strong. Hence, U.S. policymakers have long justified massive spending on military-focused R&D on grounds that the results not only helped to ensure successful national defense, but also spilled over "freely" to the commercial sector. However, as military performance requirements have become more specialized and exotic, the supporting technologies have decreased in value to the commercial sector, while their R&D has grown increasingly expensive. To be useful, explicit knowledge with potential commercial benefits must find its way to those who have tacit knowledge derived from years of learn-by-doing experience about how to adapt performance-based military technology to efficiency-focused commercial use (Teece, 1981; von Hippel, 1994). Even with tacit knowledge in place, typically, military technology must be dramatically reconfigured to have value in the commercial sector. Moreover, organizational structures and legal requirements often inhibit effective transfer of knowledge among researchers even though they may be colleagues in the same firm. DoD regulations often keep knowledge derived from military R&D projects classified for years. The U.S.-based General Electric (GE) Aircraft Engine Group and the French *Société Nationale d'Étude et de Construction de Moteurs d'Aviation (SNECMA)* joint production of the CFM-56 family of aircraft engines stands as an unusual example of transnational joint ventures in which military engine technology was used in commercial aircraft engines. Wary of giving away critical commercial and military trade secrets, the U.S. Congress required GE to protect the "hot" core of the engine (Alic, Branscomb, Brooks, Carter, and Epstein, 1992; Golich, 1989). Meanwhile, technologies developed for the commercial sector are more frequently being used, or adapted for use, by the defense sector in a process often referred to as "spin-on."

Fifth, knowledge can diffuse both intentionally and unintentionally. Just as many U.S. public policies that affect LCA production and operation are not directed at either the manufacturing or service sector of the commercial aviation industry, we often give away critical knowledge in an effort to achieve an unrelated policy—usually foreign—goal. For example, we may trade entire technologies for such outcomes as a military base or flyover privileges; airline landing, capacity and route rights; guaranteed or virtually guaranteed markets—as with the GE-SNECMA deal or more recent co-production arrangements with Japan and China. Traded technologies usually have embedded tacit and explicit knowledge that can be used by countries or corporations to develop their own economic or security base.

Sixth, and finally, all of this is complicated by the following facts related to the global marketplace: Improvements in communications and transportation systems make knowledge diffusion eminently easier than ever before. Countries around the world, no matter how economically advanced, seek knowledge that will enhance their competitiveness in the world economy. In many cases their own geopolitical resources and negotiating expertise have meant they have been able to bargain successfully for knowledge transfer in exchange for any number of trade-offs, ranging from guaranteed or potential market to military *quid pro quos*. Typically, countries other than the U.S. tend to have more carefully focused knowledge acquisition and use policies. As the National Academy of Engineering Panel pointed out in their 1985 study of the U.S. civil aviation manufacturing industry, controlling knowledge transfer requires "balancing national security or foreign policy objectives with those of strengthening the economy and preserving the U.S. position in advanced technology" (p. 92). Further, the panel noted that a delicate part of this balancing act is related to reciprocity: U.S. policy that leads to excessive restrictions on technology exchanges with other nations can impede increasingly valuable reverse flow and thus impair the aeronautics infrastructure of this country.

In summary, although policies purporting to facilitate the transfer and use of knowledge abound, we have been far more successful in producing knowledge than we have in managing its diffusion. U.S. public policy emphasizes knowledge production and the restriction of its outward flow; little attention is given either to targeting the recipients of knowledge or to acquiring knowledge created by others that might have military or commercial value for domestic commercial industries. Chapters 3 and 4 examine in greater detail the dynamics of public policy specifically concerned with the production and distribution of knowledge resulting from federally funded aeronautical R&T.

CHAPTER REFERENCES

"AIA Member Companies Reduce Waste and Improve Products While Saving Money." (1994). *AIA Newsletter* (May) 6(9): 8.

Alic, J.A.; L.M. Branscomb; H. Brooks; A.B. Carter; and G.L. Epstein. (1992). *Beyond Spinoff: Military and Commercial Technologies in a Changing World.* Boston, MA: Harvard Business School Press.

Alic, J.A. (1986). "The Federal Role in Commercial Technology Development." *Technovation* 4(4): 253-267.

Anderton, D.A. (1980). *Sixty Years of Aeronautical Research, 1917-1977.* NASA-EP-145. Washington, DC: National Aeronautics and Space Administration. (Available NTIS; 79N27089.)

Bacher, T.J. (1984). "The Economics of the Commercial Aircraft Industry." Paper presented at the conference on *The Role of South East Asia in World Airlines and Aerospace Development* sponsored by the *Financial Times Limited* in Singapore, 24-25 September.

Bailey, E. (1986). "Deregulation: Causes and Consequences." *Science* (December 5) 234(4781): 1211-1216.

Beard, M.C. (1982). Aircraft Engineering Division, Office of Airworthiness, Federal Aviation Administration. Washington, DC. Personal correspondence. (March 1).

Becker, J.V. (1980). *The High-Speed Frontier: Case Histories of Four NACA Programs, 1920-1950.* NASA SP-445. Washington, DC: National Aeronautics and Space Administration. (Available NTIS; 81N15969.)

Bergsten, C.F.; T. Horst; and T.H. Moran. (1978). *American Multinationals and American Interests.* Washington, DC: Brookings Institute.

Bluestone, B.; P. Jordan; and M. Sullivan. (1981). *Aircraft Industry Dynamics: An Analysis of Competition, Capital, and Labor.* Boston, MA: Auburn House.

Booz-Allen Applied Research, Inc. (1971a). *A Historical Study of the Benefits Derived From Application of Technical Advances to Civil Aviation. Volume 1: Summary Report and Appendix A.* NASA CR-1808. Washington, DC: National Aeronautics and Space Administration. (Available NTIS; N71-27010.)

Booz-Allen Applied Research, Inc. (1971b). *A Historical Study of the Benefits Derived From Application of Technical Advances to Civil Aviation. Volume 2: Appendices B thru I.* NASA CR-1809. Washington, DC: National Aeronautics and Space Administration. (Available NTIS; N71-27011.)

Branscomb, L.M. (ed.). (1993). *Empowering Technology: Implementing a U.S. Strategy.* Cambridge, MA: MIT Press.

Branscomb, L.M. (1992). "Information Infrastructure for the 1990s: a Public Policy Perspective." In *Building Information Infrastructure.* B. Kahin, ed. Cambridge, MA: McGraw-Hill Primis, 15-30.

Brenner, M.A. (1988). "Airline Deregulation: A Case Study in Public Policy Failure." *Transportation Law Journal* 16: 179-227.

Brenner, M.A.; L.O. Leet; and E. Schott. (1985). *Airline Deregulation.* Westport, CT: Eno Foundation.

Brown, A.E. (1987). *The Politics of Airline Deregulation.* Knoxville, TN: University of Tennessee Press.

Bush, V. (1945). *Science: The Endless Frontier.* Washington, DC: Government Printing Office.

Carey, J. and M.B. Regan. (1995). "Are Regs Bleeding the Economy?" *Business Week* (July 17) (3433): 75-76.

Carnegie Commission on Science, Technology, and Government. (1991). *Technology and Economic Performance: Organizing the Executive Branch for a Stronger National Technology Base*. New York, NY: Carnegie Corporation.

Chambers, M. (1996). "New C-17 Aircraft Blows into Langley," *Research News* (May 31) 10(11): 1-3.

Clinton, W.J. and A. Gore. (1993). *Technology for America's Economic Growth, A New Direction to Build Economic Strength*. Washington, DC: U.S. Executive Office of the President, The White House. (Available NTIS; AD-A261-553.)

Cooper, R.T. (1996). "Demise of the ICC: Is It Really 'Much Ado About Nothing'?" *Los Angeles Times* (January 3): A5.

Cordes, J.J. (1988). "The Effect of Tax Policy on the Creation of New Technical Knowledge: An Assessment of the Evidence." Chapter 11 in *The Impact of Technological Change on Employment and Economic Growth*. R.M. Cyert and D.C. Mowery, eds. Cambridge, MA: Ballinger Publishing Company, 443-480.

Council on Competitiveness. (1994). *Competitiveness Index 1994*. (July). Washington, DC: Council on Competitiveness.

Council on Competitiveness. (1988). *Picking Up the Pace: The Commercial Challenge to American Innovation*. Washington, DC: Council on Competitiveness.

Davies, R.E.G. (1964). *A History of the World's Airlines*. London, UK: Oxford University Press.

Defense Science Board. (1988). *Summer Study on the Defense Industrial And Technology Base*. Volume 1. (Final Report). Washington, DC: The Defense Science Board. (Available NTIS; AD-A202 469.)

Dempsey, P.S. and A.R. Goetz. (1992). *Airline Deregulation and Laissez-Faire Mythology*. Westport, CT: Greenwood Publishing.

Derian, J.C. (1990). *America's Struggle For Leadership in Technology*. Cambridge, MA: MIT Press.

Derthick, M. and P.J. Quirck. (1985). *The Politics of Deregulation*. Washington, DC: The Brookings Institute.

Dertouzos, M.L.; R.K. Lester; R.M. Solow. (1989). *Made in America: Regaining The Productive Edge*. Cambridge, MA: MIT Press.

Engardio, P. And D. Roberts. (1996). "The Relentless Pursuit of *Guanzi*." *Business Week* (September 30): 124-125.

Ergas, H. (1987). "Does Technology Policy Matter?" In *Technology and Global Industry: Companies and Nations in the World Economy*. B.R. Guile and H. Brooks, eds. Washington, DC: National Academy Press, 191-245.

Evenson, R.E.; P.E. Waggoner; and V.W. Ruttan. (1979). "Economic Benefits from Research: An Example from Agriculture." *Science* (September 14) 205: 1101-1107.

Flanigan, J. (1996). "Realizing Technology's Promise for the Economy." *Los Angeles Times* (March 17): D1-D4.

Fraser, R.C. and B. Maggin. (1982). *Summary and Analysis of the Role of NASA in Aeronautics Research and Development*. NASA CR-170110. Washington, DC: National Aeronautics and Space Administration. (Not available NTIS.)

Gansler, J.S. (1989). *Affording Defense*. Cambridge, MA: MIT Press.

Gellman, A.J. and J.P. Price. (1978). *Technology Transfer and Other Public Policy Implications of Multi-National Arrangements for the Production of Commercial Airframes*. NASA CR-159890. Washington, DC: National Aeronautics and Space Administration. (Available NTIS: N78-29045.)

Gidwitz, B. (1980). *The Politics of International Air Transport*. Lexington, MA: Lexington, Books.

Golich, V.L. (1992). "From Competition to Collaboration: The Challenge of Commercial-Class Aircraft Manufacturing." *International Organization* (Autumn) 46(4): 900-934.

Golich, V.L. (1990). "Liberalizing International Air Transport Services." Chapter 11 in *Privatization and Deregulation in Global Perspective*, D.J. Gayle and J.N. Goodrich, eds. New York: Quorum Books, 156-176.

Golich, V.L. (1989). *The Political Economy of International Air Safety: Design for Disaster?* London, UK: Macmillan.

Goodrich, C. (1966). "American Development Policy: The Case of Internal Improvements." Chapter 12 in *American Economic History: Essays in Interpretation*. S. Coben and F. Hill, eds. Philadelphia, PA: Lippincott, 197-207.

Gummett P. and J. Reppy. (1988). *The Relations Between Defence and Civil Technologies*. Boston, MA: Kluwer Academic Publishers.

Hall, G.R. and R.E. Johnson. (1970). "Transfers of United States Aerospace Technology to Japan." In *The Technology Factor in International Trade*. R. Vernon, ed. New York, NY: Columbia University Press, 305-358.

Hansen, J.R. (1987). *Engineer in Charge: A History of the Langley Aeronautical Laboratory, 1917-1958*. NASA SP-4305. Washington, DC: National Aeronautics and Space Administration. (Available NTIS: 87N24390.)

Hart, J.A. (1992). *Rival Capitalists: International Competitiveness in the United States, Japan, and Western Europe*. Ithaca, NY: Cornell University Press.

Hayward, K. (1986). *International Collaboration in Civil Aerospace*. New York, NY: St. Martin's Press.

Heaton, G.R., Jr. (1991). "Global Technology Policy: Is the United States Ready?" *Issues in Science and Technology* (Fall) 8(1): 36-43.

Heaton, G.R., Jr. (1989). "Commercial Technology Development: A New Paradigm of Public-Private Cooperation." *Business in the Contemporary World* (Autumn) 2(1): 87-98.

Helm, L. (1995). "Advanced Technology Program Caught in the Works of Politics." *Los Angeles Times* (November 26): D1-D3.

Hood, R. (1995). Interview (December 7) with the Acting Director, Subsonic Transportation Division, Aeronautics and Space Technology, Code R, National Aeronautics and Space Administration, Washington, DC.

Horwitch, M. (1982). *Clipped Wings: The American SST Conflict.* Cambridge, MA: MIT Press.

Hudson, C.R., Jr. (1972). *Research and Development Contribution to Aviation Progress (RADCAP). Volume 2: Military Technology, Research, and Development to Civil Aviation Programs.* NASA CR-12573. Washington, DC: National Aeronautics and Space Administration. (Available NTIS; 73N13983; AKA Project RADCAP.)

Irwin, S.M. (1993). *Technology Policy and America's Future.* New York, NY: St. Martin's Press.

Kahn, A. (1988a). "Airline Deregulation: A Mixed Bag." *Transportation Law Journal* 16: 229-251.

Kahn, A. (1988b). "Surprises of Deregulation." *American Economic Review* (May). 78(2): 316-322.

Kasper, D. (1988). *Deregulation and Globalization: Liberalizing International Trade in Air Services.* Cambridge, MA: Ballinger.

Kelly, K.; M. Oneal; G. DeGeorge; and T. Vogel. (1991). "All the Trouble Isn't in the Sky." *Business Week* (March 11) 3203: 84-85.

Kent, R.J., Jr. (1980). *Safe, Separated and Soaring: The History of Federal Civil Aviation Policy, 1961-1972.* Washington, DC: Government Printing Office.

Kerr, N.A. (1987). *The Legacy: A Centennial History of the State Agricultural Experiment Stations, 1887-1987.* Columbia, MO: Missouri Agricultural Experiment Station: University of Missouri-Columbia.

Keyworth, G.A. (1986). "Four Years of Reagan Science Policy: Notable Shifts in Priorities." In *Technological Innovation: Strategies for a New Partnership.* D.O. Gray; T. Solomon; and W. Hetzner, eds. New York, NY: North-Holland, 27-39.

Kindleberger, C.P. (1978). "Government and International Trade." *Princeton Essays in International Finance,* No. 129. (July). Camden, NJ: International Finance Section, Princeton University Press.

Kleinman, D.L. (1995). *Politics on the Endless Frontier: Postwar Research Policy in the United States.* Durham, NC: Duke University Press.

Kline, S.J. and D.E. Kash. (1992). "Government Technology Policy: What Should It Do?" *The Bridge* (Spring) 22(1): 12-18.

Knezo, G.J. (1996). "Research and Development Funding in a Constrained Budget Environment: Alternative Support Sources and Streamlined Funding Mechanisms." *CRS Report for Congress.* 96-340 SPR(April 5).

Komons, N.A. (1978). *Bonfires to Beacons: Federal Civil Aviation Policy Under the Air Commerce Act, 1926-1938.* Washington, DC: Government Printing Office.

Langford, J.S., III. (1989). *The NASA Experience in Aeronautical R&D: Three Case Studies With Analysis.* IDA Report R-319 Alexandria, VA: Institute for Defense Analyses. (Available NTIS; AD-A211,486.)

Levine, M.E. (1987a). "The Legacy of Airline Deregulation: Public Benefits, But New Problems." *Aviation Week & Space Technology* (November 9) 127(19): 161, 163.

Levine, M.E. (1987b). "Airline Competition in Deregulated Markets: Theory, Firm Strategy, and Public Policy." *Yale Journal on Regulation* (April) 4(1): 393-494.

Levine, M.E. (1981). "Revisionism Revised? Airline Deregulation and the Public Interest." *Law and Contemporary Problems* (Winter). 44(1): 179-195.

"Liberal Regulatory Environment Alters IATA's Fare-Setting Role." (1985). *Aviation Week & Space Technology* (November 11) 123(19): 102-103, 105.

Loftin, L.K. (1985). *Quest for Performance: The Evolution of Modern Aircraft.* NASA SP-468. Washington, DC: National Aeronautics and Space Administration. (Available NTIS: 85N32089.)

Logsdon, J.M. (1986). "Federal Policies Towards Civilian Research and Development: A Historical Overview." In *Technological Innovation: Strategies for a New Partnership.* D.O. Gray; T. Solomon; and W. Hetzner, eds. New York, NY: North-Holland. 9-26.

Lopez, V.C. (1994). "Reinventing Aerospace Products and Processes to Meet Environmental Goals." *AIA Facts & Perspectives* (October) 7(4): 5-12.

Lopez, V.C. and L. Yager. (1987). "An Aerospace Profile: The Industry's Role in the Economy—The Importance of R&D." *AIA Facts & Perspectives* (April): 1-4.

Lowndes, J.C. (1982). "U.S. Seeking Stronger Standards Role." *Aviation Week & Space Technology* (August 23) 117(8): 25-26.

Mann, P. (1981a). "Four Nations in Accord on Exports of Aircraft." *Aviation Week & Space Technology* (August 10): 115(6): 25.

Mann, P. (1981b). "EXIM to Eliminate Credit for Older Transports." *Aviation Week & Space Technology* (July 27) 115(4): 16.

Mansfield, E. (1986). "The R&D Tax Credit and Other Technology Policy Issues." *American Economic Review Papers and Proceedings* (May) 76: 190-194.

Mansfield, E. (1980a). "The Economics of Innovation." In *Innovation and U.S. Research.* W.N. Smith and C.F. Larson, eds. Washington, DC: ACS Symposium Series, 129, 96-97.

Mansfield, E. (1980b). "Research and Development, Productivity, and Inflation." *Science* (September 5) 209(4461): 1091-1093.

Mansfield, E. (1980c)."Basic Research and Productivity Increase in Manufacturing." *American Economic Review* (December) 70(5): 863-873.

Mansfield, E. (1968). *The Economics of Technological Change.* New York, NY: W.W. Norton.

Mansfield, E. (1965). "Rates of Return From Industrial Research and Development." *American Economic Review* (May) 55: 310-322.

Mansfield, E.; J. Rapoport; A. Romeo; E. Villani; S. Wagner; and F. Husic. (1977). *The Production and Application of New Industrial Technology.* New York, NY: W.W. Norton.

March, A. (1989). "The US Commercial Aircraft Industry and Its Foreign Competitors." Working Paper Prepared for the Commission Working Group on the Aircraft Industry. MIT Commission on Industrial Productivity.

McCraw, T.K. (1984). *Prophets of Regulation: Charles Francis Adams, Louis D. Brandeis, James M. Landis, Alfred E. Kahn.* Cambridge, MA: The Belknap Press of Harvard University Press.

Moran, T.H. (1990). "The Globalization of America's Defense Industries: Managing the Threat of Foreign Dependence." *International Security* (Summer) 15(1): 57-99.

Moran, T.H. and D.C. Mowery. (1994). "Aerospace and National Security in an Era of Globalization." Chapter 8 in *Science and Technology Policy in Interdependent Economies*. D.C. Mowery, ed. Boston, MA: Kluwer Academic Publishers, 173-198.

Mowery, D.C. (1994). "Survey of Technology Policy." Chapter 2 in *Science and Technology Policy in Interdependent Economies*. D.C. Mowery, ed. Boston, MA: Kluwer Academic Publishers, 7-56.

Mowery, D.C. (1985). "Federal Funding of R&D in Transportation: The Case of Aviation." Paper commissioned for a workshop on *The Federal Role in Research and Development* sponsored by the National Academy of Sciences, National Academy of Engineering, and Institute of Medicine in Washington, DC, 21-22 November.

Mowery, D.C. (1983). "Economic Theory and Government Technology Policy." *Policy Sciences* 16: 27-43.

Mowery, D.C. and N. Rosenberg. (1993). "The U.S. National Innovation System." Chapter 2 in *National Innovation Systems: A Comparative Analysis*. R.R. Nelson, ed. New York, NY: Oxford University Press, 29-75.

Mowery, D.C. and N. Rosenberg. (1989). "New Developments in U.S. Technology Policy: Implications for Competitiveness and International Trade Policy." *California Management Review* (Fall) 32: 107-124.

Mowery, D.C. and N. Rosenberg. (1985). "Commercial Aircraft: Cooperation and Competition Between the U.S. and Japan." *California Management Review* (Summer) 27(4): 70-92.

Mowery, D.C. and Rosenberg, N. (1982). "The Commercial Aircraft Industry." Chapter 3 in *Government and Technical Progress: A Cross-Industry Analysis*. R.R. Nelson, ed. New York, NY: Pergamon Press, 101-161.

Napier, D. (1995) "1994 Year-End Review and Forecast—Analysis," Washington, DC: Aerospace Industries Association.

National Academy of Engineering. (1987). *Strengthening U.S. Engineering Through International Cooperation: Some Recommendations for Action.* Washington, DC: National Academy Press.

National Academy of Engineering. (1985). *The Competitive Status of the U.S. Civil Aviation Manufacturing Industry: A Study of the Influences of Technology in Determining International Industrial Advantage.* Washington, DC: National Academy Press. (Available NTIS: PB88-100334.)

National Academy of Engineering. (1981). *The DoD-NASA Independent Research and Development Program: Issues and Methodology for an In-Depth Study.* Washington, DC: National Academy Press.

National Academy of Sciences, Committee on Criteria for Federal Support of Research and Development. (1995). *Allocating Federal Funds for Science and Technology.* Washington, DC: National Academy Press.

National Academy of Sciences, Transportation Research Board. (1994). *Public Investment in the Aviation Transportation Infrastructure: Summary of Workshop Proceedings.* Unpublished Report 55 of a Special Workshop sponsored by the Transportation Research Board Committee on Airfield and Airspace Capacity and Delay in Washington, DC 27 October.

National Aeronautic Association. (1996). "Boeing 777 Wins Collier Trophy." (Press Release Dated February 14.) Arlington, VA: National Aeronautic Association.

National Commission to Ensure a Strong Competitive Airline Industry. (1993). *Change, Challenge and Competition: A Report to the President and Congress.* Washington, DC: Government Printing Office.

National Research Council, Commission on Engineering and Technical Systems. (1987). *Manufacturing Technology Development: Cornerstone of a Renewed Defense Industrial Base.* Washington, DC: National Research Council. (Available NTIS: PB88-142716.)

National Research Council, Commission on Engineering and Technical Systems. (1986). *The Role of the Department of Defense in Supporting Manufacturing Technology Development.* Washington, DC: National Research Council. (Available NTIS: PB86-200672.)

National Research Council, Committee on Independent Research and Development Issues. (1981). *The DOD-NASA Independent Research And Development Program: Issues and Methodology for an In-Depth* Study. Washington, DC: National Research Council. (Available NTIS: PB82-192741.)

National Science and Technology Council. (1995). *Goals For a National Partnership in Aeronautics Research And Technology.* Washington, DC: National Science and Technology Council. (http:\\www.whitehouse.gov\wh\eop\ostp\html\aero\cvind.html)

Nelson, R.R. (1984). *High-Technology Policies: A Five-Nation Comparison.* Washington, DC: American Enterprise Institute.

Nelson, R.R. and R.N. Langlois. (1983). "Industrial Innovation Policy: Lessons Learned from American History." *Science* (February 18) 219: 814-818.

Nelson, R.R. and N. Rosenberg. (1993). "Technical Innovation and National Systems." Chapter 1 in *National Innovation Systems: A Comparative Analysis.* R.R. Nelson, ed. New York, NY: Oxford University Press, 3-21.

Newhouse, J. (1982). *The Sporty Game: The High-Risk Competitive Business of Making and Selling Commercial Airliners*. New York, NY: Alfred A. Knopf.

Paulisick, J.G. (1972). *Research and Development Contribution to Aviation Progress (RADCAP). Volume 1: Contributions of Military Technology, Research, and Development to Civil Aviation Programs*. NASA CR-129572. Washington, DC: National Aeronautics and Space Administration. (Available NTIS; 73N13982; AKA Project RADCAP.)

Pavitt, K. and W. Walker. (1976). "Government Policies Toward Industrial Innovation: A Review." *Research Policy* (January) 5(1): 11-97.

Peltz, J.F. and R. Vartabedian. (1996). "Clinton to Propose Faster Production of C-17 Cargo Jet." *Los Angeles Times* (February 23): D1-D7.

Peterson, B.S. and J. Glab. (1994). *Rapid Descent: Deregulation and the Shakeout in the Airlines*. New York, NY: Simon & Schuster.

Pinelli, T.E. (1991). *The Relationship Between the Use of U.S. Government Technical Reports by U.S. Aerospace Engineers and Scientists and Selected Institutional and Sociometric Variables*. Washington, DC: National Aeronautics and Space Administration. NASA TM-102774. (Available NTIS; 91N18898.)

President's Airport Commission. (1952). *The Airport and Its Neighbors*. Washington, DC: Government Printing Office. (AKA the Doolittle Report.)

Preston, E. (1987). *Troubled Passage: The Federal Aviation Administration During the Nixon-Ford Term, 1973-1977*. Washington, DC: Government Printing Office.

Proctor, P. (1988). "U.S., International Carriers Maneuver to Exploit Transports' Extended Range." *Aviation Week & Space Technology* (March 14) 128(11): 199, 201-202.

Rae, J.B. (1968). *Climb to Greatness: The American Aircraft Industry 1920-1960*. Cambridge, MA: MIT Press.

Ravitz, R. (1994). "Hearing on NASA Aeronautics." *AIA Newsletter* (March) 6(7): 4.

Reich, R.O. (1991). *The Work of Nations: Preparing Ourselves for 21st Century Capitalism*. New York, NY: Alfred A. Knopf.

Reich, R.O. (1990). "Who Is Us?" *Harvard Business Review* (January-February) 90(1): 53-64.

Reppy, J. (1985). "Military R&D and the Civilian Economy." *Bulletin of Atomic Scientists* (October): 10-14.

Rochester, S.I. (1976). *Takeoff at Mid-Century: Federal Civil Aviation Policy in the Eisenhower Years, 1953-1961*. Washington, DC: Government Printing Office.

Roland, A. (1985). *Model Research: The National Advisory Committee for Aeronautics 1915-1958*. Volume. 1. NASA SP-4103. Washington, DC: National Aeronautics and Space Administration. (Available NTIS; 85N23683.)

Rosenberg, N. (ed.). (1982). *Inside the Black Box: Technology in Economics*. New York, NY: Cambridge University Press.

Rosenberg, N.; A. Thompson; and S.E. Belsley. (1978). *Technological Change and Productivity Growth in the Air Transport Industry*. NASA TM-78505. Washington, DC: National Aeronautics and Space Administration. (Available NTIS: 79N10997.)

Rothwell, R. (1982). "Government Innovation Policy: Some Past Problems and Recent Trends." *Technological Forecasting and Social Change* 22: 3-30.

Sabbagh, K. (1996). *Twenty-First Century Jet: The Making and Marketing of the Boeing 777*. New York, NY: Scribner.

Schacht, W.H. (1996). "The Federal Role in Technology Development," *CRS Report for Congress*. 95-50 SPR Washington DC: U.S. Congress, Congressional Research Service.

Schwartz, B. (ed.). (1973). *The Economic Regulation of Business and Industry: A Legislative History of U.S. Regulatory Agencies*, Volume IV. New York, NY: Chelsea House Publishers.

Schwartz, J.T. (1992). "America's Economic-Technological Agenda for the 1990s." *Daedalus* (Winter) 121(1): 139-165.

Shifrin, C.A. (1987). "Competitive Airline Market Spurs IATA to Develop New Services." *Aviation Week & Space Technology* (November 16) 127(20): 45-46.

Smith, A. (1776). *The Wealth of Nations*. New York, NY: (Reissued in 1937 by) Random House Modern Library.

Smith, M.R. (ed.). (1985). *Military Enterprise and Technological Change*. Cambridge, MA: MIT Press.

Soderquist, R.J. (1996). Interview (September 20) with the Japan and Korea Program Specialist, National Science Foundation, Arlington, VA.

Solberg, C. (1979). *Conquest of the Skies: A History of Commercial Aviation in America*. Boston, MA: Little, Brown and Co.

Spitzer, R. (1996). "Opening Remarks of the Vice President for Engineering, Commercial Airplane Group, Boeing Commercial Aircraft Group." Made May 10 at the NASA Langley Research Center, Hampton, VA.

Stratzheim, M.H. (1969). *International Airline Industry*. Washington, DC: The Brookings Institute.

Taneja, N.K. (1976). *The Commercial Airline Industry: Managerial Practices and Regulatory Policies*. Lexington, MA: D.C. Heath.

Taneja, N.K. (1970). *U.S. International Aviation Policy*. Lexington, MA: Lexington Books.

Teece, D.J. (1981). "The Market for Know-How and the Efficient International Transfer of Technology." *The Annals of the American Academy of Political and Social Science* (November) 458: 81-96.

Terleckyj, N.E. (1985). "Measuring Economic Effects of Federal R&D Expenditures: Recent History With Special Emphasis on Federal R&D Performed in Industry." Paper commissioned for a workshop on *The Federal Role in Research and Development* sponsored by the National Academy of Sciences, National Academy of Engineering, and Institute of Medicine in Washington, DC, 21-22 November.

Terleckyj, N.E. (1974). *Effects of R&D on the Productivity Growth of Industries: An Exploratory Study.* Washington, DC: National Planning Association.

Thayer, F.C., Jr. (1965). *Air Transport Policy and National Security: A Political Economic and Military Analysis.* Chapel Hill, NC: University of North Carolina Press.

Thornton, R. (1970). *International Airlines and Politics.* Ann Arbor, MI: Bureau of Business Research, Graduate School of Business Administration, University of Michigan.

Tickner, J.A. (1987). *Self-Reliance Versus Power Politics: The American and Indian Experiences in Building Nation States.* New York, NY: Columbia University Press.

Tornatzky, L.G. and D. Luria. (1992). "Technology Policies and Programmes in Manufacturing: Toward Coherence and Impact." *International Journal of Technology Management* 7(1/2/3): 141-157.

Tornatzky, L.G. and M. Fleischer. (1990). *The Process of Technological Innovation.* Lexington, MA: D.C. Heath.

Tyson, L.D. (1992). *Who's Bashing Whom? Trade Conflict in High-Technology Industries.* Washington, DC: Institute for International Economics.

Tyson, L.D. (1988). "Competitiveness: An Analysis of the Problem and a Perspective on Future Policy." Chapter 3 in *Global Competitiveness: Getting the U.S. Back on Track.* M.K. Starr, ed. New York, NY: W.W. Norton, 95-120.

U.S. Congress, House Committee on Government Operations. (1966). *Creating a Department of Transportation, Part 1: Hearings Before a Subcommittee of the Committee on Government Operations.* Washington, DC: Government Printing Office.

U.S. Congress, House of Representatives. (1988a). *Aviation Safety Research Act of 1988, To Accompany H.R. 4686.* Washington, DC: Government Printing Office.

U.S. Congress, House of Representatives. (1988b). *Omnibus Trade and Competitiveness Act of 1988: Conference Report to Accompany H.R. 3.* Washington, DC: Government Printing Office.

U.S. Congress, House of Representatives. (1988c). *Effects of Federal Economic Policies on U.S. Industries.* A Report prepared by the Congressional Research Service for the use of the Subcommittee on Oversight and Investigations of the Committee on Energy and Commerce. (October). Washington, DC: Government Printing Office.

U.S. Congress, House of Representatives. (1980). *A Thorough Critique of Certification of Transport Category Aircraft by the Federal Aviation Administration.* Sixteenth Report by the Committee on Government Operations. Washington, DC: Government Printing Office.

U.S. Congress, Office of Technology Assessment. (1994). *Federal Research and Technology for Aviation.* OTA-ETI-610. Washington, DC: Government Printing Office.

U.S. Congress, Office of Technology Assessment. (1991). *Competing Economies: America, Europe, and the Pacific Rim.* OTA-ITE-498. Washington, DC: Government Printing Office. (Available NTIS; PB92-115575.)

U.S. Congress, Office of Technology Assessment. (1989a). *Holding the Edge: Maintaining the Defense Technology Base.* OTA-ISC-420. Washington, DC: Government Printing Office.

U.S. Congress, Office of Technology Assessment. (1989b). *Holding the Edge: Maintaining the Defense Technology Base: Summary.* OTA-ISC-421. Washington, DC: Government Printing Office.

U.S. Congress, Office of Technology Assessment. (1988). *The Defense Technology Base: Introduction and Overview: A Special Report.* OTA-ISC-374. Washington, DC: Government Printing Office.

U.S. Congress, Office of Technology Assessment. (1986). *Research Funding as an Investment: Can We Measure the Returns?* OTA-TM-SET-36. Washington, DC: Government Printing Office.

U.S. Department of Commerce, International Trade Administration. (1983). *An Assessment of U.S. Competitiveness in High Technology Industries.* Washington, DC: Government Printing Office.

U.S. Department of Defense. (1988). *Bolstering Defense Industrial Competitiveness: Preserving Our Heritage, Securing Our Future.* Report to the Secretary of Defense by the Undersecretary of Defense (Acquisition).

U.S. Department of Transportation, Federal Aviation Administration. (1994). *The Federal Aviation Administration Plan for Research, Engineering and Development: 1994.* Washington, DC: Government Printing Office. (Available NTIS; AD-A281 733.)

U.S. Executive Office of the President, Office of Science and Technology Policy. (1987). *National Aeronautical R&D Goals: Agenda for Achievement.* Washington, DC: Executive Office of the President, Office of Science and Technology Policy. (Available NTIS; PB90-179094.)

U.S. Executive Office of the President, Office of Science and Technology Policy. (1985). *National Aeronautical R and D Goals: Technology for America's Future.* Washington, DC: Office of Science and Technology Policy. (Available NTIS; 87N-12405.)

U.S. Executive Office of the President, Office of Science and Technology Policy. (1982a). *Aeronautical Research and Technology Policy. Volume 1. Summary Report.* Washington, DC: Executive Office of the President, Office of Science and Technology Policy. (Available NTIS; 83N-17452. AKA the Keyworth Report.)

U.S. Executive Office of the President, Office of Science and Technology Policy. (1982b). *Aeronautical Research and Technology Policy. Volume 2. Final Report.* Washington, DC: Office of Science and Technology Policy. (Available NTIS; 83N-23268; AKA the Keyworth Report.)

U. S. Executive Office of the President, Office of Science and Technology. (1966). *Alleviation of Jet Aircraft Noise Near Airports.* Washington, DC: Office of Science and Technology.

Vander Meulen, J.A. (1991). *The Politics of Aircraft: Building an American Military Industry.* Lawrence, KS: University Press of Kansas.

von Hippel, E. (1994). "'Sticky Information' and the Locus of Problem Solving: Implications for Innovation." *Management Science* (April) 40(4): 429-439.

von Hippel, E. (1976). "The Dominant Role of Users in the Scientific Instrument Innovation Process." *Research Policy* (July) 5(3): 212-239.

Wiegel, E. (1995). "Aeronautics: Developing Cutting-Edge Technology to Keep America the World's Leader in Aviation." *NASA Facts OnLine.* NASA FS-1995-08-007-HQ. Washington, DC: National Aeronautics and Space Administration.
(http://www.dfrc.nasa.gov/pais/hq/html/fs-007-hq.html)

Wilkins, B. (1984). "Airline Deregulation: Neoclassical Theory as Public Policy." *Journal of Economic Issues* (June) 18(2): 419-427.

Chapter 3

U.S. Public Policy and the Production of Federally Funded Aeronautical Research and Technology

W.D. Kay
Thomas E. Pinelli
Rebecca O. Barclay

SUMMARY

Chapter 3 provides an overview of the U.S. government's role in the production of knowledge, with a particular focus on public policy that affects aeronautical research and technology (R&T). The background section defines knowledge and explores the relationship(s) between science, technology, and technological innovation. The history of U.S. science and technology policy, both generally and specifically with respect to aeronautical R&T, is presented. Next, we define R&T and examine U.S. aeronautical R&T policy from the perspective of knowledge production, transfer, and use. The chapter concludes with implications for diffusing the results of federally funded (U.S.) aeronautical R&T.

INTRODUCTION

The Boeing 777 embodies much of what is state-of-the-art in aeronautical research and technology (R&T) (Sabbagh, 1996). Some is new—used for the first time in a U.S.-manufactured large commercial aircraft (LCA). Some is evolutionary and represents what David (1986) refers to as "advances in systemic knowledge of the useful arts" (p. 376). Much of the R&T incorporated in the B-777 is unique to the manufacturer, a result of the Boeing Company's corporate investment in research and development (R&D). Some of the "advances," including digital flight controls, the glass cockpit, quiet

engine nacelles, flight management systems, graphite-epoxy structures, and transonic supercritical airfoils, are outgrowths of the
publicly funded aeronautical R&T conducted by the National Aeronautics and Space Administration (NASA). With over 300 purchase
orders placed by airlines from around the world, the B-777 is poised
to become a commercial success (Reinhardt, Browder, and Engardio, 1996).

In many ways, the Boeing 777 is also a U.S. public policy success. With its contribution to trade, its coupling with national
security, and its symbolism of technological strength, the U.S. aircraft industry holds a unique position in the nation's industrial
and economic structure. In 1994, the worldwide sale of U.S. aircraft produced the largest trade surplus of any U.S. manufacturing
industry. The growth of the U.S. aircraft industry, especially the
LCA sector, has been the result of a strong partnership between
government, industry, and academia (U.S. Executive Office of the
President, 1982).

The tradition of U.S. public (policy) support for aeronautics
dates back to the earliest days of the aircraft industry (Mowery,
1994). U.S. public policy has continuously funded generic civil and
military aeronautical R&T, supported the dissemination of the research results, and encouraged the adoption of the technology since
1917, first through the National Advisory Committee for Aeronautics
(NACA) and later through NASA. In fact, NACA has been cited by
scholars as a model for federal involvement in civilian research and
development (Shapley and Roy, 1985). Through its research facilities and wind tunnels, NASA continues to develop precommercial
aeronautical R&T with broad application to industry and to maintain a system that distributes the documented (i.e., text and visual)
results of its aeronautical R&T. Processes used by NASA to diffuse
the results of its aeronautical R&T include informal technical
discussions; NASA technical reports; conferences, symposia, and
workshops; contracts with industry; noncontract cooperative programs; technology demonstration programs; and government and
industry personnel exchange programs.

The irony, however, is that the system at the heart of the success story is virtually unique. The U.S. federal government spends
billions of dollars annually in the area of R&D and the generally
very high quality of the American science and technology establishment is well recognized. With the exception of aeronautics, along

with agriculture and medicine, very little in the way of time, attention, or resources has been devoted to any step of the R&T process beyond the simple act of creating the knowledge itself. The atypical nature of federal aeronautical R&T policy (and its associated funding) becomes clear when it is contrasted with U.S. science and technology policy in general. However, ongoing support for publicly funded aeronautical R&T is not guaranteed. Almost since the inception of the NACA, opponents in Congress and the private sector have periodically opposed federal funding for aeronautics (Roland, 1985). In 1981, during the first term of the Reagan Administration, significant questions were raised about justifying continued support of nonmilitary aeronautical R&T (U.S. Executive Office of the President, 1982). As recently as August 1996, the Congressional Budget Office (CBO), in its review of domestic discretionary spending, revisited the arguments for federal support for producers of LCA (U.S. Congress, Congressional Budget Office, 1996).

This chapter provides an overview of the U.S. government's role in the production of knowledge, with a particular focus on public policy that affects aeronautical R&T. The background section defines knowledge and explores the relationship(s) between science, technology, and technological innovation. The history of U.S. science and technology policy, both generally and specifically with respect to aeronautical R&T, is presented. Next, we define R&T and examine U.S. aeronautical R&T policy from the perspective of knowledge production, transfer, and use. The chapter concludes with implications for diffusing the results of federally funded (U.S.) aeronautical R&T.

BACKGROUND

Knowledge is a building block, an essential ingredient of technological innovation. Innovation is necessary for creating *new* processes, products, systems, or services. Advances in knowledge are widely regarded as major sources of *improvements* in existing processes, products, systems, or services. The rate at which knowledge is created, diffused (i.e., spread, distributed, transmitted), and absorbed or utilized influences the rate of technological innovation and progress (Mansfield, 1984, 1981). Advancements in technological innovation require investments in capital, labor, and knowledge to produce tangible results that are sold in today's global markets.

A firm that produces processes, products, or systems or delivers services is deemed competitive if it can provide goods and services of superior quality or lower costs than its competitors. Countries with many competitive firms typically have high rates of economic growth and standards of living, hence the interest on the part of governments in technological innovation and technical progress.

For many economists, knowledge is the catalyst that helps allocate resources and makes a free market function. Economists now view knowledge as an engine of change and embrace it in their theoretical constructs. Many economists see knowledge living up to Daniel Bell's (1973) prediction: Knowledge will replace capital and energy as the primary wealth-creating assets, just as capital and energy replaced labor and land (Haeckel and Nolan, 1993). In an economic sense, knowledge differs from other so-called commodities or resources: (a) it is not depleted with use, it is sharable, and traditionally, it has had no intrinsic value; (b) it is difficult to distinguish between knowledge and the medium in which it is contained; (c) except for knowledge-based products and services designed to be sold, most knowledge lacks markets in which value can be determined by supply and demand; (d) unlike other so-called commodities or resources, the overwhelming importance of knowledge is as a public good (Noll, 1993); and (e) numerous individuals located at various points across the globe can possess the same knowledge, unlike other commodities or resources (Brinberg, Pinelli, and Barclay, 1995; Brinberg and Pinelli, 1993). The past 20 years have witnessed the propensity of knowledge to cross national boundaries, a phenomenon that observers have labeled the *globalization of knowledge*. The boundary–spanning propensity of knowledge is due mainly to improvements in communications (e.g., the Internet), transportation (e.g., international air travel), and the fact that developed and developing countries are spending more on creating and acquiring knowledge. The globalization of knowledge requires U.S. firms and industries involved in innovation to construct and employ strategies for exploiting extramural research and development and to develop systems for acquiring knowledge produced outside of the U.S. as a means of increasing their international reach (Ives and Jarvenpae, 1993). (See Drucker, 1993; Nelson and Winter, 1982, 1977; Nelson and Wright, 1992; Miles and Robins, 1992; Romer, 1990; Rosenberg and Birdzell, 1990; Schmookler, 1966; Schumpeter, 1951; Schwartz, 1992; and Scott, 1989 for a discussion of the role of knowledge in economic theory and growth.)

Knowledge has been variously labeled, described, and defined. It can be scientific or technical, embodied or disembodied, tacit or explicit, and product or process knowledge. *Scientific knowledge* is embodied in the laws, principles, and theorems of a specific discipline (e.g., Newton's three laws of motion in physics). It is easily codified and is unlikely to be altered by language and culture. *Technical knowledge* tends to be narrowly focused or specific; it is not always predictable, and it does not necessarily spring from scientific knowledge. Technical knowledge is not the application of scientific knowledge. It may be applicable to a particular technology like the manufacture of aircraft, but it is not easily transferred or applied to another technology. It is cumulative to an individual, groups of individuals, and organizations; it is derived from learning-by-doing (Arrow, 1962a; von Hippel and Tyre, 1995; Wright, 1936) or learning-by-using, and it is not easily or accurately codified. For example, after a particular jet engine has been in use for a decade, the cost of maintenance may have declined to only 30% of the initial level as a result of learning-by-using (Rosenberg, 1982). [See Copp and Zanella (1993, Chapter 4) and Constant (1980) for a discussion of how scientific and technical knowledge combine to make flight possible.]

Learning-by-doing and learning-by-using generate a substantial amount of what Rosenberg (1982) defines as *embodied* and *disembodied* knowledge. In the first case, early experience with a new technology leads to a better understanding of the relationship between design characteristics and performance that permits subsequent improvements, which over time lead to an optimal design of an aircraft, system, or component. Optimization may be achieved by applying advancements made in other areas like materials, manufacturing, or miniaturization. *Disembodied* knowledge results in slight but often continuing changes in design and operation that result from the experience of making or operating an aircraft. Prolonged experience with an aircraft, system, or component produces knowledge that can be used to lengthen the service life of an aircraft or reduce its operating cost. Rosenberg makes the point that disembodied knowledge is critical to aircraft design and manufacture because it is only through actual operation that the true performance (i.e., characteristics and costs) and full potential of a new aircraft can be determined. Vincenti (1992, 1990) provides excellent definitions and examples of knowledge as applied to aeronautical engineering. *Inside the Black Box—Technology and Economics*

(Rosenberg, 1982, Chapter 6) offers convincing examples of both learning-by-doing and learning-by-using within the context of aircraft production.

When a firm or organization innovates, that is, creates or improves a process, product, system, or service, it generally does so by using both *tacit* and *explicit* knowledge. Polanyi (1966) provides the following basic definitions for these two types of knowledge: *Tacit* knowledge is personal, context-specific, and therefore, hard to formalize and communicate; *explicit* knowledge is codified and refers to knowledge that is transmittable in formal, systematic language and includes grammatical statements, mathematical expressions, specifications, and manuals. Bateson (1973) offers the following distinctions between these two types of knowledge: Tacit knowledge tends to be experiential and subjective. It is derived from practice, created "here" and "now" in a specific context, and entails what Bateson refers to as an "analog" quality; whereas explicit knowledge tends to be rational and objective. It is derived from what is known and accepted, was created "there" and "then," and it is oriented toward context-free theory. Tacit knowledge cannot always be codified because it often contains an important dimension of "know-how." Individuals may know more than they are able to articulate. When knowledge has a high tacit component, it is extremely difficult to transfer without personal contact, demonstration, and involvement. Indeed, in the absence of close human contact, the diffusion of knowledge is sometimes impossible (Teece, 1981). Von Hippel (1994) argues that tacit, unlike other forms of knowledge, is often costly, difficult, and sometimes impossible to acquire, transfer, and use owing to the attributes of tacit knowledge itself. For an explanation of tacit and explicit knowledge within the context of *technical knowledge*, see Alic, Branscomb, Brooks, Carter, and Epstein (1992). Nonaka and Takeuchi (1995, Chapter 2) have proposed a theory of knowledge creation relative to the dynamics of technological innovation that contains four modes of knowledge conversion: tacit to tacit (socialization), tacit to explicit (externalization), explicit to explicit (combination), and explicit to tacit (internalization).

It is interesting to note that most of these theoretical accounts of knowledge are tightly bound to some concept of *utilization*. However it is defined and whatever its source, the assumption (particularly among economists) is that knowledge is a commodity to be used by institutions, firms, or individuals, not something that is created simply as an end in itself. It is only when considering the role of government (at least in the U.S.) that knowledge creation becomes detached from knowledge use.

Although innovation is an investment decision generally made within a firm or organization, it is also influenced, to a large extent, by public policy and the resulting laws and regulations that affect the mobilization of capital and labor (David, 1986). As suggested previously, U.S. public policy plays a major role in creating the knowledge that drives innovation through direct funding of science and technology. In addition, policy decisions *potentially* have a significant impact on knowledge diffusion. The federal government supports a range of programs, from those that simply collect knowledge and make it accessible, to those that actively seek to couple knowledge with potential beneficiaries. Finally, the adoption and utilization of knowledge and technological innovation can be influenced through a variety of programs that provide special considerations, incentives, credits, and protections affecting investments in labor and capital.

Science, Technology, and Technological Innovation

Science and technology play a major role in technological innovation through the creation of knowledge. The process of innovation, applied within a capitalist system, relies primarily on market forces and the use of human, technical, and financial resources to create new and improve existing processes, products, systems, and services. However, investments in creating and improving knowledge differ from investments in physical capital in that the results, once produced, become, in principle, free goods unless steps are taken to prevent that from happening (Matthews, 1973). This creates a basic public policy dilemma. If exclusive rights are granted to those investing in creating and improving knowledge, from a social perspective, the use of that knowledge becomes wastefully restricted. If no such rights are granted, no incentive exists to invest in creating and improving knowledge. Without knowledge, there is no innovation. Innovation begets technical progress and economic growth, and economic growth fosters technological innovation, creates jobs, and generally raises the standard of living. Therefore, from a public policy perspective, federal funding of science and technology provides considerable social benefits.

The process of technological innovation interacts with government at three essential levels (Ergas, 1987a). *First,* the government promotes the generation of this critical public good—technological innovation—through the production and purchase of goods and services that provide for the nation's defense and security. *Second,* the government facilitates the development of technological innovation and the creation of new and improved processes, products, systems, and services by funding science and technology. *Third,* the government supports the education and training of engineers and scientists, provides a legal system for defining and enforcing property rights and contracts, and maintains a uniform system for conducting commerce (i.e., weights and measures, currency values, and interest and exchange rates).

Although technical progress and economic growth and competitiveness are inextricably linked to advances in science and technology, the relationship among these variables is often sometimes stochastic. The general assumption that technology grows out of or depends upon science for its development suggests that the metamorphosis from science to technology is a continuous process (or follows a linear path) from basic research (science) through applied research (technology) to development (utilization). Indeed, the prevalence of this mindset may help explain the use of the conventional phrase "scientists and engineers."

It also explains the commonly held belief of policymakers that basic research somehow "automatically" produces technological advances and economic growth. When questioned, these same policymakers respond that "market forces" produce the aforementioned technological advances and economic growth (Mowery and Rosenberg, 1979). In fact, this belief forms the basis of much of U.S. science policy. As Logsdon (1986) points out:

> There continues to be broad acceptance of the proposition that fundamental research in science is a vital investment likely to yield good for U.S. society and that the primary responsibility of the federal government with respect to science and technology is to support long-term research. (p. 10)

In reality, however, the relationship between science and technology is not continuous, but is perhaps best described as a series of interactions that are based on need rather than a normal progression (Allen, 1977). Science and technology are embedded in a larger political, economic, and social system that constrains some and promotes other developments in each arena. Therefore, science and technology exist within "causally complex interactive systems" (Kitschelt, 1991; Kline, 1985; Shapley and Roy, 1985) characterized by dynamic reciprocity (Golich, 1992).

Science is an introverted activity. It studies problems that are usually generated internally by logical discrepancies or inconsistencies or by anomalous observations that cannot be accounted for within the present intellectual framework. Indeed, scientists are said to do their best work when investigating problems of their own selection and in a manner of their own choosing (Bush, 1945; Amabile, 1983; Amabile and Gryskiewicz, 1987). The output of science is knowledge that is regarded by scientists essentially as a free good. The expectation within the scientific community is that knowledge will be made universally available through presentations at conferences and society meetings and publication in scholarly and professional journals.

Technology, on the other hand, is an extroverted activity; it involves a search for workable solutions to problems. When technology finds solutions that are workable and effective, it does not pursue the *why* (Salomon, 1984). Moreover, the output of technology is frequently a process, product, system, or service. Technological knowledge is not easily or completely codified, nor is it freely communicated. Unlike science, the output of technology is *not* made universally available. Technology successfully functions *only* within a larger social environment that provides an effective combination of incentives and complementary inputs into the innovation process. Technology is a process dominated by engineers rather than scientists (Landau and Rosenberg, 1986).

As *social organizations*, science and technology have very different attitudes and values concerning knowledge and its ownership. Generally speaking, the scientific community tends to view knowledge as a *public consumption good*, while engineers (or, more precisely, the firms that employ them) regard it as a *private capital good* (Dasgupta, 1987). Thus, the rules of the two communities concerning the communication (i.e., disclosure) and ownership of knowledge are fundamentally different. Scientists are obligated to disclose their findings and to submit them for critical inspection to other members (i.e., peers) of the scientific community. Hence, the ability of scientists to communicate freely and openly is critical.

Moreover, knowledge production takes place in the context of two very different reward systems. In science, rewards are based on priority of discovery or the *rule of priority*. This rule acts as an incentive for scientific discovery, and serves to promote public disclosure of that discovery. Thus, scientists are compelled to take privately created knowledge and to make that knowledge accessible to the scientific community and the general public. The rule of priority also precludes a second or third place winner because from a societal point of view, there is no value added when a discovery is made a second or third time.

As is often the case, policies based on this arrangement have led to some unintended consequences, particularly when combined with other government objectives. For example, U.S. science policy supports and encourages free and open communication of federally funded science as a key factor in promoting progress. Economic and national security policies, on the other hand, favor restrictions and controls on the flow of scientific information and embrace numerous regulations that have fueled pragmatic efforts to limit or discourage the open communication of ideas and information in many scientific and technical disciplines. This has become a particularly important issue in international collaborative ventures.

The attitudes toward and values concerning knowledge and its ownership are different in the *technology* community. Disclosure is neither required nor expected in the technology community. In fact, secrecy is often the rule because the reward system is linked to "privately capturable rents" that can be earned from the production of knowledge. Technological knowledge is considered proprietary, is afforded patent protection, can be a "trade" secret, and is often the subject of industrial espionage. Its use can be licensed to those willing to pay an agreed upon price. What is interesting about knowledge in the technology community is that, although patent protection is used to make private knowledge public, it does not attempt to place a monetary or societal value on the knowledge. The worth and risk associated with obtaining privately capturable rents from knowledge are left to the marketplace. Although being first to the market often results in financial success, it is often possible for a process, product, system, or service to be imitated or improved upon. In contrast with science, there *can* be second and third place winners in technology. In that case, society usually looks for what Dasgupta (1987) refers to as a "best-practice" technique that is often subjectively determined by the consumer as a compromise between price and performance. From a societal standpoint, having more than one winner stimulates the marketplace and can produce multiple and perhaps better products for less.

The differences between science and technology and their respective attitudes toward knowledge creation can be clearly seen in the recent episode regarding so-called "cold fusion." In 1989, two chemists from the University of Utah announced the discovery of a (relatively) simple technique for creating nuclear fusion essentially in a jar of water. (Fusion is a process in which large quantities of energy are produced by "fusing" the atoms of lighter elements like hydrogen or helium. In a sense, it is the opposite of the more familiar process of nuclear fission, which "splits" heavier atoms like those of uranium or plutonium; see Kay, 1992). Had this claim been proven (it was never verified), it would have revolutionized the thinking in a number of scientific fields. Accordingly (and, as this discussion suggests, quite naturally); scientists from around the world sought to find out more about the research, and acquire the experimental data, in order to attempt to replicate the chemists' findings. This proved to be extremely difficult, however, because the discovery also had immense economic and commercial implications, and these provided a powerful incentive for the researchers (and their university) to withhold many critical details of their work for some time.

As noted previously, the U.S. federal government plays a major role in supporting basic scientific research, believing that support by itself promotes innovation, economic growth, and general social well-being. Clearly, the U.S. has been quite successful in developing and maintaining (at least for the present) a world-class scientific establishment across virtually all fields. Given the highly complex relationship—on a number of levels—between science and technology, or, more generally, between knowledge creation and knowledge diffusion and use, however, there is some question as to whether or not American science policy is actually meeting its broader objectives. It is to this question that we now turn our attention.

A BRIEF REVIEW AND EXAMINATION OF
U.S. SCIENCE AND TECHNOLOGY POLICY

Throughout its history, the United States government "has recognized the importance of technological advancement and has undertaken programs designed to stimulate the development and application of new technologies" (Irwin, 1993, p. 85). Even so, public support (i.e., funding) of science and technology is a relatively recent development. Prior to World War II, federal involvement in science and technology was limited to specific agencies, departmental programs, and functions such as national defense that were considered to be the sole responsibility of the federal government (Brooks, 1986). Apart from providing for the nation's defense, only agriculture, aviation, and medicine had successfully garnered continued federal support.

World War II and the Cold War that followed permanently transformed the federal government with respect to financial support for science and technology. According to Logsdon (1986), technological superiority was viewed as the major U.S. asset in a global geopolitical and ideological contest. For years, there was little debate in Congress over the legitimacy of the expenditure of funds to obtain or retain that superiority and to provide for the national defense. In recent years, however, the use of public money to develop generic technology not directly related to the mission of a federal agency or intended for eventual use in producing a process, product, system, or service for the marketplace has engendered considerable congressional debate. Central to the debate has been the question of the proper role of the federal government in fostering the development of technology not tied to national defense or to the mission of a specific federal agency.

Our review of U.S. science and technology policy is brief and selective. Designed to establish a contemporary (i.e., World War II to 1996) perspective, it is neither analytical nor comprehensive; we have left that task to others. (See, for example, Alic, Branscomb, Brooks, Carter, and Epstein, 1992; Averch, 1985; Bilich, 1989; Bopp, 1988; Branscomb, 1993; Dupree, 1986; England, 1982; Ergas, 1987b; Golden, 1993a, 1993b; Gray, Solomon, and Hetzner, 1986; Irwin, 1993; Kevles, 1995; Kleinman, 1995; Lambright and Rahm, 1992; Lomask, 1976; Mowery, 1994; National Academy of Engineering, 1993; Roessner, 1988; Shapley and Roy, 1985; Smith, 1990; Tassey, 1992; Tornatzky and Fleischer, 1990.)

Pre-World War II

In detailing the involvement of the United States government in science and technology, it is important to understand that the present day involvement in and expenditures for science and technology represent a radical departure from earlier arrangements. The only reference to a "science policy" in the United States Constitution relates to Congress' power to enact laws relating to patents (Article 1, Section 8, Clause 8):

> Congress shall have the power ... to promote the progress of science and useful arts, by securing for limited times to authors and inventors the exclusive right to their respective writings and discoveries.

A patent is a protection granted by a government to an inventor for a defined period of time to prevent unauthorized exploitation of an invention. (It is distinguished from trademark and copyright protection.) Patent protection allows an individual, firm, or organization to disclose an invention publicly without diluting the potential return that can be obtained from the investment. From a public policy perspective, patent protection offers a private reward for public disclosure. The reward stems from the return that is earned through the sale or use of the newly created or improved knowledge that is secured by a patent.

With regard to more general support for research and development, however, Rosenberg (1985) makes the following point:

> In spite of the permissive implications of the *general welfare* clause of the U.S. Constitution, Federal support for science and technology prior to World War II had been limited sharply by a strict interpretation of the role of the government. (p. 92, emphasis added)

To be sure, there was a modest amount of federal involvement in science and technology during the years before World War II. Examples include the establishment of the Coast and Geodetic Survey, the U.S. Geological Survey, the Weather Bureau, the National Bureau of Standards (now the National Institute of Standards and Technology), the National Institutes of Health and a host of public works projects and medical programs conducted under the auspices of the military (Dupree, 1986).

Nevertheless, such involvement often led to controversy. A case in point is the creation of the Smithsonian Institution in Washington, DC. The actual debate in Congress regarding the acceptance of James Smithson's 1829 bequest lasted for years. Congress did not actually found the Smithsonian until 1846. Support for agricultural research, perhaps the oldest federal commitment to science and technology, was also subjected to protracted congressional debate. The first Morrill Act, which established the nation's land-grant colleges, passed both houses of Congress in 1859 but was vetoed by President Buchanan. The Act became law in 1862, and the Department of Agriculture was established the same year. Still, little research was performed by the land-grant colleges until the Hatch Act, which provided federal funding for agricultural experiment stations, was passed in 1887.

Until the Second World War, agriculture continued to occupy its long-standing position of being the only sector of the economy to receive research funding support from the federal government (Mowery and Rosenberg, 1989). Although the federal agricultural research establishment, with its land-grant colleges, agricultural experiment stations, and the state agricultural extension programs that bring research results to the farmer and the farmers' agricultural problems back to the research establishment, has been criticized in recent years, the system is widely regarded as "probably the most successful government effort to date in stimulating the innovation process" (Coles, 1983, p. 36).

With the creation of NACA in 1917, the United States government established a federal research laboratory (i.e., the Langley Memorial Aeronautical Laboratory) that began to "investigate the scientific problems involved in flight and to give advice to the military air services and other aviation services of the government" (P.L. 63-271). Even considering the obvious connection to national defense, passage of this legislation did not come easily. After considerable debate, the legislation creating NACA was passed as a rider to the Naval Appropriations Act of 1916.

NACA has been described as arguably the most important and productive aeronautical research establishment in the world (Roland, 1985). Between its creation in 1917 and its demise in 1958, it published more than 16 000 technical reports that were sought after and exploited by aeronautical engineers and scientists throughout the U.S. and abroad. Many of these reports, classics in the fields of aerodynamics and aeronautics, are still used and referenced; the data contained therein are essential to understanding fundamental aerodynamics (Anderson, 1974). NACA has been cited by scholars as a model for federal involvement in science and technology (Teich, 1985), and precommercial research cooperation between the public and private sectors (Nelson, 1982).

In 1930, Congress created the National Institute of Health (NIH—the plural Institutes would come later) as part of the Public Health Service to conduct "research into chronic diseases" (Swain, 1962). As was the case with agriculture and aeronautics, extending the federal government into this area was not without controversy. NIH's funding for 1935 ($2 million) slipped through Congress as part of the Social Security Act. By the end of the decade, however, the agency had become firmly established. The National Cancer Institute was added to NIH in 1937 (Kleinman, 1995), with many more to follow (e.g., the National Institute of Mental Health, the National Heart Institute) after World War II (at which time the Institute became plural). In carrying out its funding responsibilities, NIH adopted a policy of supporting the best research early on, regardless of geographic area, and tended to interpret its "disease" mandate rather broadly. During the 1930s, for example, recognizing the medical potential of radioisotopes, it provided funding to Ernest Lawrence of Berkeley for the development of his early cyclotron (Kevles, 1995).

It is interesting to note that, despite the concerns over the proper federal role in R&D, the activities of all three of these organizations—the federal agricultural research establishment, NACA, and NIH—included science, applied science, technology, and a system for coupling the user with knowledge resulting from these programs. We will return to this point later.

World War II–1958

The entry of the United States into World War II brought dramatic changes with respect to the federal government's financial support for science and technology. From 1940 to 1945, total federal expenditures for science and technology rose from $83.2 million to $1.3 billion, with the bulk of the funds expended by the Department of Defense (Mowery and Rosenberg, 1989). After the war, this expansion continued, not only in terms of funding, but also in the number of federal agencies with science- and technology-related responsibilities: the Atomic Energy Commission (1946), the expanded (and renamed) National Institutes of Health (1948), the National Science Foundation (1950), NASA (1958), and many others.

Two noteworthy events precipitated this transformation. The *first* was the Manhattan Project, the successful completion of which ushered in the age of "big science" and helped shape the postwar imagination about the "more constructive possibilities of science when it could be applied in an organized and systematic way to the pursuit of human goals" (Rosenberg, 1985, p. 5). The *second* event was the establishment of the Office of Scientific Research and Development (OSRD) under the direction of Vannevar Bush. Prior to World War II, almost all federally funded science and technology had been performed by the government itself, that is, by civil servants in federal research centers and laboratories. Bush's approach with OSRD was to enter into contracts with the private sector (i.e., academia and industry). The success of these contractual arrangements with the private sector dramatically influenced the organization of federally funded science and technology in the post-World War II era.

In particular, the OSRD arrangement effectively split the responsibility for conducting scientific and technological research between the private and public sectors. Overall resource allocation decisions remained with the federal government while academia and industry were given considerable autonomy in terms of problem formulation and approach. As Teich (1985) has observed:

What has emerged since the Second World War is a system in which the federal government has become the dominant purchaser of R&D, but without, at the same time, becoming the dominant performer of R&D. Thus, the unique institutional development has been the manner in which the federal government has accepted a vastly broadened financial responsibility for R&D without arranging simultaneously for its in-house performance. Rather, private industry has become the main performer of federal R&D, and the university community the main performer of the basic research component. Thus, the enlarged role of the federal government in the support of R&D has been carried out within an institutional framework dominated by contractual relationships between the federal government and private performers. (p. 2)

Officially, post-World War II government funded research in science and technology was to serve as a means to improve health, defend the nation, fuel economic growth, and provide jobs in new industries (Smith, 1990). It is important to note, however, precisely how these public benefits were to be brought about. The policy of separating "government-as-user" from "government-as-sponsor" gave rise to the concepts of "spinoff" and "dual-use" technologies; that is, R&D outputs that serve the immediate purpose(s) of the government but can also be "commercialized" by the private sector. According to this view, commercialization would occur "automatically" as a by-product of federal support for basic research in the academic community and agency (e.g., DoD and NASA) support for applied research in the private sector (Alic, Branscomb, Brooks, Carter, and Epstein, 1992).

1959–1979—"The Age of Big Science"

After World War II, the government was committed to making the United States scientifically self-sufficient as a way of ensuring that the nation would not be dependent upon other countries for science as it had been on English and German scientists during and after World War II (Graham, 1985). Two key changes resulted from this commitment to fund research as a national priority. *One*, for the first time in peacetime, the U.S. government provided a major share of funds for industrial R&D with funding rising from 40% in the 1950s to 60% in the 1960s. *Two*, the government pursued a policy of increasing the number of companies performing industrial R&D, thus making industrial R&D a competitive business in the United States. Two assumptions underlie this science policy decision (Graham, 1986). *First* was the belief in "technology push," the view that technological innovation is an unidirectional transfer of knowledge from basic science through development and commercialization. *Second* was the belief in breakthrough innovations (e.g., transistor) that would lead to entirely new industries (e.g., computers), creating thousands of highly skilled, high paying jobs. Federal funding during this period focused heavily upon two industries—electronics and aeronautics—both highly dependent upon physics for their knowledge base and "seemingly the most applicable to national defense." Increased federal funding for industrial R&D created a tremendous demand for and dramatically increased the salaries of engineers and scientists. To increase the size of the pool of engineers and scientists, the government responded by passing legislation—most notably the National Defense Education Act of 1958—that provided fellowships for students pursuing advanced degrees in engineering and the physical sciences.

The "Age of Big Science" might have run its course in a decade had it not been for the Cold War and Sputnik (McDougall, 1985). (Most authorities agree that it came to an end about the mid- to late-1970s.) Events like the environmental movement, the energy crisis, the "War on Cancer," the "War on Poverty," and the Vietnam War helped change and shape United States science and technology policy during the 1960s and 1970s in two ways and created competition for funds allocated for industrial R&D (National Academy of Sciences, National Academy of Engineering, Institute of Medicine, and National Research Council, 1986). *First*, the Cold War changed the focus of federal science and technology policy to military superiority. *Second*, particularly during the earlier part of this period, federally funded science and technology came to be seen as a means of solving such pressing social problems as inadequate housing, declining environmental quality, and even poverty (Averch, 1985). The familiar refrain from this period was the question: "If we can put a man on the moon, why can't we ...?" (Nelson, 1977). Finally, beginning in the early 1960s, productivity in the U.S. began to drop and the U.S. as a nation and some industries in particular had begun to loose market shares to foreign competitors. By the mid-1960s (and continuing to today), economists and policymakers had begun to view federally funded science and technology as a force that could be used to spur the economic growth and the competitiveness of the United States in the emerging global economy.

The notion that the federal government should play a direct and active role in funding science and technology not directly associated with a specific mission (e.g., defense) or a particular sector of the economy (e.g., agriculture) was embraced by the Kennedy Administration (Averch, 1985). The goal was to design programs that would increase industrial innovation, encourage industry to do socially useful research, maintain technological leadership, and boost the nation's economic output (Logsdon, 1986). As part of their justification, the administration used the theoretical economic concept of *externalities* (Eads, 1974). In this context, "externalities" refer to a situation in which private firms underinvest in technological innovation, which constitutes a failure of the market. (For an economic explanation of *externalities* and market failures, see Baer, Johnson, and Merrow, 1977). It is the failure of the market that justifies intervention by the government in the form of financial support for science and technology. Action on the part of the government is meant to *supplement*, not *supplant*, the market through the creation (i.e., production) of knowledge.

In 1972, Richard Nixon delivered the first presidential message to Congress on science and technology. The Nixon administration's rationale for funding civilian technology was largely economic and inspired by decreasing productivity, a negative balance of trade, and rising unemployment. Nixon's message was significant for two reasons. *One*, for the first time a President declared as a matter of policy that it is appropriate for the federal government to fund private R&D to the extent that the market mechanism is not effective in bringing needed innovation into use. *Two*, Nixon called for a new partnership in science and technology that would bring together the federal, state, and local governments; the private sector; and the university research community in a coordinated effort to serve the national interest (Nixon, 1972).

The Carter Administration undertook a comprehensive policy review intended to answer the question, "What actions should the U.S. government take to encourage industrial innovation?" Two recommendations resulting from the review are noteworthy. *One*, the federal government should fund university research programs that will stimulate cooperation with industry, and *two*, funding of generic (product and process) technology should be strengthened and recognized as a federal responsibility. Some of the initiatives proposed by Carter were incorporated in the Stevenson-Wydler Technology Innovation Act of 1980 (P.L. 96-480). However, by the late 1970s, federal funding for industrial R&D ended primarily because policymakers had lost faith in the belief that R&D was one of the levers by which "government could promote technological innovation and productivity" (Nelson and Langlois, 1983, p. 815).

1980–1992

In terms of federal funding for science and technology, the Reagan Administration attempted to "return" to the consensus position, comprised of three elements, that emerged after World War II. *First*, the federal government would assume the responsibility for funding basic scientific research. *Second*, the "mission" agencies, namely Defense, Energy, and NASA, would invest heavily in both basic research and technology. *Third*, in most cases, responsibility for commercial technology development would be left to the private sector. Federal involvement in commercial technology development would be the exception, not the rule (i.e., the existence of externalities would have to be proven, rather than assumed).

Three actions of the Reagan Administration are noteworthy. *First*, defense R&D spending grew substantially, as did reliance upon spinoff and dual-use technologies (i.e., technologies having both military and commercial potential) to serve the government's immediate needs while producing technology that could also be "commercialized" by the private sector (Alic, Branscomb, Brooks, Carter, and Epstein, 1992). *Second*, grounded in the belief that research beneficial to a particular industry should not be funded by the federal government, funds for non-military aeronautics research (i.e., NASA aeronautical R&T) were sharply reduced in the President's proposed budget for fiscal year 1982. Strong congressional and industry reaction to the proposed reductions prevented their implementation. A subsequent interagency review of aeronautical R&T concluded that (a) aeronautics as a technology was still extremely important to national defense and to the economic competitiveness of the U.S. in the world economy; (b) the aircraft industry by itself would underinvest in aeronautical R&T; and (c) no other indirect or direct budgetary means for appropriate government support to aeronautics would be superior to the current system (U.S. Executive Office of the President, 1982). *Third*, statutes and regulations were used to restrict the dissemination of data, information, and knowledge resulting from federally funded science and technology (Burger, 1986; Relyea, 1994). For example, the Department of Defense Authorization Act of 1984 (P.L. 98-94) empowered the Secretary of Defense to refuse a Freedom of Information Act (FOIA) request for and to withhold from public disclosure, unclassified data and information. Kempf (1990) points out that other federal agencies have been unsuccessful in obtaining a similar exemption. He further states that the failure of policymakers to differentiate between the transfer of knowledge resulting from federally funded R&D as an end process and as an integral part of an agency's R&D program and dissemination activities has obscured efforts to develop government information policy.

Concerns about the competitive position of the United States in the global marketplace raised questions regarding the role that the federal government should play in assisting U.S. industry to develop and use new technology for competitive purposes (National Academy of Sciences, National Academy of Engineering, Institute of Medicine, and the National Research Council, 1995). During the 1980s and early 1990s, policymakers came to view federally funded R&D as an untapped national resource that could be used to (a) improve the productivity of American industries and business, (b) furnish the basis for increased automation, (c) enhance existing industries and product lines, and (d) create new business opportunities. Significant policy changes governing intellectual property (e.g., patents, patent rights, trade secrets, and copyright) created with or resulting from federally funded science and technology were enacted. (An extensive discussion of intellectual property rights and information policy may be found in U.S. Congress, Office of Technology Assessment, 1986.) These changes were part of an effort by the federal government to (a) increase the quantity and quality of technological innovation at the federal laboratories, (b) encourage its commercialization by the private sector, and (c) stimulate a sagging U.S. economy (Wessel, 1993).

The Stevenson-Wydler Technology Innovation Act of 1980 and the Patent and Trademark Laws Amendments, 1980, also known as the Bayh-Dole Act of 1980 (P.L. 99-502), formed the foundation for implementing this policy change. The Small Business Innovation Development Act of 1982 (P.L. 97-219) required all federal agencies that spend a significant amount on R&D to set aside a small proportion of those funds to support R&D projects of interest to them within small businesses. The National Cooperative Research Act of 1984 (P.L. 98-462) modified the operation of antitrust laws to encourage the formation of R&D joint ventures and provided for antitrust (both civil and criminal) immunity for joint ventures. The Federal Technology Transfer Act of 1986 (P.L. 99-502) amended the Stevenson-Wydler Innovation Act of 1980 by permitting the director of a government-owned federal laboratory to enter into cooperative research and development agreements (CRADAs) with companies and consortia of companies to pursue projects of mutual interest. P.L. 99-502 also established the Federal Laboratory Consortium for Technology Transfer and stated that technology transfer is the responsibility of each laboratory engineer and scientist.

Executive Order (E.O.) 12591, issued by President Reagan in 1987, facilitated collaboration among federal laboratories, state and local governments, universities, and the private sector for the purpose of transferring technology to the marketplace. The Omnibus Trade and Competitiveness Act of 1988 (P.L. 100-418) authorized the National Institute of Standards and Technology (NIST) to establish an Advanced Technology Program of competitive awards to firms and consortia of firms on a matching basis to support early-stage, generic technology development projects. The same act authorized what has become the Manufacturing Extension Partnerships program in NIST, which provides grants to nonprofit consortia and state and local governments for the transfer of technology and technical assistance to manufacturing firms, with an emphasis on small- and medium-sized firms. The National Competitiveness Technology Transfer Act of 1989 (P.L. 101-189) amended the Stevenson-Wydler Innovation Act of 1980 by permitting the director of any *government-owned and contractor-operated (GOCO)* federal laboratory to enter into CRADAs with companies and consortia of companies to pursue projects of mutual interest. An amendment to the Defense Authorization Act of 1993 (P.L. 102-484) permitted the DoD to fund a variety of technology development, technology deployment, and technical education and training activities in firms, consortia of firms, and nonprofit organizations. This authority was used to create the Technology Reinvestment Program of 1993 (National Academy of Sciences, National Academy of Engineering, Institute of Medicine, and the National Research Council 1995). (Keyworth, 1986 provides a view of science and technology policy during the Reagan Administration.)

The early 1990s found increased support for an enhanced federal role in the development of commercial technology and elements within the Congress clamoring for a more coherent federal technology policy. Support for an enhanced federal role in the development of commercial technology was spurred on by the end of the cold war. As Irwin (1993) points out:

> As lawmakers struggle to define new missions for America's vast defense technology infrastructure and to help the nation's defense industries convert to civilian production, the possibility of promoting greater commercial competitiveness by redirecting federal R&D funding has become increasingly politically salient. (p. 92)

The Bush Administration, however, opposed any attempts to directly subsidize the development of commercial technologies.   The funding of "critical" or "enabling" technologies was viewed by the administration as a form of industrial policy. The Bush administration had no desire to start "picking winners." In September 1990, the Bush Administration submitted a report to the Congress entitled, *U.S. Technology Policy* (U.S. Executive Office of the President, 1990). The report acknowledged that the federal government had a responsibility for participating "with the private sector in precompetitive research on generic, enabling technologies that have the potential to contribute to a broad range of government and commercial applications" (p. 5). However, the role of the federal government should be indirect and would emphasize the use of monetary and fiscal policies, provide a stable regulatory and legal environment, and enforce intellectual property rights. (See Bromley, 1993 for an overview of science and technology policy in the Bush Administration.)

1993–Present

In the policy document, *Technology for America's Economic Growth, A New Direction to Build Economic Strength* (Clinton and Gore, 1993), the Clinton Administration indicated that the traditional role of limiting government to supporting basic science and mission-oriented applied research was "not appropriate for today's profound challenges" (p. 1). The administration has shown both a willingness and a desire to shift the DoD research agenda to support dual-use technologies (i.e., those having both commercial and military potential) that might offer economic returns. Indeed, the Clinton Administration has demonstrated a desire for a more "activist" technology policy and a much greater willingness to shift the DoD and Department of Energy (DoE) research budgets into more commercially promising technologies, especially those that will help the competitive position of the U.S. automobile industry. This shift in policy has not been without controversy, however. The Department of Commerce's Advanced Technology Program (ATP), which helps to underwrite the cost of innovative, high-risk research in industry, has been under attack by Republicans in Congress since it was first established. These attacks have gained strength since the Republicans took control of the House and Senate in 1992, and ATP has twice come close to being eliminated in the last two budget cycles.

Three aspects of the administration's technology policy are noteworthy. *First* is the belief that government can and should play a key role in helping private firms develop and profit from innovations. *Second* is the emphasis placed on using communication and information technology to further economic and social development. *Third* is the emphasis on the use of information technology to promote the dissemination of federal information (e.g., economic, environmental, and technical). (Gibbons, 1994 and Smith, 1994 review science and technology policy in the Clinton Administration.)

A Critique of U.S. Science and Technology Policy

The economic vitality of the United States depends in large part on technological progress. However, U.S. science and technology policy alone will not solve the country's economic ills. Although consensus is building for involving the federal government in promoting the development of commercial technology, agreement about the role that government should play in the process remains the subject of considerable debate. Those who call for a more direct role for the federal government see technology policy as the simple and logical extension of science policy. They cite the "activist" involvement of the governments of Europe and Japan in shaping market forces as further justification. Those who promote an indirect role advocate a "market strategy" or a market-driven approach with the federal government funding knowledge production and providing the private sector with the incentives needed for the successful development of commercial technology. The complexity and dynamics of the market, they argue, require that decisions about which technologies to develop remain with the private sector.

Our primary international competitors take a slightly different position with respect to developing technology policy. If the *private* sector is likely to underinvest in knowledge production, it is also equally or more likely to underinvest in knowledge deployment and diffusion. Conversely, if societal value accrues from the public's funding of knowledge production, it is also equally or more likely that the *public* will enjoy a substantial return from an investment in knowledge deployment and diffusion.

Science and technology policy formulation represents a compromise among conflicting objectives (Mowery, 1994). There does seem to be a broad consensus on the need for supporting the creation of knowledge through the use of federal funds. As an extension of science policy, technological knowledge is viewed as a public good, the production of which should be subsidized to offset the effects of market failure (Arrow, 1962b). Hence, the use of public funds to support the production of technological knowledge is widely viewed as legitimate.

What U.S. technology policy often overlooks are the high costs of transferring and exploiting technological knowledge. These costs are often high enough to affect the utilization of that knowledge by the private sector. As the earlier theoretical discussion indicates, there is no economic return on the investment of public funds until, or unless, this knowledge is utilized. In reality, federally funded technological knowledge is both a public and a private good. Therefore, technology policy should involve more than using public funds to subsidize the creation of knowledge alone (Mowery, 1994; 1983).

In the largest context, critics contend that U.S. technology policy should be more closely tied to economic and trade policy and less to national security and defense policy. They claim that the current U.S. approach to research and technology does not take into account the factors and influences that motivate innovation. Doing so, they state, would increase the likelihood of successful technological innovation. Early on, Pavitt and Walker (1976), Nelson and Winter (1977), and, more recently, Branscomb (1991) criticized this "supply-side" policy (i.e., producing knowledge) because it encourages the creation of knowledge but not its adoption, and knowledge transfer and utilization are very inadequately served by market forces. Furthermore, they argue, this policy provides little incentive for knowledge transfer and utilization. They conclude that the government would better serve the public's interest by formulating policies that encourage the production, transfer, and use of knowledge resulting from federally funded science and technology.

Mowery (1983) states that a theoretical economic framework that ignores or does not account for the effective transmission and utilization of technological knowledge is inappropriate for developing federal technology policy. Any such framework ignores the abilities and limitations of organizations engaged in technological innovation to exploit the results of extramural research (i.e., the results of federally funded science and technology), thus ignoring knowledge production, transfer, and utilization as equally important components of the process of technological innovation. Mowery (1985) offers the following observation:

> This theoretical [economic] framework focuses primarily on the pu-
> tative undersupply of research and bases its recommendations for
> policy on this market failure. However, for policy purposes, the
> distribution and utilization of the results of research and develop-
> ment are crucial. An exclusive focus on the R&D support policies
> of the Federal government, without some cognizance of the sub-
> stantial diffusion support component of the policy structure, yields
> conclusions that differ substantially from those of an analysis that
> attempts to incorporate both the technology supply and technology
> adoption incentives operating within the overall policy framework.
> (p. 34)

Commenting on technology diffusion, public policy, and industrial competitiveness in the United States, David (1986) concludes
that promoting successful technological innovation rests more with
knowledge transfer and utilization than with knowledge production.
In making this point, he offers the following commentary:

> Innovation has become our cherished child, doted upon by all
> concerned with maintaining competitiveness and renewing failing
> industries; whereas diffusion (i.e., transfer and utilization) has
> fallen into the woeful role of Cinderella, a drudge-like creature who
> tends to be overlooked when the summons arrives to attend the
> technology policy ball. (p. 377)

The apparent success of federal involvement in agricultural,
aviation, and medical research—activities which, incidentally, have
become somewhat less controversial over time—contrasts sharply
with the results of the federal government's attempts to stimulate
technological innovation in other sectors of the civilian economy.
As shown above, in the case of agriculture, aviation, and medicine,
the government utilized a long-term, holistic, loosely coordinated
policy environment approach to technological innovation that offers
science, applied science, technology, and a system for coupling
knowledge with people who use it.

In short, our criticism of U.S. technology policy stems from our
belief that it emphasizes knowledge production, but ignores its
transfer and use. By comparison, the governments of some of our
major competitors (e.g., Germany and Japan) in the global marketplace have adopted what Branscomb (1992) and Ergas (1987a,

1987b) describe as "diffusion-oriented" technology policy that emphasizes the deployment and absorption of knowledge in addition to its production.

AN EXAMINATION OF U.S. AERONAUTICAL RESEARCH AND TECHNOLOGY POLICY

The formulation of (U.S.) public policy is a complex process that takes place within the larger context of national and party politics. According to Majchrzak (1984), public policy is not made, it accumulates; it is usually incremental, not revolutionary in character. Policy debates are shaped by a multitude of factors, assumptions, and values (Bobrow and Dryzek, 1987). Recent technology policy debates center on the appropriate relationship between the public and private sectors, the state's role in the marketplace, and the relationship of technology policy to national security, trade, domestic, and foreign policy objectives and goals. Technology policy studies are abundant. There is, however, a shortage of studies that evaluate technology policy, studies conducted to determine if the policy recommendations were implemented and the effects (outcome) of the recommendations were measured in terms of the policy objectives and goals.

Aeronautical R&T Defined

Our examination focuses on aeronautical R&T policy as a subset of U.S. technology policy. The following definitions are used to establish a context for our examination. The process of R&D is divided into three sequential phases—*R&T development, technology demonstration*, and *systems development*. The three phases represent a continuous spectrum of related activities in the R&D process in which technology is produced, demonstrated, and transferred into systems development. The key criterion in applying the definitions of R&T development and technology demonstration is the nature of the intended results, not the physical mechanisms selected to conduct the effort.

Aeronautical R&T consists of the first two phases. *R&T development* embodies activities that (a) have a high degree of uncertainty associated with their outcome; (b) are aimed at producing physical understanding, new concepts, design data, and validated design procedures for aircraft systems, subsystems, and components; (c) encompass theoretical analysis, laboratory investigations, and flight-testing experimental aircraft; (d) are conducted in DoD budget categories 6.1 (research), 6.2 (exploratory development), and some 6.3A (advanced engineering); and (e) in the NASA budget, include the R&T base and some systems technology. *Technology demonstration* consists of activities primarily aimed at demonstrating improved subsystem or system characteristics to provide the development and manufacturing decisionmaker with the confidence that the anticipated improved level of performance is indeed achievable in a new system. Technology demonstration efforts are characterized by testing configurations similar to the intended application and by a modest degree of uncertainty in outcome. These efforts are the final technology activity prior to systems development and take place before a decision is made to develop a specific system. To ensure the effective and efficient diffusion of knowledge, the aircraft industry is usually involved in some of the large-scale testing stages of technology development and during technology demonstration. *Systems development* consists of activities aimed at producing a specific aircraft or specific aircraft system for operational use. In the DoD system, development begins with budget category 6.3B.

The Scope of Aeronautical R&T Policy

Aeronautical R&T policy is operationally defined as including the entire range of government activities that support the diffusion (i.e., production, transfer, and utilization) of knowledge. These activities include (a) long-term and stable funding for research programs, the objectives of which have been agreed upon by those who will produce and use the knowledge; (b) a research infrastructure consisting of engineers, scientists, and support personnel, and facilities (e.g., wind tunnels), equipment (e.g., simulators), and tools (e.g., computational fluid dynamics); (c) funding in the form of research grants and stipends (e.g., paid tuition) to universities to educate and train engineers, engineering technologists, and scientists; (d) support for the technologies relating to and needed for product manufacture, production, and systems integration; (e) a system for assigning ownership of and protecting intellectual property; and (f) an inclusive program for managing knowledge as an intellectual asset that has both tactical and strategic value.

Aeronautical R&T Policy Examined

A variety of studies conducted over the past 15 years has examined aeronautical R&T from a number of perspectives. We selected the six studies whose scope included the manufacture of LCA (Table 3.1). Branscomb (1993) and Mowery (1994) contend that U.S. technology policy emphasizes the production of knowledge and neglects its transfer and use. We examined a subset of U.S. technology policy—aeronautical R&T policy—to determine if the criticisms of Branscomb (1992), Mowery (1994), and others are justified. Specifically, we attempted to determine if U.S. aeronautical R&T policy recommends a program or methodology for knowledge production, transfer, and use as equally important parts of the knowledge diffusion process.

Aeronautical Research and Technology Policy (1982, 1985, and 1987). Since 1917, the U.S. has provided national leadership and significant financial support for the development of aeronautical R&T, first through NACA and later through NASA. Federal support for aeronautical R&T was considered essential for U.S. preeminence in both military and civil aviation. However, during the first year of the Reagan presidency, significant questions were raised concerning the appropriateness and effectiveness of current U.S. aeronautical R&T policy and the U.S. government's role in the support of aeronautical R&T. In 1982, the Office of Science and Technology Policy (OSTP) empaneled a senior-level steering group and an interagency working group to review national aeronautical R&T policy. The purpose of the review was to determine if the potential benefits justify continued federal investment in aeronautical R&T. Furthermore, the study sought to identify the proper role of government in aeronautical R&T and determine if the existing institutional framework satisfies these roles.

The study was conducted over six months, beginning in February 1982. Six federal agencies were represented in the interagency working group; two observers, one from industry and one from academia, also participated. Background data were collected from an extensive questionnaire sent to U.S. aeronautical leaders in industry and academia. Although the working group considered the broad range of aeronautics, they concentrated on the air vehicle. To establish an historical perspective, the working group reviewed past and current government policies on aeronautical research. The

Table 3.1. U.S. Aeronautical Policy Studies and Reports

No.	Year	Title	Author/Sponsor
1	1982	*Aeronautical Research and Technology Policy*	U.S. Executive Office of the President, OSTP
2	1984	*A Competitive Assessment of the U.S. Civil Aircraft Industry*	U.S. Department of Commerce, ITA
3	1985	*National Aeronautical R and D Goals: Technology for America's Future*	U.S. Executive Office of the President, OSTP
4	1985	*The Competitive Status of the U.S. Civil Aviation Manufacturing Industry: A Study of the Influences of Technology in Determining International Industrial Competitive Advantage*	National Academy of Engineering
5	1987	*National Aeronautical R&D Goals: Agenda for Achievement*	U.S. Executive Office of the President, OSTP
6	1989	*The NASA Experience in Aeronautical R&D: Three Case Studies with Analysis*	J.S. Langford, III, Institute for Defense Analysis
7	1991	*Competing Economies: America, Europe and the Pacific Rim, Summary*	U.S. Congress, OTA
8	1992	*Aeronautical Technologies for the Twenty-First Century*	National Research Council
9	1994	*Federal Research and Technology for Aviation*	U.S. Congress, OTA
10	1994	*High-Stakes Aviation: U.S.-Japan Technology Linkages in Transport Aircraft*	National Research Council
11	1995	*Goals for a National Partnership in Aeronautics Research and Technology*	U.S. Executive Office of the President, OSTP
12	1996	*Endless Frontier, Limited Resources: U.S. R&D Policy for Competitiveness*	The Council on Competitiveness

working group examined the status and development of foreign aeronautical industries challenging the U.S. militarily and in the civil marketplace; addressed the potential for military and civil R&T and provided examples of potential benefits; examined, from an economic perspective, federal involvement in aeronautical R&T for civil applications; and summarized current NASA/DoD programs, facilities, and personnel in aeronautical R&T. Lastly, the working group focused on the development, evaluation, and selection of national (knowledge production) goals, government and agency (i.e., DoD, FAA, and NASA) roles, and policy alternatives for the operation of aeronautical facilities and the dissemination and control (knowledge transfer and use) of research results. The working group made two recommendations specific to knowledge transfer and use: the U.S. needs to collect selectively and distribute the results of foreign aeronautical R&T and to determine the extent to which the dissemination of federally funded aeronautical R&T might give foreign aeronautical firms an undue (competitive) advantage.

In March 1985, the science advisor to the President and the director of OSTP established an aeronautical policy review committee composed of government, industry, and academic experts to keep track of the implementation of the recommendations made in 1982. The committee made specific (knowledge production) goals in three areas—subsonic, supersonic, and transatmospheric—that, if accomplished, should sustain U.S. preeminence in aeronautics into the next decade. Impediments to attaining the three goals were said to be federal contracting procedures, regulatory (antitrust) policy, federal tax policy, and the need for continued (knowledge production) research and technology development activities. The review committee made no recommendations concerning knowledge transfer and use and did not comment on the two knowledge transfer and use recommendations made in 1982.

In February 1987, the science advisor to the President and the director of OSTP established an aeronautical policy review committee composed of 16 leaders from government, industry, and universities to consider the evolving international scene in aeronautics and America's future in it. The committee established three specific (knowledge production) goals—subsonic, supersonic, and transatmospheric—that would serve as the centerpiece of a national aeronautics strategy. According to the committee, the depth of foreign aeronautical resolve should not be underestimated; therefore, the committee recommended an eight-point national "call to action" plan to achieve these national goals and ensure that the U.S. aeronautical enterprise remains a viable competitor in the world aviation marketplace. The eight points contained in the national plan are as follows:

1. Increased industry investment in R&D as a means of producing the new technology required for global competitiveness.

2. Aggressive pursuit of the National Aero-Space Plane (NASP) program and the need to protect certain sensitive information for national security and U.S. competitiveness.

3. Development of the fundamental technology and design capability to support the research, development, and production (RD&P) of a long range supersonic transport.

4. Expansion of domestic R&D collaborative efforts by creating an environment that reflects the new era of global competition. The committee recommended that the federal agencies (e.g., NASA and DoD) increase the rate of technology transfer to the private sector for commercial development by balancing the early open publication of research results with the need for early domestic technology transfer. Moreover, industry and government should expand cooperative research, technology development, and licensing arrangements under the terms of the Federal Technology Transfer Act of 1986.

5. Concentration of NASA's knowledge production efforts in research areas—composite materials, propulsion, numerical simulation, and laminar flow—that promise high payoffs.

6. Government and industry cooperation to help strengthen U.S. universities as centers for basic research and as centers of excellence.

7. Improved development and integration of advanced design, processing, and computer-integrated manufacturing technologies to transform emerging R&D results into affordable U.S. products.

8. Enhanced safety and capacity of the national air transportation system through advanced automation and new vehicle concepts, including vertical and short takeoff and landing (VTOL/STOL) aircraft.

Implicit in the committee's recommendations is the recognition that the results of federally funded aeronautical R&T are an intellectual asset that should be managed as a national resource for attaining and maintaining a competitive advantage in the emerging global marketplace. The committee recommended that discoveries having commercial potential be transferred as quickly as possible to U.S. industry (early domestic dissemination) while ensuring that these same discoveries not be made available to potential foreign competitors through publication in the open literature.

The Competitive Status of the U.S. Civil Aviation Manufacturing Industry (1985). The Committee on Technology and International Economic and Trade Issues of the National Academy of Engineering conducted seven industry-specific studies, one of which focused on the U.S. civil aviation manufacturing industry. The objectives of this study were to (a) identify global shifts of industrial technological capacity on a sector-by-sector basis, (b) relate those shifts in international competitive industrial advantage to technological and other factors, and (c) assess further prospects for technological change and industrial development. The study was undertaken by a panel of experts, invited experts from outside the panel, and government agency and congressional representatives presenting current governmental views and deliberations. As part of the study, the panel developed a brief historical description of the industry and an assessment of the dynamic changes that have occurred and are anticipated in the next decade. The primary charge of the panel was to develop a series of policy options for consideration by public and private policymakers.

The panel concluded that the civil aviation manufacturing industry, which holds a unique position in the nation's industrial structure, is experiencing profound change. The implications of the change are of national importance. The panel recommended that (a) the U.S. government take a more active role in trade administration, focusing on those sectors that are significant in foreign trade; (b) the U.S. strive for balancing national security or foreign policy objectives with those of strengthening the U.S. economy and preserving the U.S. position in advanced technology; (c) DoD, FAA, and NASA reexamine the mechanisms of working with civil aircraft manufacturers to ensure that maximum advantage is taken of opportunities for dual-use capabilities in technology development; and (d) NASA's role, activities in, and resources available for technology validation be reconsidered. Lastly, the panel concurred with the OSTP (1982) recommendations that NASA collect and evaluate foreign aeronautical R&T results and distribute them to the U.S. technical community and that U.S. aircraft manufacturers and firms in the supporting infrastructure assign a higher priority to collecting and evaluating foreign technology and build the capacity to do so.

Aeronautical Technologies for the Twenty-First Century (1992). In 1991, NASA asked the Aeronautics and Space Engineering Board of the National Research Council (NRC) to assist in assessing the current status of aeronautics in the United States and to help identify the technology advances necessary to meet the challenges of the next several decades. The Aeronautics and Space Engineering Board established the Committee on Aeronautical Technologies, which defined an approach to helping NASA determine the appropriate level and focus of its near-term technology development (knowledge production) efforts to maintain a leadership role in the years 2000 to 2020. Based on these projections, the committee identified the high-leverage technologies that offer the most significant advances in aeronautics to ensure long-term competitiveness for U.S. aircraft, engines, and components, and to enhance performance and safety in the total air transportation system.

The Committee identified seven needs that must be addressed by the U.S. aeronautics community—NASA, the FAA, aircraft manufacturers, and air carriers—if the United States is to maintain or increase its share of the global aircraft market. These needs are (a) lower cost and greater convenience, (b) greater capacity to handle passengers and cargo, (c) reduced environmental impact, (d) greater aircraft and air traffic control (ATC) system safety, (e) improved aircraft performance, (f) more efficient technology transfer from NASA to industry, and (g) reduced product development times. Neither the Committee's charter nor its makeup allowed detailed consideration of the latter two needs, so they were not discussed in detail in the report.

Goals for a National Partnership in Aeronautics Research and Technology (1995). In 1995, the National Science and Technology Council (NSTC) and OSTP undertook an effort to establish national goals for aeronautical R&T. With input from academia and industry, the NSTC performed a national assessment that will be used to guide the direction and character of the federal investment in U.S. civil aviation technology for the future. The resulting document will provide a blueprint for a public-private partnership for the future. The vision for the partnership is world leadership in aircraft, engines, avionics, and air transportation system equipment for a sustainable, global aviation system. Three (knowledge production) goals are key to achieving this vision: (a) maintaining the superiority of U.S. aircraft and engines; (b) improving the safety, efficiency, and cost effectiveness of the global air transport system; and (c) ensuring the long-term environmental compatibility of the aviation system. As part of the goal of maintaining the superiority of U.S. aircraft and engines, the panel made specific (knowledge production) recommendations for subsonic aircraft, high-speed aircraft, aircraft design and manufacturing, and the aeronautics R&D infrastructure. The report contained no recommendations concerning the transfer and use of federally funded aeronautical R&T to the U.S. aircraft industry.

A Critique of U.S. Aeronautical R&T Policy

Several factors combine to affect R&D and technological innovation. The movement of capital and labor is far less subject to natural and national boundaries than at any other time in history. Increasingly, product development in many high-technology industries and sectors of the economy is the result of partnerships and collaborations involving numerous firms in several countries. The commerce associated with collaboratively developed products is occurring in a competitive, global economy. Knowledge, which is recognized as fundamental to R&D and technological innovation, is now viewed as a key intellectual asset that, if managed effectively, can help firms compete successfully. As a result of deliberate policy decisions, many national governments are taking an activist, and in some cases interventionist, role in the process of technological innovation. Government policies in such nations as Korea and Taiwan support the importation of explicit and tacit knowledge to help build an R&D knowledge base. Industrial policies in such nations as Japan call for active participation by firms in international collaborative ventures that support the acquisition and transfer of knowledge through learning-by-doing and learning-by-using (National Research Council, 1994). Industrial and trade policies of such nations as France, Germany, and Japan reflect these nations' interventionist approaches to product development and deployment as a means of ensuring employment for a skilled high-tech workforce, enhancing national prestige, and forcing U.S. manufacturers to maintain competitive prices (U.S. Congress, Office of Technology Assessment, 1991; U.S. International Trade Commission, 1993).

U.S. aeronautical R&T policy appears to recognize that the LCA portion of the U.S. aircraft industry has become global in terms of the partnerships and collaborations required for the RD&P of new aircraft. The Boeing 777 is representative of this trend, in terms of both its design and sale. U.S. aeronautical R&T policy also recognizes that other nations are committed to becoming players and competitors in the RD&P of LCA and, consequently, the governments of these nations are actively supporting efforts at various levels of involvement to ensure each nation's competitive success. These and other factors have weakened and negated many of the assumptions on which the post-World War II system of (U.S.) federal support for aeronautics is based. The emergence of civilian and commercial interests as the primary drivers of technological and product development in the U.S. has altered the assumption that defense R&D, with its promised "dual-use" technologies and "spin-off" products, would sustain technological innovation and ensure U.S. preeminence (Council on Competitiveness, 1996).

The decentralized post-World War II system of federal support for aeronautics may have been adequate when the U.S. was the unquestioned leader in LCA production and sales and when funding for science and technology was more plentiful. However, today's system is still geared to lead, "rather than to race neck and neck with highly capable rivals" (Branscomb, 1992, p. 52). Competitive advantage today has much more to do with organizational learning and development cycles that reduce the time required to deliver products with improved quality, at reduced cost, and with fewer defects. With the 777, Boeing decided to produce a "service-ready, on time" aircraft. In doing so, Boeing changed the way it builds airplanes by reducing reoccurring costs (cutting defects by 50%) and substantially decreasing the time it takes to build an individual airplane from the conventional 16 months to 12 months with the ultimate goal of reducing the time to 8 months. In effect, Boeing marketed the 777 to potential customers by allowing them to play a key role in its design and development. In simple terms, the RD&P of the 777 required Boeing to undergo a fundamental cultural change in terms of how it designs and produces LCA and how it interacts with customers, clients, subcontractors, and suppliers. No less of a change is required of the federal support system for aeronautics.

Finally, it appears that aeronautical R&T policy fails to recognize knowledge transfer and utilization (together with knowledge production) as equally important parts of the R&D process. In that sense, the criticisms of Branscomb (1992), Mowery (1994), Pinelli, Barclay, and Kennedy (1996), and others concerning federal technology policy appear to be applicable to U.S. aeronautical R&T policy, namely that greater emphasis should be placed on the diffusion and absorption of knowledge. To ensure the economic vitality of the U.S. aircraft industry, aeronautical R&T policy would include at minimum the following three objectives.

Optimize the diffusion and absorption of knowledge resulting from federally funded aeronautical R&T to enhance economic competitiveness. This requires that the knowledge base be managed as a capital asset and that a strategy, an infrastructure, and mechanisms be developed that ensure the diffusion and absorption of knowledge. The mechanisms would be user-focused and the available knowledge would be modeled to meet users' needs for solutions. Information technology and (human) intermediaries would facilitate delivery of both codified (explicit) and uncodified human (tacit) knowledge that could then be readily analyzed and absorbed by users for problem-solving purposes.

Optimize the diffusion and absorption of knowledge produced outside of the United States. This requires an understanding of the scope and value of work being done in other countries, the facilities at which the work is being conducted, and who the experts are that are conducting the work. A mechanism is needed for collecting, analyzing, and integrating foreign-produced knowledge into the U.S. aerospace knowledge base. The U.S. had a similar mechanism in the Paris (France) office of NACA prior to World War II.

Optimize the diffusion and absorption of knowledge through cooperative ventures. This requires increased use of aeronautical technology demonstration programs involving NASA, industry, and universities. Such programs are recognized as relatively successful in diffusing the results of federally funded aeronautical R&T. Greater use of personnel exchange programs among government, industry, and academic sectors would also enhance diffusion and absorption.

CONCLUSIONS

Science—The Endless Frontier (Bush, 1945) established the philosophy and the assumptions associated with post-World War II federally funded science. As envisioned by Bush, the U.S. federal government would play a major role in supporting basic scientific research, believing that such support would promote innovation, economic growth, and general social well-being. Indeed, the two decades following the war witnessed scientific discoveries that revolutionized communication and transportation and led to the creation of new industries. During that same period, U.S. corporations created new consumer goods, generated millions of high paying jobs, opened new markets, developed new manufacturing and production techniques, and instituted new approaches to industrial organization and management. From 1945 to 1965, U.S. corporations were substantially more innovative and productive than firms in Europe and Japan. The standard of living in the United States became the envy of the world.

However, from 1965 to 1985, growth in American (industrial) productivity has been surpassed by all of our major trading partners. Real hourly wages in industry have remained virtually stagnant since 1973 and have actually declined in recent years, the number of manufacturing jobs in the U.S. continues to decline, and U.S. trade deficits are near or at all-time highs. The new reality is global competition (Young, 1985). Although a visionary in many respects, Vannevar Bush could not have foreseen the emergence of an interdependent and highly competitive global economy or that 70% of the goods produced in the U.S. would compete against products made elsewhere in the world.

For such critics of U.S. post-World War II science policy as Shapley and Roy (1985), the assumption that continued funding of basic science by the federal government automatically promotes innovation, economic growth, and general social well-being is inappropriate for the reality of today's global economy.   They and others reject the notion that innovation is a linear process that grows out of basic science and conclude, instead, that, from the standpoint of supporting innovation, improving the ability of firms to absorb and exploit knowledge may be as important as creating knowledge by publicly funding basic science.   Still other critics see technology policy as distinct from science policy and more closely aligned with economic, innovation, and trade policy than with national security.   The critics do, however, agree that to remain competitive the nation needs technology policy that is based on a consensus philosophy and set of assumptions that reflect the reality of today's interdependent and highly competitive global economy. To this we add that an empirical understanding of the innovation process and the interaction and effects of technology policy with economic, innovation, trade, and national security policy is needed for sound policy formulation.   Finally, policy evaluation and program assessment is needed to determine the extent to which policy goals and objectives are being accomplished.

In recent years, global competition has led to two dramatic changes in the commercial aircraft sector of the U.S. aircraft industry. *First*, the technology-driven strategy that has underlain the RD&P of LCA for years is being replaced by one that is more market- or customer-driven. In the technology-driven strategy, knowledge was used to design LCA whose technical performance, in terms of speed, range, and fuel efficiency, significantly exceeded that of predecessors. In the market- or customer-driven strategy, knowledge takes on a different focus; technology and innovation are used to improve the reliability, dependability, safety, and overall economic performance of LCA, the parameters of which are determined by the marketplace. The RD&P of the Boeing 777 exemplifies the market- or customer-driven strategy. *Second*, the use of collaboration by LCA manufacturers to lower costs and reduce time to market (product) delivery is becoming increasingly international in scope. Boeing's 777 has a substantial international component, including the outsourcing of certain R&D-related activities.   The Japanese Aircraft Development Corporation (JADC), a consortium of the three largest Japanese manufacturers, built approximately 21% of the B-777 airframe on a risk-sharing basis, investing their own funds plus money obtained through low-interest loans from the Japanese government (Council on Competitiveness, 1996). This "program partnership" allowed Boeing to take advantage of Japan's manufacturing expertise, leverage Japanese capital, and gain access to the lucrative Japanese market. The JADC was able to expand its expertise in designing aircraft components and to improve its manufacturing processes by using the (French) Dassault computer design system. Though largely beneficial to Boeing, international collaboration raises questions about potential Japanese penetration of the U.S. aircraft industry. Japan already has world-class capabilities in manufacturing a wide array of aircraft components and composites, and the country is fostering further technical acquisition (of knowledge) through international partnerships. In fact, a recent National Research Council (1994) study suggests that U.S. aircraft manufacturers have transferred far more technology (knowledge) to Japan through such alliances than they have absorbed and exploited knowledge from Japan.

IMPLICATIONS FOR DIFFUSING THE RESULTS OF FEDERALLY FUNDED AEROSPACE R&D

The enumerated changes occurring in federally funded science and technology policy and the RD&P of LCA will undoubtedly affect federal agencies, in particular, NASA. These changes will precipitate greater sharing of resources and collaboration between academia, industry, and government and between LCA manufacturers, subcontractors, and suppliers. The enormous costs and risk associated with bringing a new LCA to market will result in increased international partnerships. Federal budgetary pressures will continue to make it difficult for NASA to maintain world-class engineers, scientists, facilities, and institutions. NASA will be faced with pressure from industry to conduct more short-term research that produces the results needed by LCA manufacturers to remain competitive while at the same time conducting the high-risk, long-term research that will help produce the technological breakthroughs needed for the next generation of LCA. Finally, pressure to maintain U.S. preeminence in LCA, to reduce reoccurring costs, and to substantially decrease the time it takes to build an individual airplane requires that NASA reexamine how it manages the results of aeronautical R&T. Wise management includes optimizing the diffusion and absorption of knowledge (a) resulting from federally funded aeronautical R&T, (b) produced outside of the United States, and (c) through cooperative ventures to enhance the economic competitiveness of the U.S. aircraft industry.

CHAPTER REFERENCES

Alic, J.A.; L.M. Branscomb; H. Brooks; A.B. Carter; and G.L. Epstein. (1992). *Beyond Spinoff: Military and Commercial Technologies in a Changing World*. Boston, MA: Harvard Business School Press.

Allen, T.J. (1977). *Managing the Flow of Technology: Technology Transfer and the Dissemination of Technological Information Within the R&D Organization*. Cambridge, MA: MIT Press.

Amabile, T.M. (1983). *The Social Psychology of Creativity*. New York, NY: Springer-Verlag.

Amabile, T.M., and S.S. Gryskiewicz (1987). *Creativity in the R&D Laboratory*. Greensboro, NC: Center for Creative Leadership.

Anderson, J. (1974). "Not a Half-Life for NACA Reports." *Sci-Tech News* (October) 28(4): 24.

Arrow, K. (1962a). "The Economics Implications of Learning by Doing." *Review of Economic Studies*. 29: 155-173.

Arrow, K. (1962b). "Economic Welfare and the Allocation of Resources For Invention." In *The Rate and Direction of Inventive Activity*. R. Nelson, ed. Princeton, NJ: Princeton University Press.

Averch, H.A. (1985). *A Strategic Analysis of Science & Technology Policy*. Baltimore, MD: Johns Hopkins University Press.

Baer, W.S.; L.L. Johnson; and E.W. Merrow. (1977). "Government-Sponsored Demonstrations of New Technologies." *Science* (May) 196(27): 950-977.

Bateson, G. (1973). *Steps to an Ecology of Mind*. London, UK: Paladin.

Bell, D. (1973). *The Coming of Post-Industrial Society: A Venture in Social Forecasting*. New York, NY: Basic Books.

Bilich, F. (1989). *Science and Technology Planning and Policy*. Amsterdam, The Netherlands: Elsevier Science Publishers.

Bobrow, D.B. and J.S. Dryzek. (1987). *Policy Analysis by Design*. Pittsburgh, PA: University of Pittsburgh Press.

Bopp, G.R. (ed.). (1988). *Federal Lab Technology Transfer: Issues and Policies*. New York, NY: Praeger.

Branscomb, L.M. (ed.). (1993). *Empowering Technology: Implementing a U.S. Strategy*. Cambridge, MA: MIT Press.

Branscomb, L.M. (1992). "U.S. Scientific and Technical Information Policy in the Context of a Diffusion-Oriented National Technology Policy." *Government Publications Review* 19(5): 469-482.

Branscomb, L.M. (1991). "Toward A U.S. Technology Policy." *Issues in Science and Technology* 7(4): 50-55.

Brinberg, H.R. and T.E. Pinelli. (1993). "A General Approach to Measuring the Value of Aerospace Information Products and Services". Paper presented at the 31st Aerospace Sciences Meeting and Exhibits of the American Institute of Aeronautics and Astronautics (AIAA), Bally's Grand Hotel, Reno, NV. (Available AIAA; 93A17511.)

Brinberg, H.R.; T.E. Pinelli; and R.O. Barclay. (1995). "Valuing Information in an Interactive Environment". Paper presented at the *International Congress on the Economics of Information*, May 18-20, Lyon, France, and sponsored by the French Higher National School of Information Science and Library (ENSSIB) and, within it, the Information Science Research Center (CERSI). (Available NTIS; 95N21977.)

Bromley, D.A. (1993). "Science and Technology in the Bush Administration." In *Science Advice to the President*. 2nd ed. W.T. Golden, ed. Washington, DC: AAAS Press, 21-66.

Brooks, H. (1986). "National Science Policy and Technological Innovation." In *The Positive Sum Strategy: Harnessing Technology for Economic Growth*. R. Landau and N. Rosenberg, eds. Washington, DC: National Academy Press, 119-167.

Burger, R.H. (Issue Editor). (1986). "Privacy, Secrecy, and National Information Policy." *Library Trends* (Summer) 35(1): entire issue.

Bush, V. (1945). *Science: The Endless Frontier*. Washington, DC: Government Printing Office.

Clinton, W.J. and A. Gore. (1993). *Technology For America's Economic Growth, A New Direction to Build Economic Strength*. Washington, DC: U.S. Executive Office of the President, The White House. (Available NTIS; AD-A261-553.)

Coles, J.S. (ed.). (1983). *Technological Innovation in the '80s*. Englewood Cliffs, NJ: Prentice-Hall.

Constant, E.W. II. (1980). *The Origins of the Turbojet Revolution*. Baltimore, MD: Johns Hopkins University Press.

Copp, N. and A. Zanella. (1993). *Discovery, Innovation, and Risk: Case Studies in Science and Technology*. Cambridge, MA: MIT Press.

Council on Competitiveness. (1996). *Endless Frontier, Limited Resources: U.S. R&D Policy for Competitiveness*. Washington, DC: Council on Competitiveness.

Dasgupta, P. (1987). "The Economic Theory of Technology Policy" Chapter 1 in *Economic Policy and Technological Performance*. P. Dasgupta and P. Stoneman, eds. Cambridge, UK: Cambridge University Press, 7-23.

David, P.A. (1986). "Technology Diffusion, Public Policy, and Industrial Competitiveness." In *The Positive Sum Strategy: Harnessing Technology for Economic Growth*. R. Landau and N. Rosenberg, eds. Washington, DC: National Academy Press, 373-391.

Drucker, P.F. (1993). *Post-Capitalist Society*. Oxford, UK: Butterworth Heinemann.

Dupree, A.H. (1986). *Science in the Federal Government: A History of Policies and Activities*. Baltimore, MD: Johns Hopkins University Press.

E.O. 12591. "Facilitating Access to Science and Technology." April 18, 1987.

Eads, G. (1974). "US Government Support for Civilian Technology: Economic Theory Versus Political Practice." *Research Policy* 3: 2-16.

England, J.M. (1982). *A Patron of Pure Science: The National Science Foundation's Formative Years, 1945-1957*. Washington, DC: U.S. Government Printing Office.

Ergas, H. (1987a). "Does Technology Policy Matter?" In *Technology and Global Industry: Companies and Nations in the World Economy*. B.G. Guile and H. Brooks, eds. Washington, DC: National Academy Press, 191-245.

Ergas, H. (1987b). "The Importance of Technology Policy." Chapter 3 in *Economic Policy and Technological Performance*. P. Dasgupta and P. Stoneman, eds. Cambridge, UK: Cambridge University Press, 50-96.

Gibbons, J.H. (1994). "Science and Technology in the Clinton Administration." In the *AAAS 1993 Science and Technology Yearbook*. A.H. Teich; S.D. Nelson; and C. McEnaney, eds. Washington, DC: AAAS Press, 5-11.

Golden, W.T. (ed.). (1993a). *Science Advice to the President*. 2nd ed. Washington, DC: AAAS Press.

Golden, W.T. (ed.). (1993b). *Science and Technology Advice to the President, Congress, and Judiciary*. Washington, DC: AAAS Press.

Golich, V.L. (1992). "From Competition to Collaboration: The Challenge of Commercial-Class Aircraft Manufacturing." *International Organization* (Autumn) 46(4): 900-934.

Graham, M.B.W. (1986). "Corporate Research and Development: The Last Transformation." In *Technology in the Modern Corporation: A Strategic Perspective*. M. Horwitch, ed. New York, NY: Pergamon Press, 86-102.

Graham, M.B.W. (1985). "Industrial Research in the Age of Big Science." In *Research on Technological Innovation, Management and Policy*. Vol. 2. R.S. Rosenbloom, ed. Greenwich, CT: JAI Press, 47-79.

Gray, D.O.; T. Solomon; and W. Hetzner. (1986). *Technological Innovation: Strategy For a New Partnership*. Amsterdam, The Netherlands: North-Holland.

Haeckel, S.H. and R.L. Nolan. (1993). "The Role of Technology in an Information Age: Transforming Symbols Into Action." In *The Knowledge Economy: The Nature of Information in the 21st Century*. Queenstown, MD: The Aspen Institute, Institute for Information Studies, 1-24.

Irwin, S.M. (1993). *Technology Policy and America's Future*. New York, NY: St. Martin's Press.

Ives, B. and S.L. Jarvenpae. (1993). "Competing With Information: Empowering Knowledge Networks With Information Technology." In *The Knowledge Economy: The Nature of Information in the 21st Century*. Queenstown, MD: The Aspen Institute, Institute for Information Studies, 53-87.

Kay, W.D. (1992). "Congressional Decision-Making and Long-Term Technological Development: The Case of Nuclear Fusion." *Science, Technology, and Politics: Policy Analysis in Congress*. G. Bryner, ed. Boulder, CO: Westview, 87-106.

Kempf, R.F. (1990). "Legal Ramifications of Intellectual Property." *Government Information Quarterly* 7(2): 197-209.

Kevles, D.J. (1995). *The Physicists: The History of a Scientific Community in Modern America*. Cambridge, MA: Harvard University Press.

Keyworth, G.A. (1986). "Four Years of Reagan Science Policy: Notable Shifts in Priorities." In *Technological Innovation: Strategies For a New Partnership*. D.O. Gray; T. Solomon; and W. Hetzner, eds. New York, NY: North-Holland, 27-39.

Kitschelt, H. (1991). "Industrial Governmental Structures, Innovation Strategies, and the Case of Japan: Sectoral or Cross-National Comparative Analysis?" *International Organization* (Autumn) 45(4): 453-493.

Kleinman, D.L. (1995). *Politics on the Endless Frontier: Postwar Research Policy in the United States*. Durham, NC: Duke University Press.

Kline, S.J. (1985). "Innovation is Not a Linear Process." *Research Management* (July/August) 27: 36-45.

Lambright, W.H. and D. Rahm. (eds.). (1992). *Technology and U.S. Competitiveness: An Institutional Focus*. New York, NY: Greenwood Press.

Landau, R. and N. Rosenberg. (eds.). (1986). *The Positive Sum Strategy: Harnessing Technology for Economic Growth*. Washington, DC: National Academy Press.

Langford, J.S., III. (1989). *The NASA Experience in Aeronautical R&D: Three Case Studies With Analysis*. IDA Report R-319 Alexandria, VA: Institute for Defense Analyses. (Available NTIS: AD-A211,486.)

Logsdon, J.M. (1986). "Federal Policies Towards Civilian Research and Development: A Historical Overview." In *Technological Innovation: Strategies for a New Partnership*. D.O. Gray; T. Solomon; and W. Hetzner, eds. New York, NY: North-Holland, 9-26.

Lomask, M. (1976). *A Minor Miracle: An Informal History of the National Science Foundation*. Washington, DC: U.S. Government Printing Office.

Majchrzak, A. (1984). *Methods for Policy Research*. Newbury Park, CA: Sage Publications.

Mansfield, E. (1984). "Economic Effects of Research and Development, the Diffusion Process, and Public Policy." Chapter 6 in *Planning For National Technology Policy*. R.A. Goodman and J. Pavon, eds. New York, NY: Praeger, 104-120.

Mansfield, E. (1981). "How Economists See R&D." *Harvard Business Review* (November-December) 59(6): 98-106.

Matthews, R.C.O. (1973). "The Contributions of Science and Technology to Economic Development." Chapter 1 in *Science and Technology in Economic Growth*. B.R. Williams, ed. New York, NY: John Wiley, 1-31.

McDougall, W. (1985). *The Heavens and the Earth*. New York, NY: Basic Books.

Miles, I. and K. Robins. (1992). "Making Sense of Information." Chapter 1 in *Understanding Information: Business, Technology and Geography*. K. Robins, ed. New York, NY: Belhaven Press, 1-16.

Mowery, D.C. (1994). *Science and Technology Policy in Interdependent Economies*. Boston, MA: Kluwer Academic Publishers.

Mowery, D.C. (1985). "Federal Funding of R&D in Transportation: The Case of Aviation." Paper commissioned for a workshop on *The Federal Role in Research and Development* sponsored by the National Academy of Sciences, National Academy of Engineering, and Institute of Medicine in Washington, DC, 21-22 November.

Mowery, D.C. (1983). "Economic Theory and Government Technology Policy." *Policy Sciences* 16: 27-43.

Mowery, D.C. and N. Rosenberg. (1989). *Technology and the Pursuit of Economic Growth*. New York, NY: Cambridge University Press.

Mowery, D.C. and N. Rosenberg. (1979). "The Influence of Market Demand Upon Innovation: A Critical Review of Some Recent Empirical Studies." *Research Policy* 8(2): 102-153.

National Academy of Engineering. (1993). *Mastering a New Role: Shaping Technology Policy for National Economic Performance*. Washington, DC: National Academy Press.

National Academy of Engineering. (1985). *The Competitive Status of the U.S. Civil Aviation Manufacturing Industry: A Study of the Influences of Technology in Determining International Industrial Competitive Advantage*. Washington, DC: National Academy Press. (Avail. NTIS; PB88-100334).

National Academy of Sciences, National Academy of Engineering, Institute of Medicine, and the National Research Council. (1995). *Allocating Federal Funds for Science and Technology*. Washington, DC: National Academy Press.

National Academy of Sciences, National Academy of Engineering, Institute of Medicine. (1986). *The Federal Role in Research and Development: Report of a Workshop*. Washington, DC: National Academy Press.

National Research Council, Aeronautics and Space Engineering Board. (1992). *Aeronautical Technologies for the Twenty-First Century*. Washington, DC: National Academy Press.

National Research Council, Committee on Japan. (1994). *High-Stakes Aviation: U.S.-Japan Technology Linkages in Transport Aircraft*. Washington, DC: National Academy Press.

Nelson, R.R. (ed.) (1982). *Government and Technical Progress: A Cross-Industrial Analysis*. New York, NY: Pergamon Press.

Nelson, R.R. (1977). *The Moon-Ghetto Metaphor: An Essay on Public Policy Analysis*. New York, NY: Norton.

Nelson, R.R. and R.N. Langlois. (1983). "Industrial Innovation Policy: Lessons Learned From American History." *Science* (February 18) 219: 814-818.

Nelson, R.R. and S.G. Winter. (1982). *An Evolutionary Theory of Economic Change*. Cambridge, MA: Harvard University Press.

Nelson, R.R. and S.G. Winter. (1977). "In Search of a Useful Theory of Innovation." *Research Policy* 6: 36-76.

Nelson, R.R. and G. Wright. (1992). "The Rise and Fall of American Technological Leadership: The Postwar Era in Historical Perspective." *Journal of Economic Literature* (December) 30(4): 1931-1964.

Nixon, R.M. (1972). *Message to the U.S. Congress on Science and Technology.*

Noll, R.G. (1993). "The Economics of Information: A User's Guide." In *The Knowledge Economy: The Nature of Information in the 21st Century.* Queenstown, MD: The Aspen Institute, Institute for Information Studies, 25-52.

Nonaka, I. and H. Takeuchi. (1995). *The Knowledge-Creating Company: How Japanese Companies Create the Dynamics of Innovation.* New York, NY: Oxford University Press.

Pavitt, K. and W. Walker. (1976). "Government Policies Toward Industrial Innovation: A Review." *Research Policy* (January) 5(1): 11-97.

Pinelli, T.E.; R.O. Barclay; and J.M. Kennedy. (1996). "U.S. Scientific and Technical Information Policy." Chapter 10 in *Federal Information Policies in the 1990's: Views and Perspectives.* Norwood, NJ: Ablex Publishing, 211-232.

Polanyi, M. (1966). *The Tacit Dimension.* Chicago, IL: University of Chicago Press.

Reinhardt, A; S. Browder; and P. Engardio. (1996). "Booming Boeing: Can it Keep Streaking Ahead While Revolutionizing the Way it Designs and Builds Planes?" *Business Week* (September 30) 3495: 119-125.

Relyea, H.C. (1994). *Silencing Science: National Security Controls and Scientific Communication.* Norwood, NJ: Ablex Publishing.

Roessner, J.D. (ed.). (1988). *Government Innovation Policy: Design, Implementation, and Evaluation.* New York, NY: St. Martin's Press.

Roland, A. (1985). *Model Research: The National Advisory Committee for Aeronautics 1915-1958.* Volume 1. NASA SP-4103. Washington, DC: National Aeronautics and Space Administration. (Available NTIS: 85N23683.)

Romer, P.M. (1990). "Endogenous Technological Change." *Journal of Political Economy* 98(5): S71-S102.

Rosenberg, N. (1985). "A Historical Overview of the Evolution of Federal Investment in Research and Development Since World War II." Paper commissioned for a workshop on *The Federal Role in Research Development* sponsored by the National Academy of Sciences, National Academy of Engineers, and Institute of Medicine in Washington, DC, 21-22 November.

Rosenberg, N. (1982). *Inside the Black Box: Technology and Economics.* London, UK: Cambridge University Press.

Rosenberg, N. and L.E. Birdzell. (1990). "Science, Technology and the Western Miracle." *Scientific American* (November) 263(5): 42-54.

Sabbagh, K. (1996). *Twenty-First Century Jet: The Making and Marketing of the Boeing 777.* New York, NY: Scribner.

Salomon, J-J. (1984). "What is Technology? The Issue of its Origin and Definitions." *History and Technology* 1(2): 113-156.

Schmookler, J. (1966). *Invention and Economic Growth.* Cambridge, MA: Harvard University Press.

Schumpeter, J.A. (1951). *The Theory of Economic Development.* Cambridge, MA: Harvard University Press.

Schwartz, J.T. (1992). "America's Economic-Technological Agenda for the 1990s." *Daedalus* (Winter) 121(1): 139-165.

Scott, M.F. (1989). *A New View of Economic Growth.* New York, NY: Oxford University Press.

Shapley, D. and R. Roy. (1985). *Lost at the Frontier: U.S. Science and Technology Policy Adrift.* Philadelphia, PA: ISI Press.

Smith, B.L.R. (1994). "The Clinton Approach: A New Linkage Between Technology and Economic Policy." In the *AAAS 1993 Science and Technology Yearbook.* A.H. Teich; S.D. Nelson; and C. McEnaney, eds. Washington, DC: AAAS Press, 19-29.

Smith, B.L.R. (1990). *American Science Policy Since World War II.* Washington, DC: The Brookings Institute.

Swain, D. (1962). "The Rise of a Research Empire: NIH, 1930-1950." *Science* 138: 1233-1237.

Tassey, G. (1992). *Technology Infrastructure and Competitive Position.* Boston, MA: Kluwer Academic Publishers.

Teece, D.J. (1981). "The Market for Know-How and the Efficient International Transfer of Technology." *The Annals of the American Academy of Political and Social Science* (November) 458: 81-96.

Teich, A.H. (1985). "Federal Support of Applied Research: A Review of the United States Experience." Paper commissioned for a workshop on *The Federal Role in Research Development* sponsored by the National Academy of Sciences, National Academy of Engineers, and Institute of Medicine in Washington, DC, 21-22 November.

Tornatzky, L.G. and M. Fleischer. (1990). *The Process of Technological Innovation.* Lexington, MA: D.C. Heath.

U.S. Congress, Congressional Budget Office. (1996). *Reducing the Deficit: Spending and Revenue Options.* (A Report to the Senate and House Committees on the Budget as Required by P.L. 93-244.) Washington, DC: Government Printing Office.

U.S. Congress, Office of Technology Assessment. (1994). *Federal Research and Technology for Aviation.* OTA-ETI-610. Washington, DC: Government Printing Office.

U.S. Congress, Office of Technology Assessment. (1991). "Government Support of the Large Commercial Aircraft Industries of Japan, Europe, and the United States." Chapter 8 in *Competing Economies: America, Europe, and the Pacific Rim.* OTA-ITE-498. Washington, DC: Government Printing Office. (Available NTIS; PB92-115575.)

U.S. Congress, Office of Technology Assessment. (1986). *Intellectual Property Rights in an Age of Electronics and Information*. OTA-CIT-302. Washington, DC: Government Printing Office.

U.S. Department of Commerce, International Trade Administration. (1984). *A Competitive Assessment of the U.S. Civil Aircraft Industry*. Washington, DC: U.S. Department of Commerce. (Available NTIS: PB84154913.)

U.S. Executive Office of the President, Office of Science and Technology Policy. (1995). *Goals for a National Partnership in Aeronautics Research and Technology*. Washington, DC: Office of Science and Technology Policy.

U.S. Executive Office of the President, Office of Science and Technology Policy. (1990). *U.S. Technology Policy*. Washington, DC: Office of Science and Technology Policy.

U.S. Executive Office of the President, Office of Science and Technology Policy. (1987). *National Aeronautical R&D Goals: Agenda for Achievement*. Washington, DC: Office of Science and Technology Policy.

U.S. Executive Office of the President, Office of Science and Technology Policy. (1985). *National Aeronautical R and D Goals: Technology for America's Future*. Washington, DC: Office of Science and Technology Policy. (Available NTIS: 87N-12405.)

U.S. Executive Office of the President, Office of Science and Technology Policy. (1982). *Aeronautical Research and Technology Policy. Volume 2*. Final Report. Washington, DC: Office of Science and Technology Policy. (Available NTIS: 83N-23268; AKA The Keyworth Report.)

U.S. International Trade Commission. (1993). *Global Competitiveness of U.S. Advanced-Technology Manufacturing Industries: Large Civil Aircraft*. Washington, DC: U.S. International Trade Commission.

Vincenti, W.G. (1992). "Engineering Knowledge, Type of Design, and Level of Hierarchy: Further Thoughts About What Engineers Know...." In *Technological Development and Science in the Industrial Age*. P. Kroes and M. Bakker, eds. Dordrecht, The Netherlands: Kluwer Academic Publishers, 17-34.

Vincenti, W.G. (1990). *What Engineers Know and How They Know It: Analytical Studies From Aeronautical History*. Baltimore, MD: Johns Hopkins University Press.

von Hippel, E. (1994). "'Sticky Information' and the Locus of Problem Solving: Implications for Innovation." *Management Science* (April) 40(4): 429-439.

von Hippel, E. and M. Tyre. (1995). "How Learning by Doing is Done: Problem Identification in Novel Process Equipment." *Research Policy* 24(1): 1-12.

Wessel, W.V. (1993). "Technology Transfer and Intellectual Property: An Analysis of the NASA Approach" *Technovation* 13(3): 133-147.

Wright, T.P. (1936). "Factors Affecting the Cost of Airplanes." *Journal of Aeronautical Science* 3: 122-128.

Young, J.A. (1985). "Global Competition: The New Reality." *California Management Review* (Spring) 27(3): 11-25.

Chapter 4

U.S. Public Policy and the Dissemination of Federally Funded Aeronautical Research and Technology

W.D. Kay
Thomas E. Pinelli
Rebecca O. Barclay

SUMMARY

Chapter 4 focuses on the system used by NASA to disseminate the results of federally funded aeronautical research and technology (R&T). Since 1917, U.S. public policy has provided funding for aeronautical R&T, maintained a system for codifying and disseminating the results, and encouraged their use and adoption, first through the National Advisory Committee for Aeronautics (NACA) and later through the National Aeronautics and Space Administration (NASA). Having established previously the criticality of knowledge to process and product innovation, the background of this chapter presents two economic models that explain the diffusion of innovation. One model regards innovation as an individual or organizational investment decision. The other stresses the importance of knowledge in reducing the uncertainty associated with innovation and provides a conceptual framework for studying the diffusion of knowledge. The effects of U.S. public policy on the dissemination of data, information, and knowledge resulting from federally funded science and technology are reviewed. Lastly, we examine the NASA system that distributes the results of federally funded aeronautical R&T and conclude with implications for diffusing the results of federally funded research and development through effective knowledge management.

INTRODUCTION

Since its founding, the government of the United States has pursued policies to encourage innovation in certain industries and to enhance the economic returns to domestic firms and citizens from endogenously developed technologies (Mowery, 1994). The primary objective of these federal policies was providing for the nation's defense. The U.S. aircraft industry is a case in point. With the emergence of global markets and international competition in the early 1980s, the focus of the policies has changed from one of military security to economic security. We lump these policies together under the general heading of (U.S.) technology policy and focus on those aspects of public policy intended to influence a firm's decision to develop, commercialize, or adopt (absorb) external knowledge. Chapter 3 featured those public policies and issues related to federal involvement in the creation or "supply" of knowledge. In this chapter, the focus shifts to those public policies and issues specifically concerned with federal involvement in the dissemination or "absorption" of knowledge resulting from federally funded aeronautical research and technology (R&T).

Sources of knowledge external to an organization are often critical to the innovation process and to the commercial success of various products, including large commercial aircraft (LCA). Cohen and Levinthal (1990) state that studies have proved this statement true for entire nations (e.g., Japan) and for entire industries (e.g., computers). At the organizational level, the results of studies by March and Simon (1958), Myers and Marquis (1969), and von Hippel (1978) suggest that most innovation results from knowledge that resides external to the organization. Ergo, the ability of organizations to exploit external knowledge is critical to technological innovation and research and development (R&D). Several factors affect an organization's capacity to absorb knowledge, assimilate it, and apply it to commercial ends. For example, organizations that conduct their own (internal) R&D are better able to absorb external knowledge than are those organizations that do not (Mowery, 1983). It appears that experience, at both the organizational and individual levels, with similar or related knowledge, determines in large part an organization's ability to evaluate, absorb, and utilize external knowledge.

The economic returns from public investment in knowledge creation depend on its utilization by the private sector. However, as Mowery (1994) points out, technology policy frequently overlooks the high cost of transferring and exploiting knowledge created at public expense. The costs are often high enough to affect the utilization of this knowledge adversely. Thus, technology policy (a) involves more than simply subsidizing the creation of knowledge; (b) should focus on the production, transfer, and use (i.e., diffusion) of knowledge; and (c) should be based on an understanding of the process by which organizations and individuals involved in technological innovation and R&D absorb, assimilate, and apply external knowledge. A wide variety of programs including demonstration projects, extension programs, and information clearinghouses has been undertaken, devised, and created in an attempt to ensure that the public benefits from using the results of its investment in knowledge creation. With the possible exception of the U.S. Department of Agriculture's research and extension programs, few comprehensive studies have been conducted to determine the effectiveness of these programs. Nor, as Mowery (1994) points out, "is very much known about the structural features of successful and unsuccessful programs" (p. 33).

Mowery (1994) has identified five broad classes of adoption-oriented (U.S.) technology policies: (a) financial subsidies, (b) government procurement, (c) government mandated technology transfer from foreign sources, (d) technical standards, and (e) information dissemination. Financial subsidies include direct payments, favorable tax treatment, and special depreciation allowances for the purchase of equipment and the adoption of new processes and technologies (i.e., solar energy). The procurement of goods and services by the public sector can include, as a cost of doing business, the requirement that new or advanced technologies be used in the process. Mowery notes that the requirement by the public sector that new technologies be used can provide the private sector with valuable "learning-by-using" (Rosenberg, 1982) experience about the behavior and performance of these technologies. Government mandated technology transfer, referred to in aerospace as "offsets," is frequently used by industrializing economies as a method of absorbing technology from foreign sources, generally from (U.S.) firms seeking to invest or sell in the domestic market of that country. (For a discussion of offsets in aerospace, see Vadas, 1996; also, see U.S. Trade Representative, 1996.) Standards development creates considerable information about the behavior and performance characteristics of a new technology, thus reducing uncertainty and costs associated with adoption. Information dissemination, the least understood or analyzed of these policies, is the focus of this chapter.

We focus specifically on the system used by NASA to disseminate the results of federally funded aeronautical R&T for the following reasons. *First*, the system used by NASA is, for all intents and purposes, a national system. *Second*, the dissemination of NASA research results is fundamental to managing the U.S. aeronautical R&T knowledge base. *Third*, little is known about the system's effectiveness and efficiency in the context of knowledge management. As background to this chapter, we present two economic models that explain the diffusion of innovation. The first considers innovation to be an individual or organizational investment decision. The second stresses the important role knowledge plays in reducing the uncertainty associated with innovation and provides a conceptual framework for studying knowledge diffusion. Next, we review how U.S. public policy affects the dissemination of knowledge resulting from federally funded science and technology. Lastly, we examine the system used by NASA to distribute the results of federally funded aeronautical R&T and conclude with implications for diffusing the results of federally funded research and development through effective knowledge management.

BACKGROUND

Economists acknowledge that technological innovation gives rise to new products, greater productivity, lower unit costs, and increasing profitability. Technological innovation at the organizational or industry level does not occur in isolation, however, and may create pressures affecting existing relationships among labor, capital, and resources. Public (policy) interests in technological innovation include, but are not limited to, economic viability, natural resources and the environment, employment, education and training, health and safety, national security, trade, competition, taxes, and business and industry regulation. Because each issue-area is embedded in a "complex interactive system," government policies designed to influence (increase) the rate of innovation can trigger unanticipated consequences in another issue-area (Kitschelt, 1991). Recent years have witnessed growing interest on the part of policymakers in the diffusion of knowledge as a means of (a) understanding technological innovation, (b) analyzing changes in the rate of innovation, and (c) predicting the effects of technological innovation.

A useful characterization of knowledge diffusion includes the general acceptance (i.e., adoption) over time of a specific idea, concept, or practice by an individual, group of individuals, or other adopting agents or units. The adopting agents are linked through specific channels of communication to a social system and to a given system of values or culture. This "model" of diffusion more accurately reflects how an anthropologist or sociologist, rather than an economist, would view knowledge diffusion. However, knowledge diffusion as an interdisciplinary research agenda is a recent undertaking (within the past 30 years). Early examples of research into knowledge diffusion include investigations by rural sociologists of how agricultural knowledge is diffused among farmers, and studies by psychologists of how new educational theories and teaching methodologies are diffused into school districts and among classroom teachers. A common thread in this work is the desire to understand the process by which knowledge is communicated through certain channels over time among the members of a social system. Notable work in knowledge diffusion has been conducted by Brown (1981), Hägerstand (1967), and Havelock (1973). Rogers' (1995) contribution is recognized by many as the seminal work in knowledge diffusion.

Analytical Framework

Anthropologists and sociologists are interested in the informational, communicative, and interpersonal aspects of acceptance or adoption as well as the social and cultural factors that create resistance to acceptance or adoption. Economists are interested in the following: (a) the speed with which an innovation spreads from firm to firm or industry to industry in an effort to relate the rate of diffusion to such factors as the structure and character of the industry and the size and nature of the organization; (b) the growth of new products and industries stemming from major innovations; and (c) the extent to which major innovations affect established industries by replacing or displacing predecessor techniques or practices (Rosegger, 1977). Notable work in knowledge diffusion has been done by such economists as Baker (1979), Gold (1981), Hill and Utterback (1979), Mansfield (1968), Mansfield, et al., (1982), Mowery and Rosenberg (1989), Nabseth and Ray (1974), Rosenberg (1986, 1982, 1976), and Sahal (1982). Collectively, this work includes studies of the role of knowledge in technological progress, the diffusion of innovations in such industries as steel, and the diffusion of specific innovations within and between countries.

Knowledge diffusion is a multifaceted phenomenon that inspires examination from a variety of perspectives. For example, a sociologist is likely to focus on the process of knowledge diffusion. Of particular interest are the factors that accelerate or impede the diffusion process. Sociologists would examine who delivered and who received the knowledge, the speed with which it was disseminated, the channels and sources which were or were not effective, and the roles played by various participants in the process. (See Rogers, 1995, for a detailed discussion.) Economists are more interested in the role played by the firm or organization as an innovator or producer. They contend that, although knowledge is an essential part of technological innovation, decisions to innovate are largely investment decisions influenced by various economic considerations like potential profit and loss.

From an economic standpoint, innovation involves the search for and the discovery, experimentation, development, imitation, and adoption of *new* or *improved* processes, products, systems, or services (Dosi, 1988). Innovation activities undertaken by profit-motivated agents involve the perception of unexploited technical and or economic opportunities. However, such perceptions rarely include detailed or "perfect" knowledge of the events that will shape the outcome or, for that matter, knowledge of the outcome itself. Decisions to innovate involve *uncertainty*, which is usually *not* the lack of relevant knowledge about what is known but rather a lack of knowledge about what is *not known* or cannot be predicted with a high degree of certainty (i.e., the likelihood of technical problems for which acceptable solutions do not exist or can be developed). Economists consider that *scientific* knowledge and a large portion of what passes as *technical* knowledge exist as a public good. Much of this knowledge is well known, well articulated, and in some cases well documented. For example, most scientific and technical publications are open and freely available to the public. Furthermore, economists consider that a substantial amount of *tacit* knowledge is also well known in the sense that "any good engineer or technician" knows and understands how something is done.

However, acquiring much of the knowledge necessary to innovate involves several costs (Gold, 1977), chief among them being the cost of acquiring proprietary information (i.e., patent rights, licensing agreements, royalty fees), capital equipment, and worker training and education. Some costs can be borne by third party groups; for example, companies selling or leasing the capital equipment will often include "free" training as part of the purchase or lease plan. Occasionally, job training and education are paid for by the U.S. government. Even so, decisions by a firm or industry to innovate are usually made incrementally and are reversible. The process of innovation consists of a course of "events" in a series of decision cycles that include several "evaluation" points. The results of the evaluation can lead to a decision to terminate the process. Although knowledge is crucial to technological innovation, technical progress, and economic competitiveness, the linkages among the various sectors of the technology infrastructure appear to be weak and poorly defined in the U.S. Nevertheless, the importance of knowledge to the process of innovation cannot be discounted or dismissed. (For a discussion of the theoretical contributions to the study of diffusion, see Mowery, 1988.)

Two Economic Models

Two models—an investment-centered one developed by David (1986) and an uncertainty-centered one developed by Mansfield (1968)—can be used to explain the diffusion and adoption of innovation. David's model assumes that (a) diffusion is an equilibrium process, (b) all potential adopters are equally well-informed concerning the costs and profitability of an innovation, and (c) the diffusion and adoption of an innovation is the outcome of a series of individual investment decisions. Knowledge plays no role in David's model. Unlike Mansfield, David assumes that profitability and the costs associated with innovation, not uncertainty and the lack of information, influence the diffusion and adoption of innovation. Mansfield's model emphasizes uncertainty and the role of knowledge in reducing the uncertainty associated with the diffusion and adoption of an innovation. Profitability and the costs associated with innovation are constant over time and are the same for all members of the population in the Mansfield model. The critical element in the diffusion and adoption of an innovation is the level of knowledge among the members of the population concerning the characteristics of the innovation. The fact that such knowledge is not available to all the members of the population at the same time is fundamental to determining or predicting the diffusion and adoption of an innovation. Rogers (1982) has stated that technological innovation is a process that involves considerable risk and grappling with unknowns that may be technical, economic, or merely the manifestation of personal and social variables. When faced with uncertainty and complex tasks and decisions, individuals seek information, which is why information (communication) behavior cannot be ignored when studying technological innovation.

In summary, knowledge diffusion is neither a simple process nor solely an economic one wherein market forces and expected profits determine or explain the rate of diffusion and adoption of innovation. In taking this position, we agree with Rubbert (1994) that the technology-driven strategy underlying the RD&P of a new LCA is being replaced by one that is more market- or customer-driven, as evidenced by the B-777. For the purposes of this study, knowledge diffusion in the U.S. aircraft industry is viewed as a process in which a number of economic, political, and social factors combine to influence the rate of diffusion and the introduction and adoption of innovation. Technological innovation occurs in an economy that Derian (1990) describes as *sheltered* (in contrast to *exposed*); the industry is subject to a unique set of *externalities* that result from U.S. government intervention (a) as a provider (i.e., funding agent) of precommercial aeronautical R&T, (b) through a system maintained by an agency of the U.S. government that couples the results of publicly funded R&T with the people who would use it in the field, and (c) as a purchaser of the goods produced by the industry. As noted in Chapter 2, such public policy concerns as the environment (e.g., emissions and noise pollution) will definitely influence the technology developed for and the commercial development of high-speed civil transports that will fly at speeds up to Mach 2.4 and could enter into service by 2006.

Increasing collaboration with foreign producers will result in a more international manufacturing environment that will also influence the rate at which U.S. (publicly) funded aeronautical R&T is diffused globally. The ability of the U.S. to remain competitive in the world LCA market will affect the domestic economy as well as political considerations associated with the system that couples the results of publicly funded aeronautical R&T with the people who would use it in the field. Ensuring that the U.S. remains competitive depends on improving and maintaining the professional competency of U.S. aerospace engineers and scientists, and enhancing technological innovation and productivity, all of which require judicious management of the NASA aeronautical R&T knowledge base. Our research indicates that U.S. aerospace engineers and scientists spend more time using and producing data, information, and knowledge than they devote to any other engineering or scientific activity. However, an empirical understanding of the knowledge diffusion process within the U.S. aircraft industry is limited by a lack of understanding of the channels, products, and sources used by U.S. aerospace engineers and scientists to communicate and of the information-seeking behaviors of the members of the social system that comprises the U.S. aircraft industry. Even less is known about the efficiency and effectiveness of the system that couples the aeronautical R&T knowledge base with potential users. Understanding how knowledge is diffused in the U.S. aircraft industry should contribute to our ability to manage the aeronautical R&T knowledge base prudently.

A BRIEF REVIEW AND ANALYSIS OF U.S. PUBLIC POLICY AFFECTING THE DISSEMINATION OF DATA, INFORMATION, AND KNOWLEDGE RESULTING FROM FEDERALLY FUNDED SCIENCE AND TECHNOLOGY

Policymakers are beginning to realize that knowledge is as vital to technological innovation as technological innovation is to economic growth, global competitiveness, and international trade. In the U.S., economic growth, measured in terms of real wage earnings and worker productivity, has slowed relative to economic growth in such countries as Japan. To try to explain this phenomenon, economists and politicians have begun to examine the relationships among product design, product quality, and product and process innovation together with the factors affecting them as components of a national system. As Nelson (1993) points out, the term "national system" is justified because: (a) there is a connection between technological innovation and a nation's competitive performance and long-term economic growth; (b) technological innovation, commerce, and trade are increasingly international in scope; (c) national governments frequently provide direct and indirect support for knowledge production by funding science and technology; and (d) the policies of national governments contribute to and often target the efficient development, introduction, and exploitation of *new* processes, products, systems, or services as well as *improvements* in existing processes, products, systems, or services.

The U.S. national system is enormous compared to the size of other national systems. The role, the relative importance of, and the relationships among three key "actors"—the federal government, industry, and the universities as producers of knowledge—differ somewhat from the role of these actors in other national systems. Two specific public policies—antitrust and national defense—affect the structure and performance of the U.S. system and differ somewhat from their role and application in the national innovation systems of other countries (Mowery and Rosenberg, 1990). Finally, since World War II, the U.S. government has continuously supported the national system by investing in (i.e., funding) science and technology and underwriting the cost of educating and training engineers and scientists. The U.S. government appropriated more than $70 billion in fiscal year (F.Y.) 1997 for science and technology (U.S. Executive Office of the President, Office of Management and Budget, 1996, p. 117). This expenditure is generally considered to represent an investment in the country's future. However, Mowery and Rosenberg (1990) make the point that this investment is not based on any economic strategy.

In recent years, there has been growing concern that the data, information, and knowledge resulting from the government's investment in science and technology may not be well-utilized or that the public may not be receiving an "adequate" return on their investment. Attempts to measure the return on the federal investment in science and technology in a quantitative sense have been largely unsuccessful for two reasons. *First*, "economic payoff is only one of many criteria and often a secondary consideration for the federal investment in science and technology" (U.S. Congress, Office of Technology Assessment, 1986, p. 61). The federal investment in science and technology is shaped by and must be responsive to many groups and constituencies. The goal of federally funded R&D is not profitability, but rather a means to achieving various social and economic objectives and the enhancement of knowledge and education. *Second* is that the output or product of this investment is usually *knowledge*. As Mowery and Rosenberg (1989) point out:

> The product of research is rarely, if ever, a final product to which
> the marketplace can attach a price tag. Rather, the output is a pe-
> culiar kind of intermediate good that may be used, not to *produce*

a final good, but to play some further role in the *invention* of a new final good. (p. 11)

Nevertheless, there is general agreement among policymakers that the results of the federal investment in science and technology can and do enhance technological innovation and improve the economic competitiveness of the United States. These policymakers, however, are concerned that inadequacies in the federal system that is responsible for managing the massive amounts of data, information, and knowledge resulting from federally funded science and technology may contribute to the inefficient utilization of these intangible products. This concern was supported by *Helping America Compete—The Role of Federal Scientific and Technical Information* (U.S. Congress, Office of Technology Assessment, 1990), one of the latest in a series of studies stretching over 30 years, which concluded that the United States could and should make better use of the results of the federal investment in science and technology (Wood, 1991). In congressional hearings on federal scientific and technical information policy, Representative Doug Walgren (U.S. Congress, House of Representatives 1989) stated the following:

> Our government is the largest single source of scientific and technical information in the world. Each year this committee [science, space, and technology] authorizes billions of scarce taxpayer dollars for scientific research and development and without efficient management and dissemination of the results of this activity, these funds are wasted. ... Without an overall federal policy, agency efforts to collect and disseminate scientific and technical information will remain at best fragmented and at worst redundant. ... The Office of Science and Technology Policy is the logical place to pull all the elements together and fashion a coherent policy which will meet the complex needs of both the government and the public. (pp. 3-5)

This section contains a brief review and analysis of U.S. public policy affecting the dissemination of data, information, and knowledge resulting from federally funded science and technology. An historical review of the federal system that collects, stores, and otherwise makes accessible the results of this investment is followed by a critique of U.S. policy intended to coordinate and manage the results of the nation's investment in science and technology. Again, the review and analysis is neither analytical nor comprehensive; we have left that task to others. (See, for example, Adkinson, 1978; Aines, 1984; Averch, 1985; Ballard, et al., 1989; Bikson, Quint, and Johnson, 1984; Hernon and McClure, 1988, 1987; Hernon, McClure, and Relyea, 1996; McClure and Hernon, 1989; McClure, Hernon, and Relyea, 1989; Pinelli, Henderson, Bishop, and Doty, 1992; Wooster, 1987.)

Background

In Vannevar Bush's (1945) post-World War II plan for science and technology, basic research was assumed to be a federal responsibility because the benefits would accrue too far into the future to make industrial support feasible. The Bush plan effectively split the responsibility for science between the public and private sectors. Overall resource allocation decisions remained with the federal government, although academia and industry had considerable autonomy in terms of problem formulation, approach, and solution. Bush also firmly believed that the free and open exchange of ideas through the presentation, publication, and wide dissemination of new ideas, discoveries, and experimental results was essential to the advancement of science and progress and that the traditional means of science communication (conferences and journals) was the appropriate mechanism for handling the results of federally funded science. U.S. public policy would support the traditional (established) means of science communication by allowing journal "page charges" and conference registration fees as legitimate expenses against federal research grants and contracts.

Although his ideas and plan were truly visionary, Bush (1945) did not foresee the Cold War and the corresponding U.S. investment in science and technology that followed. His idea of a system of science run by scientists operating in the public's trust was no match for the events and politics of the 1960s, 1970s, and 1980s. Furthermore, Bush could not have fathomed that these events and the tremendous amount of data, information, and knowledge resulting from this investment would overwhelm the traditional means of science communication and severely test his belief that the free and open exchange of ideas and wide dissemination of new theories, discoveries, and experimental results were essential for the advancement of science and technical progress.

The Federal System

Massive amounts of data, information, and knowledge, most of which resided in technical reports produced during World War II and the literal trainloads of Axis (i.e., German) reports captured by the Allies, created the impetus for U.S. public policy with respect to making available the results of federally funded science and technology. (See Richards, 1994, for a discussion of scientific information in World War II.) President Truman's issuance of E.O. 9568 and E.O. 9604, the creation of the Publications Board and the Office of Technical Services (eventually to become the National Technical Information Service), and the Baker Report submitted by the President's (Eisenhower) Science Advisory Committee (1958) shaped the early debate concerning federal involvement in coordinating, managing, and making accessible the results of the U.S. government's investment in science and technology.

This investment increased enormously with the Cold War. The creation of NASA and the resulting Mercury, Gemini, and Apollo programs produced tremendous quantities of data, information, and knowledge. Furthermore, the corresponding number of contracts and grants generated by this investment ostensibly helped to create specialized groups within the communities of science and technology that are best identified by the phrase "DoD, DoE, and NASA and their contractors." The foregoing also signaled the beginnings of an era in which it was assumed that, as a matter of public policy, the United States government had an obligation to coordinate and make available the results of the country's investment in science and technology to the nation's scientific community, industry, and the public. The so-called "mission agencies"—DoD, DoE, and NASA— created information clearinghouses and used large mainframe computers to create bibliographic databases and related information products designed to facilitate information awareness and access within the DoD, DoE, and NASA communities and within other communities of science and technology. (The historical development of the federal system is presented by Adkinson, 1978. For an overview of the DoD, DoE, and NASA systems, see Coyne and Fornwell, 1989; Molholm, Fox, Klinefelter, McCauley, and Thompson, 1988; Pinelli, 1990.)

During this period, the technical report became synonymous with documenting and reporting the results of federally funded science and technology (Pinelli, Barclay, and Kennedy, 1993; Pinelli and Henderson, 1989). According to Adkinson (1978), distribution of printed results of federally funded science and technology changed "from almost complete reliance on traditional journals and monographs to widespread use of the government technical report" (p. 29). The proliferation of the U.S. government technical report coincides with the expanding role of DoD, DoE, and NASA after World War II and three military necessities—classification for reasons of national security, the need to control dissemination, and the need to distribute data, information, and knowledge selectively and quickly to individuals, groups of individuals, and organizations having a "need to know." However, the origin of the [U.S. government] technical report dates back to 1941 as does the creation of the OSRD (Redman, 1965/1966; Godfrey and Redman, 1973). Auger (1975) states that the use of the technical report in the U.S. coincides with the development of aeronautics, the aviation industry, and the creation of the NACA.

By the end of the 1970s, U.S. public policy had recognized the social value inherent in the country's investment in science and technology and, therefore, the responsibility of the federal government to coordinate the collection and storage of these data, information, and knowledge and to make them accessible (Ballard, et al., 1989). Through the years, the U.S. government has attempted to increase the "availability" of the results of federally funded science and technology. A recent example, the American Technology Preeminence Act of 1991 (P.L. 102-245), requires the head of each federal executive department or agency to:

> transfer in a timely manner to the National Technical Information Service (NTIS) unclassified scientific, technical, and engineering information which results from federally funded research and development activities for dissemination to the private sector, academia, state and local government, and federal agencies. Only information which would otherwise be available for public dissemination shall be transferred to NTIS. [Section 108 (a)]

Early enthusiasm for increased financial support for science and technology also carried over into creating a system for making the results of federally funded science and technology accessible. However, that initial enthusiasm has degenerated, producing a system that is passive, fragmented, unfocused, and largely unresponsive in a user context (Bikson, Quint, and Johnson, 1984). In part, this can be explained by the fact that, unlike the former Soviet Union, the United States made a conscious decision *not* to build a single, centralized system or facility to store the results of federally funded science and technology (Smith, 1990). It did, however, intend for the federal enterprise to be coordinated, purposeful, and responsive.

The Committee on Scientific and Technical Information (COSATI) was created in 1964 under the auspices of the Federal Council for Science and Technology to help coordinate the collection, storage, and dissemination of data, information, and knowledge resulting from federally funded science and technology. COSATI was transferred in 1971 from the Office of Science and Technology (OST) to the Office of Science Information (OSI) in the National Science Foundation (NSF), where it was abolished the following year. This essentially ended federal efforts to provide leadership and to coordinate federal activities related to the collection and storage of data, information, and knowledge resulting from federally funded science and technology.

The National Science and Technology Policy, Organization and Priorities Act of 1976 (P.L. 94-282) gave the Office of Science and Technology Policy (OSTP) the responsibility for "coordinating and unifying [the nation's] scientific and technical information systems." However, Brown (1987) states that although OSTP has the legislative mandate to provide policy leadership, it has shown neither the interest nor the will to carry it out. (See Aines, 1984, for a discussion of the "precipitous" retreat from the overall planning, coordination, and management at the federal level.) These events contributed to creating the existing federal system that has no coherent, policy-oriented, systematically designed approach to managing the results of federally funded science and technology. Much of the reason for the fragmented and unfocused nature of the existing federal system can be attributed to the lack of Executive Branch leadership, the idiosyncratic missions and mandates of federal agencies that are more or less responsive to shifting political alliances and imperatives, reduced federal spending for information resource management, and recently to various attempts to make the results of federally funded science and technology less accessible to foreign competitors and hostile nations (Shattuck and Spence, 1989).

Through the years and despite continuing attempts to improve the efficiency of government operations, the roles of the various government agencies involved in science and technology have become increasingly passive and blurred. The basic infrastructure created by the federal government to collect and manage the results of federally funded science and technology has changed little since its creation, although the Cold War has ended, dramatic advances in computer and information technology have occurred, and the ability of the United States to compete effectively in a global market has weakened. One effort to remedy the federal leadership void has been the formation of CENDI (Commerce, Energy, NASA, and Defense Information), a loose federation of federal agencies working together to develop standards and solve problems common to CENDI agencies (McClure, 1988). Although the efforts of CENDI are admirable, they fall short of the leadership and coordinated policy that is needed to fill the policy void that exists at the federal level (Gold, 1993).

Two models—*appropriability* and *dissemination*—have dominated the U.S. government's approach to managing the data, information, and knowledge resulting from federally funded science and technology (Ballard, et al., 1989; Williams and Gibson, 1990). The *appropriability model* emphasizes the production of data, information, and knowledge by the federal government, which would not otherwise be produced by the private sector, and competitive market pressures to promote the use of that knowledge. This model emphasizes the production of basic research results as the driving force behind technological development and economic growth, assumes that the federal provision of science and technology will be rapidly assimilated (i.e., sought after) by the private sector, and approximates the model for federally funded science advocated by Vannevar Bush (1945). Deliberate transfer mechanisms and intervention by information intermediaries are viewed as unnecessary. Appropriability stresses the supply (production) of knowledge in sufficient quantity to attract potential users. Good science, according to this model, sells itself and offers clear policy recommendations regarding federal priorities for improving technological development and economic growth. Federal activities are limited to providing access to government (public) information through such mechanisms as the Government Printing Office (GPO) depository library program and the Government Information Locator Service (GILS). This model assumes that the results of federally funded science will be acquired and used by the private sector, ignores the fact that most basic research is irrelevant to technological innovation, dismisses the process of technological innovation within the firm, and disregards the fact that innovation-adaption decisions are seldom made on the basis of "advances in systemic knowledge of the useful arts" (David, 1986, p. 374).

The *dissemination model* attempts to compensate for the deficiencies of the appropriability model. Principally, this model recognizes that the results of federally funded science and technology will not necessarily be sought after, the supply (production) of data, information, and knowledge is not sufficient to ensure its utilization, and intervention at the producer level is required to provide potential users with the access linkages. (Linkage mechanisms include various information products and services, as well as intermediaries.) This *one-way, producer-to-user* approach assumes that these mechanisms, in and of themselves, are sufficient to ensure that the results of federally funded science and technology will be utilized because they provide opportunities for users to determine what knowledge is available, acquire it, and apply it to their needs. The strength of this model rests on the recognition that transfer and use (in addition to production) are critical elements of the process of technological innovation. Its weakness lies in the fact that the one-way, producer-to-user approach is passive in that users are considered only when they interact with or contact the system for assistance. The existing federal (i.e., DoD, DoE, and NASA) system is based on a dissemination model and employs one-way, producer-to-user procedures that are seldom responsive in the user context. User requirements and behaviors are not known or considered in the design of linkage mechanisms. This model does not take into account the process of technological innovation within the firm, nor does it acknowledge that small, medium, and large firms interact differently with the external environment. Lastly, this model fails to recognize that the willingness and ability of firms to absorb extramurally produced research results vary from industry to industry.

A Critique of U.S. Public Policy

Data, information, and knowledge are central to the process of technological innovation and its management (Fischer, 1980). However, the United States has no coherent, centrally organized, or systematically designed policy with respect to the results of federally funded science and technology (Bikson, Quint, and Johnson, 1984). What the U.S. has are numerous policies and programs that cut across political jurisdictions and the idiosyncratic missions and mandates of multiple agencies that are more or less responsive to a series of shifting political alliances and imperatives. This assessment is hardly new. Bishop and Fellows (1989) analyzed 12 major policy studies concerned with the coordination and management of data, information, and knowledge resulting from the U.S. government's investment in science and technology. From their analysis, these researchers concluded that today's imperfect federal system is not due to a lack of policy studies but rather the failure of the studies' producers to influence policymakers. Perhaps this conclusion is a bit harsh considering that the individuals who conducted the studies were hired to analyze the present state of federal policy and policymakers have yet to act upon the recommendations found in the earliest studies. Instead, much of the blame for the current state of the federal enterprise rests with inadequate direction from the Executive Branch, the lack of coordination among federal agencies and departments, and the fact that federal science and technology policy emphasizes knowledge production, not its transfer and use (Gold, 1993).

With respect to the results of federally funded science and technology, U.S. public policy, like many broad areas of government concern, is not articulated in a comprehensive form, nor is it part of national information policy (Gould, 1986). Federal policy relative to the data, information, and knowledge resulting from federally funded science and technology is embodied in various policy instruments derived from statutes, legislative intent and histories, Executive Order 12958 "Classified National Security Information," National Security Decision Directives, and agency regulations and management instructions. (An overview and analysis of U.S. policy instruments affecting the data, information, and knowledge resulting from federally funded science and technology is provided by Doty and Erdelez, 1989.) Gould (1986) accurately summarized U.S. public policy with respect to the results of federally funded science and technology in the following statement:

> Government shall place no restrictions on the free and open exchange of ideas through the presentation, publication, and dissemination of new ideas, discoveries, and experimental results except as provided for in the applicable provisions of law. (p. 66)

The provisions of law are broad and varied (Hernon and Relyea, 1991). In a commercial sense, they include patents, trade secrets, and proprietary data. Perhaps the strongest statutory basis for restricting "free and open exchange" relates to national security and data, information, and knowledge having military applications. (See Relyea, 1994, for a discussion of "free vs. restricted access" to the results of federally funded science and technology and national security.) Applicable legislation includes the Invention Secrecy Act of 1951 (P.L. 82-256), the Atomic Energy Act of 1954 (P.L. 83-703) as amended, the Arms Export Control Act of 1976 (P.L. 95-92), the Export Administration Act of 1985 (P.L. 99-64), the Omnibus Trade and Competitiveness Act of 1988 (P.L. 100-418), the National Defense Authorization Act of 1991 (P.L. 101-510), and Executive Order 12928 "Classified National Security Information." Technical reports issued by federal agencies (e.g., DoD, DoE, and NASA) that are subject to these statutes carry a legend indicating that the information contained therein is or may be subject to control under the Export Administration Regulations (EAR), International Traffic in Arms Regulations (ITAR), Military Critical Technologies List (MCTL), or Nuclear Export Controls (NEC). (For additional information on the legal basis for controls on the "export" of data, information, and knowledge, see Cinquegrana and Sheperd, 1984; Finkler, Boezer, Foss, Jorstad, and Ramsbotham, 1990; Monahan, 1983; Morillo, 1994; U.S. Congress, Office of Technology Assessment, 1994.)

Concerns with U.S. competitiveness have resulted in the passage of legislation and the issuance of Executive Orders that, in effect, determine the ownership of data, information, and knowledge resulting from federally funded science and technology. Applicable legislation includes the Stevenson-Wydler Technology Innovation Act of 1980 (P.L. 96-480), the Patent and Trademark Act of 1980 (P.L. 96-517), the Small Business Innovation Development Act of 1982 (P.L. 97-219), the National Cooperative Research Act of 1984 (P.L. 98-462), the Trademark Clarification Act of 1984 (P.L. 98-620), the Federal Technology Transfer Act of 1986 (P.L. 99-502), the Omnibus Trade and Competitiveness Act of 1988 (P.L. 100-418), the National Competitiveness Technology Transfer Act of 1989 (P.L. 101-189), E.O. 12591 "Facilitating Access to Science and Technology," and E.O. 12618 "Uniform Treatment of Federally Funded Inventions." Here, the U.S. government takes the position that the best hope for commercialization rests with private ownership of publicly funded knowledge. According to Goldstone (1986), the existing federal policy of requiring contractors and grantees to submit reports that document the results of government-funded research has gradually changed. The change was made in the belief that contractors and grantees could do a better job of commercializing new technology than could the federal agencies supporting those efforts. Although this change in policy might have a positive influence on the commercialization of new technology, the absence of a report denies or restricts access to others who might be interested in or who could use the results of publicly funded science and technology. Goldstone (1986) speaks to a U.S. technology policy in which the results of publicly funded science and technology are equally available for commercialization by any and all firms. The argument against this "evenhanded" policy is that it negates the profit incentive that drives commercialization. Without proprietary rights (ownership) and patent protection, few, if any, companies will invest the resources required to translate research results into practical products. (See Berghel, 1996; Cohen and Noll, 1994; and Schriesheim, 1990-91 for a variety of views of this topic.)

Furthermore, there is the situation that White (1990) describes as the "The 26-Mile, 380-Yard Marathon." Every year, the U.S. government invests billions of dollars in science and technology. Although the resulting data, information, and knowledge are very valuable, many potential users either do not know that they exist or do not know how to access them. The U.S. government spends considerable sums of money annually to create public awareness of and to provide access to these data, information, and knowledge. White points out that the federal government expects the public to "again pay for what they have already paid for" (p. 51). By charging the public to obtain the results of federally funded science and technology, White asserts that the government quits the race five yards short of the finish line. President Clinton (Clinton and Gore, 1993) illustrates what White sees as an apparent contradiction in U.S. policy:

> We are committed to using new computer and networking technology to make this information [the information collected and processed by the federal government] more available to the taxpayers who paid for it. In addition, it will require consistent federal information policies designed to ensure that federal information is made available at *a fair price* [emphasis added] to as many users as possible while encouraging growth of the information industry. (p. 17)

Facing continuing budget cuts, federal agencies are rethinking user fees for information products. However, the Paperwork Reduction Act of 1995 (P.L. 104-13) provides for user fees no higher than the cost of information dissemination. Federal agencies are enjoined from making a profit on their information products and permitted only to recover costs (Sprehe, 1996). (For additional views on the dissemination of government information, see U.S. Congress, House of Representatives, 1995, and U.S. Senate, 1995 reports on the Paperwork Reduction Act of 1995.)

Public policy concerned with the results of federally funded science and technology is not tied to U.S. science and technology policy. Rather, it is derived from the generic federal information policy that is binding on Executive Branch agencies through the Paperwork Reduction Act of 1995. This law is implemented by the Office of Management and Budget (OMB) through OMB Circular No. A-130, "The Management of Federal Information Resources." The Paperwork Reduction Act of 1995 has several purposes, including minimizing the paperwork burden resulting from information collected by or for the federal government; minimizing the cost to the federal government for the creation, collection, use, and dissemination of information retained by or for the federal government; and providing for the dissemination of public information on a timely basis, on equitable terms, and in a manner that promotes the utility of the information to the public.

OMB Circular A-130 implements the Paperwork Reduction Act of 1995 and sets standards and guidelines as to how federal agencies should manage their information. Sprehe (1995) has argued that OMB should take the lead within the Executive Branch in coordinating and managing the results of federally funded science and technology using the authority granted to it by the Paperwork Reduction Act of 1995. Critics contend, however, that as a policy instrument, OMB Circular A-130 does not provide the coherent, centrally organized, or systematically designed policy approach that is needed for coordinating and managing the results of federally funded science and technology. Further, they argue that Circular A-130 is too generic; pays precious little attention to data, information, and knowledge as content; places far too much emphasis on the physical package (i.e., records management) in which data, information, and knowledge reside; attempts to apply "life cycle" management to data, information, and knowledge; and is largely ignored by Executive Branch departments and agencies.

Finally, whereas variations of the dissemination and the appropriability models have been tried within the Executive Branch, at the federal level, management continues to be driven by a supply-side, dissemination model. Scholars like Branscomb (1991) argue, however, that this approach and the *trickle-down* benefits associated with the funding of basic research and mission-oriented (i.e., applied) research are inadequate in terms of stimulating technological innovation and competitiveness. Branscomb (1992) advocates the adoption of a *diffusion-oriented* policy for U.S. science and technology to gain a competitive advantage in the emerging global economy.

The *knowledge diffusion model* is grounded in theory associated with planned change research, the practices of the agricultural extension program, and the clinical models of social research and public health. Knowledge diffusion emphasizes "active" intervention in conjunction with dissemination and access; it stresses reliance on interpersonal communication as a means of identifying and removing barriers between users and producers, and relies on human intercession to help bring about a desired outcome or planned change. The knowledge diffusion model substitutes an interactive, *two-way* approach to communicating for the one-way, producer-to-user approach found in the dissemination model. It also assumes that knowledge production, transfer, and use are equally important components of the process of technological innovation. This model emphasizes the link between producers, transfer agents, and users, and relies on user-oriented mechanisms specifically tailored according to the needs, circumstances, and behaviors of the user. The interactive, two-way approach, which is at the heart of this model, allows for feedback, which is a vital component of program evaluation. The knowledge diffusion model assumes that the results of federally funded science and technology will be under-utilized unless they are organized for users in a "problem relevant" manner and ongoing (interpersonal) relationships are developed among users and producers. Because the knowledge diffusion model emphasizes "active" intervention in conjunction with dissemination and access, it requires a substantial human presence, and it runs contrary to many of the dominant assumptions of established federal science and technology policy. Although the U.S. relies on a "dissemination-oriented" model, other industrialized nations, notably Germany and Japan, have adopted "diffusion-oriented" models that increase the power of the user to absorb and employ new technologies productively (Branscomb, 1992, 1991).

THE SYSTEM FOR DISSEMINATING THE RESULTS OF FEDERALLY FUNDED AERONAUTICAL R&T

The U.S. aircraft industry is virtually unique among U.S. manufacturing industries in that a federal research organization, the NACA (later NASA) has for many years conducted and funded research on airframe and propulsion technologies (Mowery, 1985). The tradition of public support (i.e., funding) in the U.S. for aeronautical research goes back to the earliest days of the industry and the creation of the NACA. As stated in its enabling legislation, the Naval Appropriations Act of 1916 (P.L. 63-271), the NACA was created to "supervise and direct the scientific study of the problems of flight with a view to their practical solutions and to give advice to the military air services and other aviation services of government."

Throughout most of its history, the NACA was arguably the most important and productive aeronautical research establishment in the world (Roland, 1985). It developed wind tunnels and the tools and techniques necessary to conduct aeronautical research. In its wind tunnels and laboratories, the NACA worked on problems of aerodynamics and aeronautics common to both military and civil aviation. The NACA was responsible for a series of aeronautical innovations that helped foster the establishment of a U.S. aircraft industry. The development of an engine cowl reduced wind drag, research in aerodynamic efficiency assisted determination of optimal engine placement, and the creation of a family of airfoils by the NACA allowed engineers to develop new wing shapes and designs. (See Gray, 1948; Hansen, 1987; Hunsaker, 1955; Roland, 1985, for a history of the NACA. See Hudson, 1972; Paulisick, 1972, for the contributions of federally funded aeronautical R&D to commercial aviation.)

Within the NACA, a coordinated approach to innovation and adoption was the norm. The NACA provided a rich supply of knowledge and went to great lengths to encourage its diffusion and adoption, including annual aircraft engineering conferences that highlighted recent NACA research accomplishments and identified potential research problems. In addition to conducting and co-ordinating fundamental research, the NACA devoted a substantial portion of its technical expertise and talents to solving problems for the military and the aircraft industry. It also encouraged aircraft companies with comparable technical abilities and interests to participate in collaborative NACA-directed research and to share the findings among all of the participants (Shapley and Roy, 1985). Furthermore, the NACA also collected, evaluated, and distributed the results of European aero-nautical research. However, the creation of NASA (P.L. 85-568) in 1958 permanently altered the *de facto* innovation policy established by the NACA.

NASA aeronautical R&T can be viewed as two parts of a whole— the *focused programs* [e.g., high-speed research (HSR)] and the *R&T base*. Focused programs are collaborative in nature and typically involve NASA and U.S. commercial aircraft companies. Much of this research is conducted by the aircraft companies and the results tend to be proprietary or, at minimum, have a limited (i.e., restricted) distribution. [Note: Section 203 (c)(5) and (6) of The National Aeronautics and Space Act of 1958 was amended in October 1992 to provide an exception to previous laws regarding the public disclosure of sensitive data and information. Pursuant to Section 303 of this Act, data and information generated under (Space Act) agreements between NASA and nonfederal parties can be protected by this amendment from Freedom of Information Act (FOIA) requests, thus benefiting the competitiveness of U.S. industry.] The R&T base can be thought of as the NASA in-house (generic) research program.

Examining the NASA System

The framework for our examination of the NASA system incorporates the following: the communications processes employed by individuals and organizations involved in R&D and technological innovation, what is known about knowledge diffusion in theory and practice, and the extent to which aeronautical R&T policy recommendations concerning knowledge diffusion have been implemented. Knowledge is critical for the competitive, sustainable advantage of the high-technology LCA industry. The NASA system is a major source of the *external* knowledge used by the LCA portion of the U.S. aircraft industry. It is also a national system, as defined by Mowery (1992) and Nelson (1993). Thus, examining NASA's system for codifying and disseminating the results of federally funded aeronautical R&T should prove useful for policymakers, R&D managers, and individuals charged with operating the NASA system.

Communications processes. Technological innovation is, in large measure, a communication and information processing activity (Fischer, 1980). Problem solving is an integral part of the innovation process (Ebadi and Utterback, 1984). Formal (written) and informal (oral) communications alike play a key role in identifying, acquiring, and transferring the *explicit* and *tacit* knowledge needed for problem solving and innovation. Although formal channels exist for the exchange of knowledge, the most valuable data, information, and knowledge are typically communicated through informal channels (Rogers, 1982). As Allen (1977) pointed out, engineers and scientists both require large amounts of information to perform their work; however, the outputs of engineers and scientists differ significantly. Scientists use information to produce explicit knowledge; thus, the inputs and the outputs of the information-processing system in science are inherently compatible. Engineers, however, consume information and transform it into tacit knowledge that is embedded in products, processes, and designs. The inputs and outputs of the information-processing system in engineering differ, with engineering outputs typically having both physically encoded and verbally encoded parts. Human intervention, often in the form of person-to-person communication, is needed to articulate, interpret, and supplement the tacit knowledge embedded in engineering process and product development.

LCA manufacturers are knowledge-intensive organizations that function as open systems (Galbraith, 1973) that must deal with complexity and sources of work-related uncertainty. LCA manufacturers must cope with technical and market uncertainty from outside the organization as well as uncertainty concerning problem solving within the organization. In the research, development, and production (RD&P) of an LCA, uncertainty increases as a problem becomes more complicated, the environment becomes more dynamic, or task interdependence becomes more complex. LCA manufacturers use business and technical data, information, and knowledge obtained largely from the *external* environment to reduce complexity and uncertainty. The greater the complexity and uncertainty, the greater the information-processing requirements and the greater the need for external data, information, and knowledge. However, it is the nature of organizations involved in R&D and technological innovation to isolate themselves from the external environment. Such behavior is due, in large part, to the need for organizations to maintain stability and control, and because the organizations are involved in activities of a proprietary nature that entail trade secrets, patent protection, and intellectual property. Organizations use a variety of "boundary-spanning" techniques or linkages to maintain contact with the external environment and to identify, acquire, and absorb data, information, and knowledge. The three primary boundary-spanning linkages fall into two groups—the *informal* that relies on collegial or peer group contacts and gatekeepers (Davis and Wilkof, 1988) or key communicators (Chakrabarti and O'Keefe, 1977) and the *formal* that relies on such information intermediaries as librarians.

NASA employs both informal and formal communications to help ensure that the results of its aeronautical R&T are made accessible to the U.S. aerospace community. Informal communications methods include collegial contacts, academic and industry visits to NASA facilities, and visits by NASA personnel to academia and industry; personnel exchange programs; and collaborative research programs involving universities and aircraft companies. Formal communications methods include the publication (e.g., conference-meeting papers, journal articles, and technical reports) and presentation of the results of NASA-sponsored (i.e., contracts and grants) and performed research at the meetings of technical and professional societies. The results of NASA aeronautical R&T are also distributed through NASA-sponsored symposia and workshops. NASA technical reports are formally distributed (mailed) to foreign and domestic aerospace organizations, usually the library. The results are, therefore, "available" to individuals or organizations choosing to identify, acquire, and absorb this U.S. publicly supported (i.e., funded) data, information, and knowledge. Citations and abstracts of publications that contain the results of federally funded aeronautical R&D appear in the NASA *RECON* (REsearch CONnection) and AIAA *Aerospace* databases. To increase access to the results of NASA aeronautical R&T, the full-text versions of many NASA technical reports are accessible in electronic form via the Internet and the World Wide Web (Nelson, Gottlich, and Bianco, 1994; Roper, et al., 1994). Information intermediaries working in the technical libraries of each of the NASA aeronautical centers identify and acquire data, information, and knowledge from the external environment, although the primary purpose of the technical library is to serve as the repository for data, information, and knowledge generated by that center.

The RD&P of a (subsonic) LCA requires a large investment of capital—$8 billion to $10 billion by some estimates—and time. To reduce the risks involved, LCA manufacturers have begun to partner with foreign and domestic firms. Similarly, LCA manufacturers find it too costly and cumbersome to develop, on their own, all of the knowledge, capabilities, and expertise needed to introduce a new LCA successfully. In response, LCA manufacturers have created tactical and strategic alliances and linkages that give them access to knowledge, expertise, capabilities, and facilities and allow them to work with other organizations to develop new knowledge, expertise, and capabilities. U.S. LCA manufacturers have a long-established knowledge link with NASA and its predecessor, the NACA. However, the effectiveness of this alliance is unknown because the actual linkages are unknown, and a methodology for assessing their effectiveness has not been established. Actual linkages would, of necessity, encourage person-to-person (oral) communication; targeted participation in selected conferences and meetings; specialized personnel exchange programs; the fostering of gatekeepers to serve as linking agents within and outside the organization; and the development of mechanisms that link knowledge seekers to those who can provide the needed knowledge. In addition, the data, information, and knowledge created by and for NASA would be organized in a problem-solution oriented context and access points to it moderated to ensure effective utilization by U.S. LCA manufacturers. Lastly, actual linkages would exist between the information intermediaries (librarians) of U.S. LCA manufacturers and those at the NASA aeronautical research centers.

Knowledge diffusion in theory and practice. The NASA system is actually a subset of the federal system that makes the results of federally funded R&D accessible to academia, industry, state and local governments, and other federal agencies. Within the system, the NASA technical report is used as a primary product for knowledge transfer. A model depicting the dissemination of NASA aeronautical R&T via the U.S. government technical report appears as Figure 4.1. The model is composed of two parts—the *informal* that relies on collegial contacts between aerospace engineers and scientists and the *formal* that relies on surrogates, producers, and information intermediaries to complete the "producer-to-user" dis-

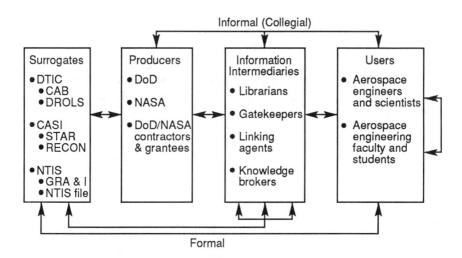

Figure 4.1. A Model Depicting the Dissemination of NASA Aeronautical R&T via the U.S. Government Technical Report.

semination process. When U.S. government (i.e., NASA) technical reports are published, the initial or primary distribution is made to institutions and organizations composed almost entirely of libraries and technical information centers in academia, industry, and other federal agencies. The addresses are both domestic (U.S.) and foreign. Copies are sent to surrogates for secondary and tertiary distribution. A limited number are set aside as "author" copies for "scientist-to-scientist" or collegial exchange (Hernon and Pinelli, 1991).

The producers include NASA and NASA contractors and grantees. Within NASA, three research centers (i.e., Ames, Langley, and Lewis) produce the preponderance of technical reports containing the results of aeronautical R&T. Surrogates serve as technical report repositories or clearinghouses for the producers and include the Defense Technical Information Center (DTIC), the NASA Center for AeroSpace Information (CASI), and the National Technical Information Service (NTIS), an organizational element of the U.S. Department of Commerce. The surrogates have created a variety of print products—*Current Awareness Bibliographies (CAB), Scientific and Technical Aerospace Reports (STAR),* and *Government Reports An-*

nouncement and Index (GRA&I)—and computerized information systems—*Defense RDT&E Online System (DROLS)*, *NASA RECON*, and *NTIS Online*—designed to facilitate awareness of and access to U.S. government technical reports. (Note: The print-based *CAB* and *STAR* are no longer published.)

Information intermediaries are, in large part, librarians and technical information specialists in academia, government, and industry. Those representing the producers serve as what McGowan and Loveless (1981) describe as "knowledge brokers" or "linking agents." Information intermediaries connected with users act, according to Allen (1977), as "technological entrepreneurs" or "gatekeepers." The more "active" the intermediary, the more effective the transfer process becomes (Goldhor and Lund, 1983). Active intermediaries move information from the producer to users, often utilizing interpersonal (i.e., face-to-face) communication in the process. Passive information intermediaries, on the other hand, "simply array information for the taking, relying on the initiative of the user to request or search out the information that may be needed" (Eveland, 1987, p.7).

The system for disseminating the results of federally funded R&D, of which the NASA system is a part, is "passive, fragmented, and unfocused" (Ballard, et al., 1989, p. 37). Effective knowledge transfer is hindered because the federal government "has no coherent or systematically designed approach to transferring the results of federally funded research to the user" (Bikson, Quint, and Johnson, 1984, p. v). In their study of federal scientific and technical information activities, Bikson, Quint, and Johnson found that many of the individuals they interviewed believed that "dissemination activities were afterthoughts, undertaken without serious commitment by federal agencies whose primary concerns were with [knowledge] production and not with knowledge transfer and use"; therefore, "much of what has been learned about knowledge diffusion has not been incorporated into activities designed to transfer the results of federally funded R&D from producers to users" (p. 23).

Problematic to the *informal* part of the federal (and the NASA) system is that knowledge users can learn from collegial contacts only what those contacts happen to know. Ample evidence supports the claim that no one researcher can know about or keep up with all the research in his or her area(s) of interest. Like other members of the scientific community, engineers and scientists are faced with the problem of too much data, information, and knowledge to know about, to keep up with, and to screen. Furthermore, information is becoming more interdisciplinary in nature and more international in scope. Two problems exist with the formal part of the system. *First*, the formal part of the system employs one-way, source-to-user transmission. The problem with this kind of transmission is that such formal one-way, "supply side" transfer procedures do not seem to be responsive to the user context (Bikson, Quint, and Johnson, 1984). Rather, these efforts appear to start with a system into which the users' requirements are retrofitted (Adam, 1975). The consensus of the findings from the empirical research is that interactive, two-way communications are required to transfer data, information, and knowledge effectively from producers to users (Bikson, Quint, and Johnson, 1984).

Second, the formal part relies heavily on information intermediaries to complete the producer-to-user process. However, a strong methodological base for measuring or assessing the effectiveness of the information intermediary is lacking (Beyer and Trice, 1982). In addition, empirical data on the effectiveness of information intermediaries and the role(s) they play in knowledge transfer are sparse and inconclusive. The impact of information intermediaries is likely to be strongly conditional and limited to a specific institutional context (Pinelli, Barclay, Hannah, Lawrence, and Kennedy, 1992).

Most federal information dissemination activities have been ineffective in stimulating technological innovation and in transferring technology (Alic, 1986; Tornatzky and Luria, 1992; U.S. Congress, General Accounting Office, 1991). The numerous federal programs designed to do so are "highest in frequency and expense yet lowest in impact," and federal "information dissemination activities have led to few documented successes" (Roberts and Frohman, 1978, p. 36). Roberts and Frohman also note that "governmental programs start to encourage knowledge utilization only after the research results have been generated" (p. 36), rather than during the idea development phase of the innovation process. David (1986), Mowery (1983), and Mowery and Rosenberg (1979) conclude that successful technological innovation rests more with the transfer and utilization of data, information, and knowledge than with their production.

Implementation of aeronautical R&T policy recommendations. Two previous policy studies, the Office of Science and Technology Policy (OSTP) aeronautical R&T policy study (U.S. Executive Office of the President, Office of Science and Technology Policy, 1982) and the National Academy of Engineering (1985) study on the competitive status of the U.S. civil aviation manufacturing industry include recommendations for knowledge diffusion specific to the NASA system. Both studies concluded that the unrestrained transfer of data, information, and knowledge resulting from NASA aeronautical R&T could unduly aid foreign competitors and that the competitive status of the U.S. LCA aircraft industry could be increased by exploiting the results of foreign aeronautical research results. The two studies made the following recommendations:

- NASA policy and practice relative to the dissemination of Agency technical reports should be reviewed to determine if, and to what extent, dissemination activities unduly aid foreign competitors.

- NASA should collect, analyze, and distribute the results of foreign aeronautical R&T to the U.S. technical community.

The group responsible for distributing NASA technical reports is organizationally disassociated (separated) from the NASA aeronautics program. General guidance is derived from federal statutes (e.g., FOIA), E.O. 12958, and OMB Circular A-130. However, NASA policy and practice, with respect to the dissemination of data, information, and knowledge resulting from federally funded aeronautical R&T is not articulated in a comprehensive, written form. Finally, it appears that NASA policy and practice relative to the dissemination of Agency technical reports has not been evaluated to determine if, and to what extent, dissemination activities unduly aid foreign competitors.

A 1995 examination of the distribution list for unclassified NASA technical reports in Category 01-Aeronautics-General revealed that 109 (62%) of the addressees were non-U.S. organizations and institutions. Mailing these reports directly to research organizations in such countries as France, Germany, and Japan would appear to ignore the studies' recommendations and concerns that NASA policy and practice might unduly aid foreign competitors. The distribution of unclassified NASA technical reports outside of the U.S. is an agreed on condition, however, in a variety of bilateral and multinational agreements to which NASA is a signatory. As Sprehe (1994) points out, in the case of bilateral and multinational agreements, federal agencies frequently find themselves bound up in the larger context of U.S. foreign policy and often give more (information) than they receive.

Lastly, the NASA system was examined to determine the extent to which aeronautical data, information, and knowledge from countries outside of the U.S. are collected, analyzed, and distributed to the U.S. technical community. The OSTP and the National Academy of Engineering studies recognized that significant aeronautical research capabilities exist in many foreign countries and that the U.S. stands to benefit from collecting, translating, analyzing, and distributing the results of foreign aeronautical R&D. Recognizing that the U.S. lacks a centralized organization to collect, translate, analyze, and distribute the best unclassified non-U.S. aeronautical R&D, a role formerly performed by the NACA, both study groups recommended that NASA develop and implement a plan for doing this. To date, NASA has neither developed nor implemented such a plan. Instead, NASA appears to rely on the publications received from its bilateral and multinational agreements as the primary source of non-U.S. aeronautical R&D. Specifically, most of the non-U.S. aeronautical R&D publications obtained by NASA come from a bilateral agreement with the European Space Agency (ESA). It is unclear how many of these reports focus on aeronautical R&D, the nature of ESA's work being space-related, not aeronautical. Finally, NASA makes virtually no attempt to analyze and distribute the results of foreign aeronautical R&T to the U.S. technical community.

CONCLUSIONS

Knowledge is a building block of technological innovation. The rate at which knowledge is created, diffused (i.e., spread, distributed, transmitted), and adopted (i.e., applied or utilized) influences the speed of technological innovation and progress. Policymakers are beginning to realize that knowledge is as vital to technological innovation as technological innovation is to economic growth, global competitiveness, and international trade. Recent years have witnessed growing interest on the part of policymakers in the diffusion of knowledge as a means of (a) understanding technological innovation, (b) analyzing changes in the rate of innovation, and (c) predicting the effects of technological innovation. Since World War II, the U.S. government has continuously invested in (i.e., funded) science and technology. This expenditure is generally considered to represent an investment in the country's future; it also serves as a means of improving the health of the nation's citizens, defending the nation, fueling economic growth, providing jobs in new industries, and generally raising the standard of living in the U.S (Gray, Solomon, and Hetzner, 1986).

During the quarter century following World War II, the United States was the world's most productive economy by virtually any measure (Nelson and Wright, 1992). No longer. Although the LCA portion of the aircraft industry is a notable exception, the U.S. technological lead has eroded in many industries, and in some industries the U.S. now lags behind other nations. Numerous reasons have been cited for the decline, and a variety of measures has been posited as solutions. There appears to be general agreement that (a) the U.S. is now competing in a transnational (i.e., global), not a national economy; (b) trade, business, and commerce have become internationalized; (c) national borders mean far less now to the flow of science and technology than they used to; and (d) the results of the nation's investment in science and technology may not be well utilized.

Three reasons are frequently given to support this last opinion. *First*, existing U.S. technology policy is inappropriate for the current global competitive economy because it neglects the effective transmission and utilization of knowledge and does not take into account the relationship among knowledge production, transfer, and utilization as equally important components of the process of knowledge diffusion. *Second*, the United States has no coherent, centrally organized, or systematically designed public policy with respect to managing the results of federally funded science and technology. Existing public policy pays far too much attention to the creation of knowledge and not enough to its diffusion and adoption. *Third*, the federal system that is responsible for managing the massive amounts of data, information, and knowledge resulting from federally funded science and technology is passive, fragmented, and unfocused. The *dissemination model*, on which the system is based, utilizes a one-way, producer-to-user approach that is passive. Users are considered only when they interact with or contact the system for assistance. Branscomb (1992) advocates replacing the dissemination model with a *diffusion-oriented* technology policy and model for managing the data, information, and knowledge resulting from federally funded science and technology. A *diffusion-oriented* technology policy and model offer the promise of stimulating technological innovation and competitiveness, thus permitting the United States to gain a competitive advantage in a global economy.

Federal involvement in agriculture and aviation is widely cited as a model for public (government) support for civilian R&D. Both agriculture and aviation employ systems that effectively produce knowledge and include "total delivery" systems designed to encourage the dissemination and adoption of knowledge. The American tradition of publicly funded agriculture and aviation has contributed to long-term increases in productivity and has yielded high rates of return to the public (Feller, Madden, Kaltreider, Moore, and Sims, 1987). Since 1917, the U.S. government has provided national leadership and significant financial support for the development of aeronautical R&T. In recent years, policymakers and critics have questioned the appropriateness and effectiveness of current U.S. aeronautical R&T policies and the U.S. government's role in the support of aeronautical R&T. With respect to knowledge diffusion, existing U.S. aeronautical R&T policy may not be appropriate for the current competitive global economy for four reasons. *First*, current policy does not recognize knowledge production, transfer, and utilization as equally important components of the R&D process. *Second*, the producer-oriented dissemination model on which the NASA system for transferring knowledge is based is inappropriate. *Third*, specific policy recommendations for enlarging and enriching the U.S. aeronautical R&T knowledge base have been virtually ignored. *Fourth*, specific policy concerns that foreign competitors might benefit unduly from receiving the results of U.S. aeronautical R&T have not been addressed.

IMPLICATIONS FOR DIFFUSING THE RESULTS OF FEDERALLY FUNDED AEROSPACE R&D

Although the LCA sector of the U.S. aerospace industry continues as a leading positive contributor to the U.S. balance of trade, the sector is experiencing significant changes whose implications may not be well understood. Increasing collaboration between U.S. and foreign producers is creating a more international manufacturing and marketing environment and continuously altering the structure of the LCA sector (Sabbagh, 1996).   International alliances are forcing U.S. companies to push ahead with new technological developments and, at the same time, position themselves to develop the next generation of LCA.   International alliances are also fostering the creation of knowledge linkages—alliances that give U.S. LCA manufacturers access to the knowledge, skills, and capabilities of other U.S. and non-U.S. organizations and permit the collaborative development of new knowledge, skills, and capabilities. These factors combine to make the results of federally funded R&D a "knowledge asset" that has to be managed both tactically and strategically. The production, transfer, and use of knowledge are becoming equally important parts of the aerospace R&D process. To ensure effective diffusion of aerospace knowledge, the current dissemination policy and model for transferring federally funded aerospace knowledge would need to be replaced with a *diffusion-oriented* policy and model. Furthermore, knowledge linkages between NASA and U.S. LCA manufacturers would, of necessity, have to become more proactive and user-oriented.

CHAPTER REFERENCES

Adam, R. (1975). "Pulling the Minds of Social Scientists Together: Towards a Science Information System." *International Social Science Journal* 27(3): 519-531.

Adkinson, B.W. (1978). *Two Centuries of Federal Information*. Stroudsburg, PA: Dowden, Hutchinson and Ross.

Aines, A.A. (1984). "A Visit to the Wasteland of Federal Scientific and Technical Information Policy." *Journal of the American Society for Information Science* (May) 35(3): 179-184.

Alic, J.A. (1986). "The Federal Role in Commercial Technology Development." *Technovation* 4(4): 253-267.

Allen, T.J. (1977). *Managing the Flow of Technology: Technology Transfer and the Dissemination of Technological Information Within the R&D Organization*. Cambridge, MA: MIT Press.

Auger, C.P. (1975). *Use of Technical Reports Literature*. Hamden, CT: Archon Books.

Averch, H.A. (1985). *A Strategic Analysis of Science & Technology Policy*. Baltimore, MD: Johns Hopkins University Press.

Baker, M.J. (1979). *Industrial Innovation: Technology, Policy, Diffusion*. London: Macmillan Press.

Ballard, S.; T.E. James, Jr.; T.I. Adams; M.D. Devine; L.L. Malysa; and M. Meo. (1989). *Innovation Through Technical and Scientific Information: Government and Industry Cooperation.* New York, NY: Quorum Books.

Berghel, H. (1996). "U.S. Technology Policy in the Information Age." *Communications of the ACM* (June) 39(6): 15-18.

Beyer, J.M. and H.M. Trice. (1982). "The Utilization Process: A Conceptual Framework and Synthesis of Empirical Findings." *Administrative Science Quarterly* 27: 591-622.

Bikson, T.K.; B.E. Quint; and L.L. Johnson. (1984). *Scientific and Technical Information Transfer: Issues and Options.* Washington, DC: National Science Foundation. (Available NTIS; PB-85-150357.)

Bishop, A.P. and M.O. Fellows. (1989). "Descriptive Analysis of Major Federal Scientific and Technical Information Policy Studies." Chapter 1 in *U.S. Scientific and Technical Information Policies: Views and Perspectives.* C.R. McClure and P. Hernon, eds. Norwood, NJ: Ablex, 3-55.

Branscomb, L.M. (1992). "U.S. Scientific and Technical Information Policy in the Context of a Diffusion-Oriented National Technology Policy." *Government Publications Review* 19(5): 469-482.

Branscomb, L.M. (1991). "Toward a U.S. Technology Policy." *Issues in Science and Technology* (Summer) 7(4): 50-55.

Brown, G.E. (1987). "Federal Information Policy: Protecting the Free Flow of Information." *Government Information Quarterly* 4(4): 349-358.

Brown, L.A. (1981). *Innovation Diffusion: A New Perspective.* London: Methuen & Co.

Bush, V. (1945). *Science: The Endless Frontier.* Washington, DC: Government Printing Office.

Chakrabarti, A.K. and R.D. O'Keefe. (1977). "A Study of Key Communicators in Research and Development Laboratories." *Group & Organization Studies* (September) 2(3): 336-346.

Cinquegrana, A.R. and J.M. Shepherd. (1984). "The Current Legal Basis for Controls on the 'Export' of Technical Information." *Boston College International & Comparative Law Review* 7(2): 285-302.

Clinton, W.J. and A. Gore. (1993). *Technology For America's Economic Growth, A New Direction to Build Economic Strength.* Washington, DC: U.S. Executive Office of the President, The White House. (Available NTIS; AD-A261-553.)

Cohen, L.R. and R.G. Noll. (1994). "Privatizing Public Research." *Scientific American* (September) 271(3): 72-77.

Cohen, W.M. and D.A. Levinthal. (1990). "Absorptive Capacity: A New Perspective on Learning and Innovation." *Administrative Science Quarterly* (March) 35: 128-152.

Coyne, J.G. and M.D. Fornwell. (1989). "Managing Scientific and Technical Information in DoE: A Departmental Approach." Chapter 11 in *United States Scientific and Technical Information Policies: Views and Perspectives.* C.R. McClure and P. Hernon, eds. Norwood, NJ: Ablex, 299-315.

David, P.A. (1986). "Technology Diffusion, Public Policy, and Industrial Competitiveness." In *The Positive Sum Strategy: Harnessing Technology for Economic Growth*. R. Landau and N. Rosenberg, eds. Washington, DC: National Academy Press, 373-391.

Davis, P. and M. Wilkof. (1988). "Scientific and Technical Information Transfer: Keeping the Figure in its Ground." *R&D Management* 18(1): 45-58.

Derian, J.C. (1990). *America's Struggle For Leadership in Technology*. Cambridge, MA: MIT Press.

Dosi, G. (1988). "The Nature of the Innovative Process." Chapter 10 in *Technical Change and Economic Theory*. G. Dosi; C. Freeman; R. Nelson; G. Silverberg; and L. Soete, eds. New York, NY: Pinter Publishers, 221-238.

Doty, P. and S. Erdelez. (1989). "Overview and Analysis of Selected Federal Scientific and Technical Information (STI) Policy Instruments, 1945-1987." Chapter 2 in *United States Scientific and Technical Information Policies: Views and Perspectives*. C.R. McClure and P. Hernon, eds. Norwood, NJ: Ablex, 56-83.

Ebadi, Y.M. and J.M. Utterback. (1984). "The Effects of Communication on Technological Innovation." *Management Science* (May) 30(5): 572-585.

E.O. 9568 "Providing for the Release of Scientific Information." June 8, 1945.

E.O. 9604 "Providing for the Release of Scientific Information Extension and Amendment of E.O. 9568." August 25, 1945.

E.O. 12591 "Facilitating Access to Science and Technology." April 10, 1987.

E.O. 12618 "Uniform Treatment of Federally Funded Inventions." December 22, 1987.

E.O. 12958 "Classified National Security Information." April 17, 1995.

Eveland, J.D. (1987). *Scientific and Technical Information Exchange: Issues and Findings*. Washington, DC: National Science Foundation. (Not available NTIS.)

Feller, I.; P. Madden; L. Kaltreider; D. Moore; and L. Sims. (1987). "The New Agricultural Research and Technology Transfer Policy Agenda." *Research Policy* 16: 315-325.

Finkler, R.A.; G.L. Boezer; E.J. Foss; N.D. Jorstad; and A.J. Ramsbotham. (1990). *Technology Transfer in a Changing National Security Environment*. IDA Paper P-2409. Alexandria, VA: Institute for Defense Analysis. (Available DTIC; AD-A232,458.)

Fischer, W.A. (1980). "Scientific and Technical Information and the Performance of R&D Groups." In *Management of Research and Innovation*. B.V. Dean and J.L. Goldhar, eds. New York, NY: North-Holland, 67-89.

Galbraith, J.K. (1973). *Designing Complex Organizations*. Reading, MA: Addison-Wesley.

Godfrey, L.E. and H.F. Redman. (1973). *Dictionary of Report Series Codes*. 2nd ed. New York, NY: Special Libraries Association.

Gold, B. (1981). "Technological Diffusion in Industry: Research Needs and Shortcomings." *Journal of Industrial Economics* (March) 29(3): 247-269.

Gold, B. (ed.). (1977). *Research, Technological Change, and Economic Analysis*. Lexington, MA: Lexington Books.

Gold, D. (1993). "Improving the Impact of Federal Scientific and Technical Information: A Call for Action." *Government Information Quarterly* 10 (2): 221-234.

Goldhor, R.S. and R.T. Lund (1983). "University-to-Industry Advanced Technology Transfer: A Case Study." *Research Policy* 12: 121-152.

Goldstone, N.J. (1986). "How Not to Promote Technology Transfer." *Technology Review* (July) 89: 22-23.

Gould, S.B. (1986). "Secrecy: Its Role in National Scientific and Technical Information Policy." *Library Trends* (Summer) 35(1): 61-81.

Gray, D.O.; T. Solomon; and W. Hetzner. (1986). *Technological Innovation: Strategy For a New Partnership*. Amsterdam, The Netherlands: North-Holland.

Gray, G.W. (1948). *Frontiers of Flight: The Story of NACA Research*. New York, NY: A.A. Knopf.

Hägerstand, T. (1967). *Innovation Diffusion As a Spatial Process*. Chicago, IL: The University of Chicago Press.

Hansen, J.R. (1987). *Engineer in Charge: A History of the Langley Aeronautical Laboratory, 1917-1958*. NASA SP-4305. Washington, DC: National Aeronautics and Space Administration. (Available NTIS; 87N24390.)

Havelock, R.G. (1973). *Planning for Innovation Through Dissemination and Utilization of Knowledge*. Ann Arbor, MI: Institute for Social Research.

Hernon, P. and C.R. McClure. (1988). *Public Access to Government Information: Issues, Trends, and Strategies*. 2nd. ed. Norwood, NJ: Ablex.

Hernon, P. and C.R. McClure. (1987). *Federal Information Policies in the 1980's: Conflicts and Issues*. Norwood, NJ: Ablex.

Hernon, P.; C.R. McClure; and H.C. Relyea. (1996). *Federal Information Policies in the 1990's: Views and Perspectives*. Norwood, NJ: Ablex.

Hernon, P. and T.E. Pinelli. (1991). "Scientific and Technical Information (STI) Policy and the Competitiveness Position of the U.S. Aerospace Industry." Paper presented at the *30th Aerospace Sciences Meeting and Exhibits of the American Institute of Aeronautics and Astronautics* (AIAA), Bally's Grand Hotel, Reno, NV, 9 January. AIAA-92-0796. (Available AIAA; 92A28233.)

Hernon, P. and H.C. Relyea. (1991). "Information Policy." In *Encyclopedia of Library and Information Science*. A. Kent and C.M. Hall, eds., 48:11. New York, NY: Marcel Dekker, 176-204.

Hill, C.T. and J.M. Utterback. (eds.). (1979). *Technological Innovation for a Dynamic Economy*. New York, NY: Pergamon Press.

Hudson, C.R., Jr. (1972). *Research and Development Contribution to Aviation Progress (RADCAP). Volume 2: Military Technology, Research, and Development to Civil Aviation Programs*. NASA CR-129573. Washington, DC: National Aeronautics and Space Administration. (Available NTIS: 73N13983; AKA Project RADCAP.)

Hunsaker, J.C. (1955). "Forty Years of Aeronautical Research." In the *Annual Report of the Smithsonian Institution*. Washington, DC: The Smithsonian Institution, 241-271.

Kitschelt, H. (1991). "Industrial Governmental Structures, Innovation Strategies, and the Case of Japan: Sectoral or Cross-National Comparative Analysis?" *International Organization* (Autumn) 45(4): 453-493.

Mansfield, E. (1968). *Industrial Research and Technological Innovation: An Econometric Analysis*. New York, NY: W.W. Norton.

Mansfield, E.; A. Romeo; M. Schwartz; D.J. Teece; S. Wagner; and P. Brach. (1982). *Technology Transfer, Productivity, and Economic Policy*. New York, NY: W.W. Norton.

March, J.G. and H.A. Simon. (1958). *Organizations*. New York, NY: John Wiley.

McClure, C.R. (1988). "The Federal Technical Report Literature: Research Needs and Issues." *Government Information Quarterly* 5(1): 27-44.

McClure, C.R. and P. Hernon. (eds.). (1989). *United States Scientific and Technical Information Policies: Views and Perspectives*. Norwood, NJ: Ablex.

McClure, C.R.; P. Hernon; and H.C. Relyea. (eds.). (1989). *United States Government Information Policies: Views and Perspectives*. Norwood, NJ: Ablex.

McGowan, R.P. and S. Loveless. (1981). "Strategies for Information Management: The Administrator's Perspective." *Public Administration Review* 41(3): 331-339.

Molholm, K.N.; B.L. Fox; P.M. Klinefelter; E.V. McCauley; and W.M. Thompson. (1988). "The Defense Technical Information Center: Acquiring Information: Imparting Knowledge." *Government Information Quarterly* 5(2): 323-340.

Monahan, P.J. (1983). "The Regulation of Technical Data Under the Arms Export Control Act of 1976 and the Export Administration Act of 1979: A Matter of Executive Discretion." *Boston College International & Comparative Law Review* 6(1): 169-197.

Morillo, J.P. (1994). "The Clinton Administration's New National Export Strategy." *Law & Policy in International Business* 25(3): 1113-1128.

Mowery, D.C. (1994). *Science and Technology Policy in Interdependent Economies*. Boston, MA: Kluwer Academic Publishers.

Mowery, D.C. (1992). "The U.S. National Innovation System: Origins and Prospects for Change." *Research Policy* (Fall) 21: 125-144.

Mowery, D.C. (1988). "The Diffusion of New Manufacturing Technologies." Chapter 12 in *The Impact of Technological Change on Employment and Economic Growth*. R.M. Cyert and D.C. Mowery, eds. Cambridge, MA: Ballinger Publishing Company, 481-509.

Mowery, D.C. (1985). "Federal Funding of R&D in Transportation: The Case of Aviation." Paper commissioned for a workshop on *The Federal Role in Research and Development* sponsored by the National Academy of Sciences, National Academy of Engineering, and Institute of Medicine in Washington, DC, 21-22 November.

Mowery, D.C. (1983). "Economic Theory and Government Technology Policy." *Policy Sciences* 16: 27-43.

Mowery, D.C. and N. Rosenberg. (1990). *The U.S. National Innovation System*. Consortium on Competitiveness & Cooperation Working Paper No. 90-3. Berkeley, CA: Center for Research in Management, University of California at Berkeley.

Mowery, D.C. and N. Rosenberg. (1989). *Technology and the Pursuit of Economic Growth*. New York, NY: Cambridge University Press.

Mowery, D.C. and N. Rosenberg. (1979). "The Influence of Market Demand Upon Innovation: A Critical Review of Some Recent Empirical Studies." *Research Policy* 8(2): 102-153.

Myers, S. and D.C. Marquis. (1969). *Successful Industrial Innovation: A Study of the Factors Underlying Innovation in Selected Firms*. NSF 69-17. Washington, DC: National Science Foundation.

Nabseth, L. and G.F. Ray. (eds.). (1974). *The Diffusion of New Industrial Processes: An International Study*. London: Cambridge University Press.

National Academy of Engineering. (1985). *The Competitive Status of the U.S. Civil Aviation Manufacturing Industry: A Study of the Influences of Technology in Determining International Industrial Competitive Advantage*. Washington, DC: National Academy Press. (Available NTIS; PB88-100334.)

Nelson, M.L.; G.L. Gottlich; and D.J. Bianco. (1994). *World Wide Web Implementation of the Langley Technical Report Server*. NASA TM-109162. Washington, DC: National Aeronautics and Space Administration. (Available NTIS; 95N13198.)

Nelson, R.R. (ed.). (1993). *National Innovation Systems: A Comparative Analysis*. New York, NY: Oxford University Press.

Nelson, R.R. and G. Wright. (1992). "The Rise and Fall of American Technological Leadership: The Postwar Era in Historical Perspective." *Journal of Economic Literature* (December) 30(4): 1931-1964.

Paulisick, J.G. (1972). *Research and Development Contribution to Aviation Progress (RADCAP): Contributions of Military Technology, Research, and Development to Civil Aviation Programs. Volume 1*. Washington, DC: National Aeronautics and Space Administration. NASA CR-129573. (Available NTIS; 73N13982.)

Pinelli, T.E. (Issue Editor). (1990). "National Aeronautics and Space Administration Scientific and Technical Information Programs." *Government Information Quarterly* 7(2): entire issue.

Pinelli, T.E.; R.O. Barclay; and J.M. Kennedy. (1993). "The U.S. Government Technical Report and the Transfer of Federally Funded Aerospace R&D." *Government Publications Review* 20(3): 393-411.

Pinelli, T.E.; R.O. Barclay; S. Hannah; B. Lawrence; and J.M. Kennedy. (1992). "Knowledge Diffusion and U.S. Government Technology Policy: Issues and Opportunities for Sci/Tech Librarians." *Science & Technology Libraries* 13(1): 33-55.

Pinelli, T.E. and M.M. Henderson. (1989). "Access to Federal Scientific and Technical Information Through U.S. Government Technical Reports." Chapter 4 in *United States Scientific and Technical Information Policies: Views and Perspectives.* C.R. McClure and P. Hernon, eds. Norwood, NJ: Ablex, 109-138.

Pinelli, T.E.; M.M. Henderson; A.P. Bishop; and P. Doty. (1992). *Chronology of Selected Literature, Reports, Policy Instruments, and Significant Events Affecting Federal Scientific and Technical Information (STI) in the United States: 1945-1990.* Washington, DC: National Aeronautics and Space Administration. NASA TM-107658. (Available NTIS; 92N27170.)

President's Science Advisory Committee. (1958). *Improving the Availability of Scientific and Technical Information in the United States.* Washington, DC: U.S. Government Printing Office. (AKA the Baker Report.)

Redman, H.F. (1965/1966). "Technical Reports: Problems and Predictions." *Arizona Librarian* (Winter) 23: 11-17.

Relyea, H.C. (1994). *Silencing Science: National Security Controls and Scientific Communication.* Norwood, NJ: Ablex Publishing.

Richards, P.S. (1994). *Scientific Information in Wartime: The Allied-German Rivalry, 1939-1945.* Westport, CT: Greenwood Press.

Roberts, E.B. and A.L. Frohman. (1978). "Strategies for Improving Research Utilization." *Technology Review* (March/April) 80(5): 32-39.

Rogers, E.M. (1995). *Diffusion of Innovations.* 4th ed. New York, NY: The Free Press.

Rogers, E.M. (1982). "Information Exchange and Technological Innovation." Chapter 5 in *The Transfer and Utilization of Technical Knowledge.* D. Sahal, ed. Lexington, MA: D.C. Heath, 105-123.

Roland, A. (1985). *Model Research: The National Advisory Committee for Aeronautics 1915-1958. Volume 1.* NASA SP-4103. Washington, DC: National Aeronautics and Space Administration. (Available NTIS; 85N23683.)

Roper, D.G.; M.K. McCaskill; S.D. Holland; J.L. Walsh; M.L. Nelson; S.L. Adkins; M. Ambur; and B.A. Campbell. (1994). *A Strategy For Electronic Dissemination of NASA Langley Technical Publications.* Washington, DC: National Aeronautics and Space Administration. NASA TM-109172. (Available NTIS; 95N18936.)

Rosegger, G. (1977). "Diffusion of Technology in Industry." Chapter 4 in *Research, Technology Change, and Economic Analysis*. B. Gold, ed. Lexington, MA: Lexington Books.

Rosenberg, N. (1986). *Perspectives on Technology*.(2nd ed.) London, UK: Cambridge University Press.

Rosenberg, N. (1982). *Inside the Black Box: Technology and Economics*. London, UK: Cambridge University Press.

Rosenberg, N. (1976). *Perspectives on Technology*. London, UK: Cambridge University Press.

Rubbert, P.E. (1994). "CFD and the Changing World of Airplane Design." Paper presented at the *19th Congress of the International Council of the Aeronautical Sciences*, 18-23 September, held in Anaheim, CA. In ICAS Proceedings of 1994, Volume I, LVII-LXXXIII.

Sabbagh, K. (1996). *Twenty-First Century Jet: The Making and Marketing of the Boeing 777*. New York, NY: Scribner.

Sahal, D. (ed.). (1982). *The Transfer and Utilization of Technological Knowledge*. Lexington, MA: Lexington Books.

Schriesheim, A. (1990-91). "Toward a Golden Age of Technology Transfer." *Issues in Science and Technology* (Winter) 7(2): 52-58.

Shapley, D. and R. Roy. (1985). *Lost at the Frontier: U.S. Science and Technology Policy Adrift*. Philadelphia, PA: ISI Press.

Shattuck, J. and M.M. Spence. (1989). *A Presidential Initiative on Information Policy, Number 7*. Washington, DC: The Benton Foundation. (Available ERIC; ED-324 022.)

Smith, B.L.R. (1990). *American Science Policy Since World War II*. Washington, DC: The Brookings Institute.

Sprehe, J.T. (1996). "Ways to Think About User Fees for Federal Information Products." *Government Information Quarterly* 13(2): 175-186.

Sprehe, J.T. (1995). "Does the Federal Government Need an A-130 for STI?" *Government Information Quarterly* 12(2): 213-223.

Sprehe, J.T. (1994). *A Study of Policy Issues Affecting the Defense Technical Information Center*. Alexandria, VA: Defense Technical Information Center. (Available DTIC; AD-A286 629.)

Tornatzky, L.G. and D. Luria. (1992). "Technology Policies and Programmes in Manufacturing: Toward Coherence and Impact." *International Journal of Technology Management* 7(1/2/3): 141-157.

U.S. Congress, General Accounting Office. (1991). *Technology Transfer: Federal Efforts to Enhance the Competitiveness of Small Manufacturers*. GAO/RCED-92-30. Washington, DC: General Accounting Office.

U.S. Congress, House of Representatives, 104th Congress. (1995). House Report 104-37. *Paperwork Reduction Act of 1995 Together With Additional Views on the Information Dissemination Provision of H.R. 830*. Washington, DC: Government Printing Office.

U.S. Congress, House of Representatives, 101st Congress. (1989). *Federal Scientific and Technical Information Policy*. Hearing Before the Committee on Science, Space, and Technology. 12 October. Washington, DC: Government Printing Office.

U.S. Congress, Office of Technology Assessment. (1994). *Export Controls and Nonproliferation Policy*. OTA-ISS-596. Washington, DC: Government Printing Office. (Available NTIS; PB94-179975.)

U.S. Congress, Office of Technology Assessment. (1990). *Helping America Compete: The Role of Federal Scientific & Technical Information*. OTA-CIT-454. Washington, DC: Government Printing Office.

U.S. Congress, Office of Technology Assessment. (1986). *Research Funding as an Investment: Can We Measure the Returns?* OTA-TM-SET-36. Washington, DC: Government Printing Office.

U.S. Congress, Senate, 104th Congress. (1995). Senate Report 104-8. *Paperwork Reduction Act of 1995, Report of the Committee on Governmental Affairs, United States Senate, Together With Additional Views to Accompany S. 244*. Washington, DC: Government Printing Office.

U.S. Executive Office of the President, Office of Management and Budget. (1996). *Budget of the United States Government: Analytical Perspectives*. Washington, DC: Government Printing Office.

U.S. Executive Office of the President, Office of Science and Technology Policy. (1982). *Aeronautical Research and Technology Policy. Volume 2. Final Report*. Washington, DC: Office of Science and Technology Policy. (Available NTIS; 83N-23268; AKA the Keyworth Report.)

U.S. Trade Representative, Trade Promotion Coordinating Committee. (1996). *National Export Strategy—Toward the Next American Century: A U.S. Strategic Response to Foreign Competitive Practices*. (Fourth Annual Report to the U.S. Congress.) Washington, DC: Government Printing Office.

Vadas, D.H. (1996). "Tools of the Trade: Offsets, Outsourcing, and Joint Ventures." *Aerospace America* (September) 34(9): 10-12.

von Hippel, E. (1978). *The Sources of Innovation*. New York, NY: Oxford University Press.

White, H.S. (1990). "The 26-Mile, 380-Yard Marathon." *Library Journal* (November) 15(115): 51.

Williams, F. and D.V. Gibson. (1990). *Technology Transfer: A Communication Perspective*. Newbury Park, CA: Sage Publications.

Wood, F.B. (1991). "Helping America Compete Through More Effective Use of Scientific and Technical Information: An Opportunity for Office of Science and Technology Policy Leadership." *Government Information Quarterly* 8(1): 105-112.

Wooster, H. (1987). "Historical Note: Shining Palaces, Shifting Sands: National Information Systems." *Journal of the American Society for Information Science* (September) 38(5): 321-335.

Chapter 5

Distinguishing Engineers from Scientists—The Case for an Engineering Knowledge Community

John M. Kennedy
Thomas E. Pinelli
Rebecca O. Barclay
Ann P. Bishop

SUMMARY

Chapter 5 makes the case for an engineering knowledge community. We begin by discussing the differences between science and technology. We next discuss the similarities and differences between engineers and scientists. Finally, we analyze previous research into the information use behaviors of engineers. The goal of this chapter is to show that engineers are not scientists and that knowledge production and use differ in engineering and science. We believe that the existing system for disseminating the results of federally funded aerospace research and development (R&D) is based on the information-seeking behaviors of scientists, not engineers. The system assumes that engineers and scientists are essentially the same and that there are few differences in their information-seeking behaviors. The distinctions between engineering and science, engineers and scientists, and the information-seeking behaviors of engineers and scientists have multiple implications for diffusing the results of federally funded aerospace R&D.

INTRODUCTION

The relationship between science and technology is often expressed as a continuous process or normal progression from basic research (science) through applied research (technology) to development (utilization). This relationship assumes that technology grows out of or is dependent upon science for its development. This "assumed" relationship is the foundation upon which U.S. science pol-

177

icy is based and may help to explain the use of the conventional phrase "scientists and engineers." It also helps us to understand why information dissemination practices in aerospace are aimed toward scientists. The assumption that all technology has its ultimate roots in science perpetuates a dissemination system that assumes all technology will have a science base.

However, the belief that technological change is somehow based on scientific advances has been challenged in recent years. Technological change has been increasingly seen as the adaptation of existing technological concepts in response to demand (Langrish, Gibbons, Evans, and Jevons, 1972). Moreover, several years of study that attempted to trace the flow of information from science to technology have produced little empirical evidence to support the relationship (Illinois Institute of Technology, 1968; U.S. Department of Defense, 1969). Price (1965), for example, claimed that most technological advances are derived immediately from the technology that preceded them, not from science or applied science. Price concluded that science and technology progress independently of one another. Technology builds upon its own prior developments and advances in a manner independent of any link with the current scientific frontier and often without any necessity for an understanding of the basic science underlying it. Shapley and Roy (1985) contend that a normal progression from science to technology does not exist, nor is there direct communication between science and technology. Rather, both are directly and indirectly supported by each other.

Science and Technology

Many researchers have questioned the classic distinctions between science and technology and between scientists and engineers in the past few years. Many current theories of science and technology (e.g., Latour, 1987) argue that if researchers make their observations at either the actor level or the societal level, the distinctions between science and technology appear to fade. Some theorists of technology studies believe that the structures of societies determine the technologies that will be developed (Law, 1987; Weingart, 1984). Law and Callon (1988), for example, argue that engineers are social activists who design societies and social institutions to fit technologies (p. 284). Rip (1992) argues that the perspective of the researcher indicates the interpretation of where to place activities or actors in science or technology. He further argues that "the dancing partnership of science and technology [is] a relation between activities oriented to different reference points and groups, rather than a matter of combining different cognitive-technical repertoires" (p. 257). That is, science and technology, scientists and engineers do many of the same activities but in different ways.

The distinction is further clouded when one looks closely at the varieties of actors and organizations that constitute technology. For example, in aerospace some engineers and scientists are working on methods to explore the edge of the universe and others on how to best design an aircraft for passenger comfort. Some deal with very abstract ideas and others with difficult technological, economic, or management issues. Much research that attempts to understand the differences between science and engineering has examined what Constant (1980) termed radical science or technology. That is, much research focuses on changes in paradigms or fundamental ways of thinking about a phenomenon or artifact. For example, Constant (1980) examined the role of presumptive anomalies in technology to understand fundamental changes. His best example is the adoption of the jet engine. Little research focuses on the day-to-day activities of scientists and engineers where science and technology are maintained through routinized activities.

Allen's (1977) study of the transfer of technology and the dissemination of technological information in R&D organizations found little evidence to support the relationship between science and technology as a continuous relationship. Allen concluded that the relationship between science and engineering is best described as a series of interactions that are based on need rather than on a normal progression. According to Allen, the results of science do progress to technology in the sense that some sciences such as physics are more closely connected to technologies such as electronics, but overall, a wide variation exists between science and technology. A direct communication system between science and technology does not exist. The most direct communication between science and engineering takes place through the process of education.

In summarizing the differences between science and technology, Price (1965) made the following 12 points:

1. Science has a cumulating, close-knit structure; that is, new knowledge seems to flow from highly related and rather recent pieces of old knowledge, as displayed in the literature.

2. This property is what distinguishes science from technology and from humanistic scholarship.

3. This property accounts for many known social phenomena in science and also for its surefootedness and high rate of exponential growth.

4. Technology shares with science the same high growth rate, but it shows quite complementary social phenomena, particularly in its attitude to the literature.

5. Technology therefore may have a similar, cumulating, closeknit structure to that of science, but the structure is of the state of the art rather than of the literature.

6. Science and technology therefore have their own separate cumulating structures.

7. A direct flow from the research front of science to that of technology, or vice versa, occurs only in special and traumatic cases since the structures are separate.

8. It is probable that research-front technology is strongly related only to that part of scientific knowledge that has been packed down as part of ambient learning and education, not to research-front science.

9. Research-front science is similarly related only to the ambient technological knowledge of the previous generation of students, not to the research front of the technological state of the art and its innovation.

10. This reciprocal relationship between science and technology, involving the research front of one and the accrued archive of the other, is nevertheless sufficient to keep the two in phase in their separate growths within each one's otherwise independent cumulation.

11. It is naive to regard technology as applied science.

12. Because of this, one should be aware of any claims that a par-
    ticular scientific research is needed for particular technological
    breakthroughs, and vice versa. Both cumulations are only
    supported for their own separate ends (Price, 1965, pp. 557-
    563).

Allen (1977) also stated that the independent nature of science
and technology and the different functions performed by engineers
and scientists directly influence the flow of information in science
and technology. Science and technology are ardent consumers of
information. Both engineers and scientists require large quantities
of information to perform their work. At this level, there is a strong
similarity between the information input needs of engineers and
scientists. However, the difference between engineers and scientists
in terms of information processing becomes apparent upon examination of their outputs. Scientists use information to produce
information. From a system standpoint, the input and output, both
of which are verbal, are compatible. The output from one stage is
in a form required for the next stage. Engineers use information to
produce some physical change in the world. Engineers consume
information, transform it, and produce a product that is infor-
mation-bearing; however, the information is no longer in verbal
form. Whereas scientists consume and produce information in the
form of human language, engineers transform information from a
verbal (or often, visual or tacit) format to a physically encoded form.
Verbal information is produced only as a by-product to document
the hardware and other physical products produced.

Allen finds an inherent compatibility between the inputs and
outputs of the information-processing system of science. Because
both are in verbal formats, the output of one stage is in the format
required for the next stage. The problem of supplying information
to the scientist is a matter of collecting and organizing these
outputs and making them accessible. Since science operates for
the most part on the premise of free and open access to information, the problem of collecting outputs is made easier.

In technology, however, there is an inherent incompatibility
between inputs and outputs. Since outputs typically differ in form
from inputs, they usually cannot serve as inputs for the next stage.

Further, the outputs are usually in two parts, one physically encoded and the other verbally encoded. The verbally encoded part does not serve as input for the next stage because it is a by-product of the process and is itself incomplete. Those unacquainted with the development of the hardware or physical product therefore require some human intervention to supplement and interpret the information contained in the documentation. Since technology operates to a large extent on the premise of restricted access to information, the problem of collecting the documentation and obtaining the necessary human intervention becomes difficult.

Allen and others used a somewhat restricted definition of technology in that they assume that it is always a physical product. Engineers in aerospace and in other industries often create systems and products that are verbally encoded, such as management systems and software. These differences do not alter the basic premise that substantial differences exist between the goals of engineers and scientists as they produce the different types of outputs in their daily activities. The connection between science and technology, in aerospace and elsewhere, is tenuous, vague, and sporadic. The processes used in science and technology to produce their respective outputs create parallel and weakly connected systems. A clear recognition of these differences is needed to establish a context for and to understand aerospace knowledge diffusion (i.e., production, transfer, and use).

Engineers and Scientists

For our purposes, we define the essential difference between engineers and scientists based on the primary goal of the output of their work—scientists produce knowledge (facts) and engineers produce designs, products, and processes (artifacts). Engineers and scientists exhibit many other important differences in education, technical discipline, and type of work activities. These differences point to differences in their information-seeking behaviors and information needs. In this section, we describe many of the differences, starting with differences in characteristics and proceeding to differences in their outputs.

Differences between engineers and scientists are difficult to determine from either self-classification or the analysis of their tasks. Citro and Kalton (1989, pp. 26-56) describe differences based on analyses of tasks, job descriptions, education, and self-identification. Their analysis indicated that even using multiple indicators did not reduce the error in classifications into engineering and science. We suspect that the increasing bureaucratization of these professions makes it more difficult to accurately differentiate them. Kintner (1993) attempted to determine who was an engineer based on the job classification, education, and job history. The results of a multivariate statistical procedure indicated that at least 15% of those who were doing engineering work would be missed using various classification schemes.

Latour (1987) used the term "technoscience" to describe the relationship between engineering and science. Using a network actor perspective, he described the daily activities of both scientists and engineers. He found that personal success in technoscience did not depend primarily on how well engineers and scientists performed their jobs, but on how well they were able to recruit others into believing in the value of what they did. For those in technoscience, recruiting others included writing proposals, looking for funding for projects, doing research, and other activities that would not be considered either science or engineering. That is, success in engineering and science does not depend so much on what is made (engineers) or on the development of new knowledge (scientists) but rather on how well the engineers and scientists are able to recruit others into the process of technoscience.

When one examines engineers and scientists over the course of their careers, it becomes increasingly difficult to distinguish them. When each does those activities that we traditionally consider the activities of engineers and scientists (making new products and new knowledge, respectively), each group appears to behave quite differently. Yet many of their activities, such as management, are the same. Contradictions based on the various views of the differences between the groups contribute to the misunderstanding that engineers are the same as scientists.

Differences

Despite the changes in engineering and science over the past 20 years, many differences noted by Ritti (1971) still distinguish the two groups. In his study of engineers in industry, Ritti found striking contrast between the goals of engineers and scientists—(a) the goals of engineers in industry are very much in line with meeting schedules, developing products that will be successful in the mar-ketplace, and helping the company expand its activities; (b) al-though both engineers and scientists desire career advancement or development, advancement for the engineer is tied to activities within the organization, whereas advancement for the scientist is dependent upon the reputation established outside the organi-zation; and (c) whereas publication of results and professional autonomy are clearly valued goals of the Ph.D. scientist, they are clearly the least valued goals of the baccalaureate engineer (Ritti cited in Allen 1977, p. 5).

Blade (1963) states that engineers and scientists differ in train-ing, values, and methods of thought. In particular, in their individ-ual creative processes and in their creative products—(a) scientists are concerned with discovering and explaining nature; engineers use and exploit nature; (b) scientists search for theories and prin-ciples; engineers seek to develop and make things; (c) scientists seek a result for its own end; engineers are engaged in solving a problem for the practical operating results; and (d) scientists create new unities of thought; engineers invent things and solve problems. Danielson (1960) found that engineers and scientists are funda-mentally different in terms of how they approach their jobs, the type and amount of supervision they require, the type of recognition they desire, and their personality traits.

Allen (1977) stated that the type of person who is attracted to a career in engineering is fundamentally different from the type of person who pursues a career as a scientist. He wrote that:

> Perhaps the single most important difference between the two is the level of education. Engineers are generally educated to the bacca-laureate level; some have a master's degree, while some have no college degree. The research scientist is usually assumed to have a doctorate. The long, complex process of academic socialization involved in obtaining the Ph.D. is bound to result in persons who differ considerably in their lifeviews. (p. 5)

According to Allen (1988), these differences in values and attitudes toward work will almost certainly be reflected in the behavior of the individual, especially in the use and production of information.

Much of the research on the differences between engineers and scientists is dated and does not reflect the impact of changes in post-World War II engineering curricula. During World War II and throughout the era of Sputnik, government and industry leaders recognized that engineering training in the U.S. was not adequate to meet military and industrial challenges (Grayson, 1993). The Grinter Report, prepared by a committee of the American Society for Engineering Education (ASEE), urged the inclusion of more science and liberal arts into engineering education. This 1955 report transformed engineering education over the subsequent two decades from "hands-on" training to a more theoretical perspective resembling other types of academic disciplines, particularly the sciences. In his history of engineering education in the U.S., Grayson (1993) terms the period from World War II through 1970 the "scientific" period. Engineering education since the 1960s has tended to blur the distinction between the training of engineers and scientists. In addition, the types of work that they do in the large bureaucratic organizations that employ them makes it increasingly difficult to differentiate them by title alone. From a research perspective, it is difficult to observe a clear difference between engineers and scientists in many settings. Later in this chapter, we demonstrate the differences between engineers and scientists that are clearly reflected in their daily activities.

Engineering is defined as the creation or improvement of technology. As such, it clearly encompasses both intellectual and physical tasks (i.e., both knowing and doing). Engineering work is fundamentally both a social and a technical activity. It is a social activity in that it often involves teamwork, as individuals are required to coordinate and integrate their work. It is also a social activity in that the production of the final product depends on the ability to maintain successful social relationships (e.g., negotiate with vendors, maintain smooth personal relations among members of a work group). Membership in a community is important for the effective functioning of current engineering and engineers. Engineers do their work in an embedded set of contextual relationships. Science, on the other hand, allows scientists to conduct their daily activities with only a vague reference to others doing similar work.

Similarities

A number of writers note that engineers behave very similarly to scientists. At times, they adopt the methods used by scientists to generate knowledge. For example, according to Ritti (1971), engineering work consists of scientific experimentation, mathematical analysis, design and drafting, building and testing of prototypes, technical writing, marketing, and project management. Kemper (1990), too, noted that the typical engineer is likely to define problems, come up with new ideas, produce designs, solve problems, manage the work of others, produce reports, perform calculations, and conduct experiments. Florman (1987) described engineering work as encompassing both theory and empiricism. Ziman (1984) wrote that:

> Technological development itself has become "scientific". It is no longer satisfactory, in the design of a new automobile, say, to rely on rule of thumb, cut and fit, or simple trial and error. Data are collected, phenomena are observed, hypotheses are proposed, and theories are tested in the true spirit of the hypothetico-deductive method. (p. 130)

Constant (1980) also described the similarities between engineering and science in his detailed history of the origin of the modern jet engine. He defined a "variation-retention" model to describe how engineers and scientists create technological change. Change, in technology, results from random variation and selective retention. Technological conjecture, which can occur as a result of knowledge gained from either scientific theory or engineering practice, yields potential variations to existing technologies. For example, in the case of the turbojet revolution, technological conjecture was based on engineers' knowledge of scientific theories. In contrast, in their writings, scientists usually describe their methods as following the hypothetico-deductive method. However, in many of their daily research activities, they use methods similar to those used by engineers—particularly the variation-retention method.

Convergence

We expect that many of the differences found by earlier researchers between engineers and scientists should decrease over time. The previous research on the differences and similarities between engineers and scientists has been vague and often based on small samples. Many studies that focused on the differences included engineers who received training before the impact of the Grinter Report. In addition, undergraduate engineering curricula are continually changing and are very similar to undergraduate curricula for scientists in that engineers receive more humanities, liberal arts, and business training. As a result, there may be a convergence of the training of each group.

During this century, and especially since World War II, engineers and scientists have been increasingly employed in such large bureaucratic organizations (Florman, 1987; Layton, 1974; Meiksins and Smith, 1993) as the major corporations and the federal government. The integration of engineers and scientists into these organizations has significantly reduced their autonomy. Both groups have increased attempts to maintain their autonomy by defining and controlling separate spheres of knowledge. Yet, in most organizations, the opportunity for upward mobility is limited to management. Both engineers and scientists tend to move into management during their careers. If we look at both groups over their careers, we see that they tend to converge in their daily activities. Although they may consider themselves engineers or scientists based on education or professional orientation, in reality they become managers and behave alike in bureaucratic organizations.

Scientific and Technological Communities

There are other differences between engineers and scientists, in addition to their daily activities, that affect our understanding of their production and use of knowledge. Both engineers and scientists work in structured social settings that influence their behavior. An analysis of the differences in the settings is critical to understanding the production, transfer, and use of aerospace knowledge. Engineers and scientists interact within each professional group in systematic ways that produce differences in the methods they use to generate and access knowledge.

Each group conducts its work activities in somewhat different environments. For the most part, scientists work within organizations, but they are primarily "outside oriented." Although scientists in industry may create proprietary information as part of their duties, they expect that their work will become known to others outside the organization. In academia and government, most scientific work is aimed at others outside the organization. Engineers' work is usually more "inside oriented"; that is, they work for the benefit of their organizations. Yet engineers, too, interact regularly with others outside their organizations (Kennedy, Pinelli, and Barclay, 1995).

But, as in other ways, the differences between the professions are not entirely distinct. It is generally accepted that scientists have an "invisible college" that they use to share information. This "college" extends far past the boundaries of their work organizations. Within the "college," information and ideas are shared among its members. It is assumed that engineers do not have this college. What engineers have might better be described as having a "knowledge community" with which they share information and ideas (Vincenti, 1990). A community is a group of people who maintain social contact with each other and who have common goals, behavioral norms, and knowledge. As members of a profession, engineers share common knowledge and a set of espoused values. The profession prescribes its own approach to work behavior. Engineering is also a social activity; most work is accomplished as a result of group effort and requires extensive interpersonal communication both within and outside the organization.

Studies of scientific communities look at the values, norms, knowledge, methods, reward system, and culture shared by community members and frequently underscore the role of interpersonal communication in defining the community and holding it together (see Barber, 1962; Doty, Bishop, and McClure, 1991; Kuhn, 1970). This type of investigation has not often been performed in relation to engineering knowledge communities. Gaston (1980) noted that ["the problem of the internal workings of the engineering community"] is virtually unexplored.... In contrast to the sociology of the scientific community, "little is known about the sociology of the technological [engineering] community." (p. 495). Constant (1984) also noted the lack of research on engineering communities. He wrote that "while extensive research has been done on 'invisible colleges', research fronts, and the community structure of science, there has been little analogous, sociological or historical investigation of [engineering] practice" (p. 8).

Rothstein (1969), pointing to the diversity inherent in engineering, warned that defining the entire profession of engineering as a single knowledge community provides a model that is inadequate to describe engineering behavior. He argued that the huge variety of occupations and disciplines in engineering demonstrates that there is no such thing as a single engineering knowledge community. Further, he contended that most discussions of professional communities fail to direct enough attention to the nature of professional knowledge and its influence on behavior. He further contended that the heterogeneity, rate of change, and degree of specialization of engineering knowledge also led to the emergence of specific communities in engineering.

Some work has begun to explore the extent to which members of engineering knowledge communities share similar work tasks, goals, and methods; are governed by shared social and technical norms; and engage in extensive informal information exchange among themselves. Laudan (1984) found justification for this approach in that:

> Cognitive change in technology is the result of the purposeful problem-solving activities of members of relatively small communities of practitioners, just as cognitive change in science is the product of the problem-solving activities of the members of scientific communities. (p. 3)

Layton (1974) also contended that "the ideas of technologists cannot be understood in isolation; they must be seen in the context of a community of technologists" (p. 41). Donovan (1986) noted that "the study of engineering knowledge must not be divorced from the social context of engineering" and suggests that "the interplay of social values and theoretical understanding in the evolution of scientific disciplines certainly has its analogues in engineering, although the values and knowledge involved are often quite different" (p. 678).

The nature of engineering work suggests that engineers require access to a variety of tools and information resources. In addition, the use of these tools and resources and the way they are integrated into engineering work may be planned in some cases and ad hoc in other situations. The engineering knowledge community, although it has received minimal attention from researchers, clearly plays an important role in the conduct of engineering work and the production, transfer, and use of engineering knowledge.

Yet the notion of community can be pushed too far and can cause problems for understanding how engineers behave in their daily activities. As Allen (1977) points out, there is a vague limit on the amount of interpersonal communication engineers can use in problem-solving. It is acceptable for engineers to ask for help from other engineers when they have a technical problem, but it is also expected that engineers will not seek help from other engineers for every problem they encounter. The normative system in engineering prevents them from asking for all the help they might need because to do so would indicate a lack of engineering competence. When the engineers reach the point where interpersonal communications can no longer be used, the alternatives—tacit knowledge, testing, or finding the information—become more important.

Differences between Engineers and Scientists in Knowledge Diffusion

Most research on knowledge production, transfer, and use has not focused on the differences between engineers and scientists. Rather, the research has focused on how scientists create and use knowledge, or it has assumed that both groups are similar. Engineers and scientists are similar in that knowledge production and use is critical to the performance of their jobs, but there are major differences in how and when they use it. To thoroughly understand the need for effective knowledge diffusion, research needs to be focused more closely on why and how engineers and scientists differ in their production and use of knowledge.

Scientists use knowledge as part of the process of generating new knowledge. Latour (1987) described how scientists "recruit" their intellectual and research predecessors to demonstrate the importance of their current research. They use knowledge to show how their research differs from or improves upon previous research in the field. The intellectual context of the research must be established if the priority and importance of a finding or fact are to be established. In most instances, scientists gather most of their knowledge before beginning research or at least before writing their research results.

Engineers, on the other hand, use knowledge to help make decisions. They care more about the ability of the research to provide guidance on their particular problem than about its intellectual history. They use knowledge throughout the research, design, development, and manufacturing processes. When engineers produce new knowledge, they often do so solely to provide guidance to others in their organization who might face a similar issue in the development of another product. In these instances, the intellectual history of previous research is not as important as documenting the procedures and results.

Scientists tend to use hypothesis testing (at least as they describe their research) in gaining new knowledge. Engineers are more likely to use iterative parameter variation and selective retention (Vincenti, 1990) to generate new knowledge. Each technique produces different types of documentation. Journal articles are appropriate for scientists to describe the development and testing of one idea. In contrast, technical reports are more appropriate for engineers to document engineering outcomes. When preparing to do research, scientists will search for knowledge that may not be directly related to the research but can be used to place the research in context. The information needs of engineers are more immediate, at least in critical phases of design. They select information because it directly relates to solving a problem.

For example, according to Allen (1977):

> Engineers read less than scientists, they use literature and libraries less, and they seldom use information services which are directly oriented to them. They are more likely to use specific forms of literature such as handbooks, standards, specifications, and technical reports. (p. 80)

What an engineer usually wants, according to Cairns and Compton (1970), is "a specific answer, in terms and format, that is intelligible to him—not a collection of documents that he must sift, evaluate, and translate before he can apply them" (pp. 375-376). Young and Harriott (1979) report that:

> The engineer's search for information seems to be based more on a need for specific problem solving than around a search for general opportunity. When engineers use the library, it is more in a personal-search mode, generally not involving the professional (but nontechnical) librarian. (p. 24)

Young and Harriott conclude by saying:

> When engineers need technical information, they usually use the most accessible sources rather than searching for the highest quality sources. These accessible sources are respected colleagues, vendors, a familiar but possibly outdated text, and internal company [technical] reports. He [the engineer] prefers informal information networks to the more formal search of publicly available and cataloged information. (p. 24)

We are not convinced that there is a neat dichotomy between engineers and scientists in their production, transfer, and use of knowledge. Rather, there is a continuum of activities and behaviors that each group uses in differing amounts in their daily activities. Included among these are knowledge production, transfer, and use activities that appear to be similar to those of the other profession. Because of the variety of tasks that engineers and scientists perform, it is difficult to assume that any model of knowledge diffusion can be simple and meet the needs of all engineers and scientists.

The Nature of Engineering Knowledge

Vincenti (1990) proposed a schema for engineering knowledge that categorizes knowledge as descriptive (factual knowledge), prescriptive (knowledge of the desired end), or tacit (knowledge that cannot be expressed in words or pictures but is embodied in judgment and skills). Both descriptive and tacit knowledge are embedded in the daily activities and understandings of engineers. These types of knowledge guide their everyday behaviors for creating and designing products and services. Descriptive knowledge, for the most part, resides in commonly shared knowledge. The community of engineers uses the three sources of knowledge as its form of collective identification as a profession.

Vincenti (1990) traced five "normal" (as opposed to revolutionary) developments in the history of aerospace engineering to detail what he called "the anatomy of engineering design knowledge" (p. 9). His examples reveal that technological developments require a range of scientific, technical, and practical knowledge as well as information about social, economic, military, and environmental issues. Vincenti conducted three important analyses of engineering knowledge. One analysis is his own elaboration on the variation-selection model of the growth of technological knowledge. Vincenti concluded, after examining numerous examples from history, that the mechanisms for producing variations in engineering design include three types of cognitive activities (p. 246)—(a) searching past experience to find knowledge that has proved useful, including the identification of variations that have not worked; (b) incorporating novel features thought to have a chance of working; and (c) "winnowing" the conceived variations to choose those most likely to work. Vincenti noted that these activities occur in an interactive and disorderly fashion. Selection occurs through physical trials such as everyday use, experiments (e.g., the use of wind tunnels), simulations, analytical tests such as the production of sketches of proposed designs, calculations, and other means of imagining the outcome of selecting a proposed variation (pp. 247-248).

Descriptive and prescriptive knowledge are explicit; tacit knowledge is implicit. Both tacit and prescriptive knowledge are procedural and reflect a "knowing how" (p. 197-198). Vincenti defined specific engineering knowledge categories: fundamental design concepts, criteria and specifications, theoretical tools (i.e., mathematical methods and theories and intellectual concepts), quantitative data, practical considerations, and design instrumentalities (i.e., procedural knowledge and judgmental skills) (pp. 208-222). He identified several sources of engineering knowledge that include transfer from science or generation by engineers during invention, theoretical and experimental engineering research, design practice, production, or direct trial and operation (p. 235).

Vincenti (1992) and others use a "demarcationist" approach to engineering knowledge (Downey and Lucena, 1995); that is, they identify engineering knowledge as different from scientific knowledge. Science has the same three categories of knowledge that Vincenti attributed to engineering knowledge, but the two professions might differ in the value placed on each. For example, scientists may not assign as much importance to tacit knowledge as engineers do. Using another model, the network actor model, to understand the differences between engineers and scientists would place an important consideration on the differences between the groups in their production and use of knowledge. Both groups use substantial amounts of knowledge, but each uses it in different ways to recruit others into accepting the merit of their designs (engineers) or ideas (scientists). This difference implies that the knowledge diffusion system must meet differing needs if it is to be used successfully by both groups.

Engineers as Information Processors

The ultimate goal of engineering is to produce a design, product, or process. It is informative to view engineering as an information-processing system that uses knowledge to reduce work-related uncertainty. That is, engineers are heavy information processors. The concept of engineering as an information-processing activity represents an extension of the arguments developed by Tushman and

Nadler (1980) and has its roots in open systems theory developed by Katz and Kahn (1966). Throughout the engineering process, data, information, and knowledge are acquired, produced, transferred, and used. The fact that these data, information, and knowledge may be physically or hardware encoded should not detract from the observation that the process of engineering is fundamentally an information-processing activity.

Uncertainty, defined as the difference between the information possessed and the information required to complete a task, is central to the concept of engineering as an information-processing activity. Rogers (1982) stated that coping with uncertainty is the central concept in information use behavior. The process of engineering is one of grappling with the unknown. These unknowns or uncertainties may be technical, economic, or merely the manifestations of personal and social variables. When faced with uncertainty, engineers typically seek data, information, and knowledge. In other words, data, information, and knowledge are used by engineers to moderate technical uncertainty. Because engineering generally entails coping with a relatively high degree of uncertainty, engineering can certainly be viewed as an informational process. Consequently, information-seeking behavior and patterns of technical communication cannot be ignored when studying engineers.

In Orr's (1970) conceptual framework, the engineer is an information processor. This framework focuses on information-seeking behavior and assumes that an internal, consistent logic governs the information-seeking behavior of engineers, individual differences notwithstanding. A project, task, or problem that precipitates a need for information is central to the conceptual framework. This need for information may be internally or externally induced and is referred to by Orr as inputs or outputs, respectively. Orr, citing the work of Voight (1960), Menzel (1964), Storer (1966), and Hagstrom (1965), stated that inputs originate within the mind of the individual engineer and include the data, information, and knowledge needed to keep up with advances in one's profession; to perform one's professional duties; to interact with peers, colleagues, and coworkers; and to obtain stimulation and feedback from them. Outputs frequently result from an external stimulus or impetus and serve a variety of functions, including responding to a request for information from a supervisor, a coworker, peer, or colleague; reporting progress; providing advice; reacting to inquiries; defending; advocating; and proposing. Inputs and outputs require the use of specific kinds and types of data, information, and knowledge.

The conceptual framework for our research assumes that, to address a project, task, or problem successfully, specific kinds or types of data, information, and knowledge are needed. In response, engineers are confronted with two basic alternatives: They can create the information through experimentation, observation, or other accepted engineering procedures; alternatively, they can search the existing information sources to determine if the information is currently available and usable. If they act rationally, the decision to "make or buy" the information is influenced by three factors—(a) their subjective perception of the relative likelihood of success in acquiring the desired information in the time allocated for addressing the project, task, or problem; (b) their perception of the relative cost (money and/or effort) of these alternatives; and (c) the anticipated acceptance of their resolution of the problems by others (e.g., managers, other engineers, contractors).

If a decision is made to search the existing information, engineers must choose between two information channels. They can choose to gather the information through *informal* methods such as interpersonal (oral) communications with peers, coworkers, colleagues, gatekeepers, vendors, consultants, "key" personnel, and supervisors, or the use of their personal collections of information. They might also choose to use the *formal* information system, which includes libraries, technical information centers, librarians and technical information specialists, information products and services, and information storage and retrieval systems. It is assumed that the decision to choose a particular information channel is influenced by personal and institutional characteristics. Other factors, such as the previous successes and failures of each channel, will further influence the channel selection (Gerstberger and Allen, 1968; Orr, 1970; Rosenberg, 1967).

More recent work highlights the value of exploring contextual and situational factors related to information seeking and use. Taylor's (1991) theoretical investigation of information use environments emphasizes the importance of understanding the context in which information is sought, conveyed, and applied. Context for professional groups, including engineers, is defined as a combination of the nature of work problems, solutions, and settings associated with particular types of jobs. Context for engineers also includes the engineering community (i.e., the methods that other engineers would use to solve a problem). We might add that engineers also consider their community when making these decisions; that is, they consider how the engineering community would approach the problem and the perceived acceptability of their choice of action in the community. Taylor assumes, in other words, that members of a profession share tasks, goals, and needs in a way that influences their use of information. Taylor's analysis recognizes that information-seeking behavior and use are determined by the nature of the particular project, task, or problem at hand (pp. 217-255).

A shift in emphasis toward the study of cognitive and situational factors surrounding information seeking and use, and away from users' personal characteristics and specific systems features, has been advocated by a number of communications and information science researchers, most notably Dervin and Nilan (1986). They have devoted special attention to understanding what there is about a particular situation that encourages an individual to use networks in fulfilling an information need. The subjective perception of cost, time, and likelihood of success may often be situationally driven.

The data, information, and knowledge that result from an engineer's search are evaluated subjectively. The engineer as an information processor faces three possible courses of action: *First*, if the created or available data, information, and knowledge used to complete the project or task or solve the problem are sufficient, the process is terminated; *second*, if the created or available data, information, and knowledge are useful but only partially sufficient to complete the project or task or to solve the problem, a decision is made either to continue the process by reevaluating the information source selected or to terminate the process; and *third*, if the created or available data, information, and knowledge are not applicable to or do not complete the project or task or solve the problem, a decision is made either to continue the process by redefining the project, task, or problem or to terminate the process. Throughout the process, the engineer evaluates both process and outcomes in light of what others in the engineering community would do and also in light of the anticipated acceptance by others within the engineering community and the employing organization. The complexities of the decision processes used by engineers to evaluate knowledge require an understanding of the personal, situational, contextual, and community characteristics in which the engineer works.

Major Empirical Studies of Engineering Information-Seeking Behavior

Pinelli (1991), Pinelli, Bishop, Barclay, and Kennedy (1993), and King, Casto, and Jones (1994) reviewed many of the studies of information-seeking behavior by engineers and scientists. For the most part, except for other papers and reports produced as part of this project, the previous studies either focused on the production and use of knowledge by scientists, or the studies do not distinguish between engineers and scientists. Because the following studies focused primarily on the information use behavior of engineers, we review them at length. Overall, there is a remarkable consistency in their findings. Engineers use information to solve a problem. They use only as much as needed when it is needed. They choose informal sources of information because they believe these sources are effective and efficient.

Herner

Herner's (1954) work was one of the first "user" studies specifically concerned with "differences" in information-seeking behavior. He reported significant differences in terms of researchers performing "basic and applied" research, researchers performing "academic and industry" type duties, and their information-seeking behavior. Herner stated that researchers performing "basic research" or "academic" duties made greater use of formal information channels or sources, depended mainly on the library for their published material, and maintained a significant number of contacts outside of the organization. Researchers performing "applied research" or "industry" duties made greater use of informal channels or sources, depended on their personal collections of information and colleagues for information, made significantly less use of the library than did their counterparts, and maintained fewer contacts outside of the organization. Applied or industry researchers made substantial use of handbooks, standards, and technical reports. They also read less and did less of their reading in the library than did their counterparts in basic research.

Rosenbloom and Wolek

In 1970, Rosenbloom and Wolek published the results of one of the first large-scale industry studies that was specifically concerned with the diffusion of knowledge within R&D organizations. They reported three significant and fundamental differences between engineers and scientists: (a) engineers tend to make substantially greater use of information sources within the organization than do scientists; (b) scientists make considerably greater use of the professional (formal) literature than do engineers; and (c) scientists are more likely than are engineers to acquire information as a consequence of activities directed toward general competence rather than at a specific task.

Considering interpersonal communication, the engineers in the Rosenbloom and Wolek study recorded a higher incidence of interpersonal communication with people in other parts of their own corporation, whereas scientists recorded a greater incidence of interpersonal communication with individuals employed outside their own corporation. When using the literature, engineers tended to consult in-house technical reports or trade publications, while scientists made greater use of the professional (formal) literature.

Rosenbloom and Wolek also reported certain similarities between engineers and scientists. The propensity to use alternative types of technical information sources is related to the purposes that give meaning to the use of that information. Work that had a professional focus drew heavily on sources of information external to the user's organization. Work that had an operational focus seldom drew on external sources, and relied heavily on information that was available within the employing organization. Those engineers and scientists engaged in professional work commonly emphasized the simplicity, precision, and analytical or empirical rigor of the information source. Conversely, those engineers and scientists engaged in operational work typically emphasized the value of communication with others who understood and were experienced in the same real context of work.

Allen

Allen's (1977) study of technology transfer and the dissemination of technological information within the R&D organization was the result of a 10-year investigation. Allen described the study, which began as a "user study," as a systems-level approach to the problem of communication in technology. Many information professionals consider his work to be the seminal research on the flow of technical information within R&D organizations. Allen was among the first to produce evidence supporting different information-seeking behavior for engineers and scientists. These differences, Allen noted, led to different philosophies and habits regarding the use of the technical literature and other sources of information by engineers. The most significant of his findings were the relative lack of importance of the technical literature in terms of generating new ideas and in problem definition, the importance of personal contacts and discussions among engineers, the existence of technological "gatekeepers," and the importance of the technical report. Allen stated that "the unpublished report is the single most important informal literature source; it is the principal written vehicle for transferring information in technology" (p. 91).

Kremer

Kremer's (1980) study was undertaken to gain insight into how technical information flows through formal and informal channels among engineers in a design company. The need to solve a problem was the reason given most frequently to search for information. Kremer found that colleagues within the company were contacted first for needed information; then colleagues outside of the company were contacted. In terms of the technical literature, handbooks were most important, followed by standards and specifications. Libraries were not important sources of information and were used infrequently by company engineers. Regardless of age and work experience, design engineers demonstrated a decided preference for internal sources of information. They consulted personal files for needed information. The perceived accessibility, ease of use, technical quality, and amount of experience a design engineer had with an information source strongly influenced the selection of an information source. Technological gatekeepers appeared among design engineers. In general, they were high technical performers, and a high percentage were first line supervisors.

Shuchman

Shuchman's (1981) study was a broad-based investigation of information transfer in engineering. The respondents represented

14 industries and the following major engineering disciplines: civil, electrical, mechanical, industrial, chemical and environmental, and aeronautical. The engineers, regardless of discipline, displayed a strong preference for informal sources of information. Further, these engineers rarely found all the information they needed for solving technical problems in one source. The major difficulty engineers encountered in finding the information they needed was identifying a specific piece of missing data and then learning who had it. In terms of information sources and solving technical problems, Shuchman reported that engineers first consulted their personal stores of technical information, followed in order by informal discussions with colleagues, discussions with supervisors, use of internal technical reports, and contact with a "key" person in the organization who usually knew where the needed information was located. A small proportion of the engineering profession used technical libraries and librarians. In general, Shuchman found that engineers do not regard information technology as an important adjunct to the process of knowledge production, transfer, and use. Although technological gatekeepers appear to exist across the broad range of engineering disciplines, their function and significance are not uniform. Considering the totality of engineering, gatekeepers accounted for only a small part of the information transfer process.

Kaufman

Kaufman's (1983) study was concerned with the factors relating to the use of technical information by engineers in solving problems. The study reported that, in terms of information sources, engineers consulted their personal collections first, followed by colleagues, and then by formal literature sources. In terms of the formal literature sources used for technical problem solving, engineers used technical reports, followed in order by text books and technical handbooks. Most sources of information, according to Kaufman (1983), were found primarily through an intentional search of written information, followed by relying on personal knowledge and then by asking someone. The criteria used in selecting all information sources, in descending order of frequency, were accessibility, familiarity or experience, technical quality, relevance, comprehensiveness, ease of use, and expense. Engineers used various information sources for specific purposes. They primarily utilized librarians and information specialists to find leads to information sources. Engineers used online computer searches primarily to define the problem, and technical literature to learn techniques applicable to dealing with the problem. They relied primarily on personal experience to find solutions to the problem.

Kaufman (1983) reported that the criteria used in selecting the most useful information sources, in descending order of frequency, were technical quality or reliability, relevance, accessibility, familiarity or experience, comprehensiveness, ease of use, and expense. In terms of the effectiveness, efficiency, and usefulness of the various information sources, personal experience was rated the most effective source of information. Librarians and information specialists received the lowest rating for efficiency and effectiveness. Most engineers used several different types of information sources in problem solving; however, engineers depended on their personal experience more often than on any other single specific information source.

Differences among Types of Engineers

Just as a careful analysis of engineers and scientists requires that we examine them as separate groups when analyzing some issues, it is equally important to recognize the differences among types of engineers when we look at engineers' attitudes and behaviors related to knowledge production, transfer, and use. Researchers rarely recognize that engineers perform such varied duties as engineering research, design, and production engineering. The daily activities of some engineers are not readily distinguishable from those of scientists. For other engineers, there are few similarities in their daily activities to those of scientists. A critical factor in the analysis of engineering information needs is the possibility of differences among types of engineers and the varied duties that engineers perform.

It would be almost impossible to answer the question completely of who an engineer is and what an engineer does. For the most part, studies of engineers and scientists are based on self-identification of professional status. The descriptions of the myriad of engineering behaviors and practices contained in earlier research indicate that the activities performed by engineers are diverse and multifaceted. However, we found no studies that contained a detailed analysis of the daily activities of engineers.

The diversity of engineers implies a diversity in information needs. Our research not only assumes that scientists and engineers have different information needs but also attempts to determine where all types of engineers and scientists differ and in what ways. For example, do the information needs of research engineers differ from those of production engineers? All engineers and scientists are heavy information users, but they might not use the same kinds or the same mix of information. A knowledge diffusion system recognizes the differences and meet the needs of each.

The more "science-oriented" model of engineering education created a need for additional training in using information resources as part of engineers' skills training. Two conflicting forces, the professionalization of engineering and the bureaucratization of engineering, have significantly affected engineering during this century. One, engineering has evolved into a more professional occupation—similar to medicine and law. Increasing numbers of engineers were academically trained, especially at the technical and the land-grant universities. Baccalaureate degrees replaced apprenticeship as the method of entering the profession, although Kintner (1993) found that about 15% of the engineers in her study did not have bachelor's degrees. Academic training, too, changed from a "hands-on" approach to more theoretical and professional training. At the same time, the major aerospace, automobile, and chemical firms and other large bureaucratic organizations began employing larger numbers of engineers. Two, these organizations attempted to constrain the professional activities and independence of engineers by treating them as technical rather than professional employees (Meiksins and Smith, 1993; Whalley, 1986; Zussman, 1985).

The desire by the leaders of engineering societies to professionalize engineering and engineers' feelings that they were merely highly-paid technical workers led to a "revolt of the engineers" during the first half of this century (Layton, 1974). Engineers became more oriented toward their profession than to their employers. Increasingly, engineers felt allegiance to a profession rather than an organization. That is, engineering became more of a "stand-alone" profession whose skills could be transferred from one job setting to another. The independence of engineers increased their reliance on shared information as critical to their professional identity. Increased reliance on knowledge, along with the changes in academic curricula, contributed to an increased recognition by engineers that the ability to use knowledge was essential to their professional careers, perhaps more so than hands-on training, because it made them less dependent on the training provided by their employers. The sharing of engineering knowledge, as it is contained in information, is the key link among engineers as they define themselves as professionals.

At the same time, the increased concentration of engineers in large bureaucratic organizations, along with the increasing amount of information in engineering knowledge, created the need to develop specialties within engineering that went beyond those tied to academic curricula. In U.S. industries, and in particular the U.S. aerospace industry, multiple types of engineers perform a variety of engineering tasks that range from basic science through engineering science to design and development and to manufacturing and production engineering. We expected to find differences among these subdisciplines in the types of information engineers used, the sources used to find information, and the amount of information they needed to perform their duties. These differences are important to understanding of effective diffusion of aerospace knowledge.

The Aerospace Engineering Community

Our research is unique among engineering studies in that we focus on a single engineering discipline and its many subdisciplines, and on one industry. Previous research on engineers and scientists has generally focused on one organization, or it was widespread across engineering and science disciplines. The focus on an organization can confuse the organizational characteristics with the characteristics of the engineers and scientists. The large multidisciplinary engineering studies do not provide sufficient focus for understanding one engineering discipline thoroughly (e.g., aerospace). We contend that a more effective analysis of engineers must consider the various technological tasks performed by aerospace engineers and scientists, which we describe as a continuum from applied (engineering) science through engineering research to design and development, production engineering, and marketing and sales. At each step, the tasks become more directly related to the production of an aerospace product. At each step along this continuum, the production, transfer, and use of knowledge may differ. This lack of attention to differences has probably contributed to the lack of an effective aerospace knowledge diffusion system.

Aerospace employs a wide range of engineers and scientists who represent many engineering specialties and scientific disciplines. Most of the scientists in aerospace are employed in the national labs and in universities, but some are employed in major aerospace firms. In contrast, engineers are employed in private companies that range from major firms like Boeing to small engineering firms with a few employees. Aerospace is further unique in that most of the engineers in one discipline—aerospace engineering—are employed in large bureaucracies. As such, aerospace engineers have unique work environments that affect their production, transfer, and use of knowledge.

The notion of an aerospace engineering community was addressed by Vincenti (1990). He described informal communities of practitioners as the most important source of knowledge generation and means of knowledge transfer in aerospace. Vincenti defined a community as those involved in work on a particular aerospace development or problem (e.g., fasteners, airfoils, or propellers), and he attributed several functions to these engineering communities. Competition among members supplies motivation, while cooperation provides mutual support. The exchange of knowledge and experience generates further knowledge, which is disseminated by word of mouth, publication, and teaching, and is also incorporated into the tradition of practice. The community also plays a significant role in providing recognition and reward.

Vincenti (1990) described the particular roles of important types of aerospace engineering institutions, such as government research organizations, university departments, aircraft manufacturers, military services, airlines, professional societies, government regulatory agencies, and equipment and component suppliers (pp. 238-240). He concluded, however, that:

> Formal institutions do a complex multitude of things that promote and channel the generation of engineering knowledge. They do not, however, constitute the locus for that generation in the crucial way that informal communities do. Their role ... is to supply support and resources for such communities. (p. 240)

Constant (1980) described aerospace engineering communities as the central locus of technological cognition. He noted that the aeronautical community is, in fact, composed of a multilevel, overlapping hierarchy of subcommunities; he argued that technological change is better studied at the community level than at the individual, organizational, national, or industry level. Constant described the community as the embodiment of traditions of practice. Technological traditions of practice define an accepted mode of operation, the conventional system of accomplishing a specified technical task. Such traditions encompass aspects of relevant scientific theory, engineering design formulas, accepted procedures and methods, specialized instrumentation, and, often, elements of ideological rationale. A tradition of technological practice is proximately tautological with a community that embodies it; each serves to define the other. Traditions of practice are passed on in the preparation of aspirants to community membership. A technological tradition of practice has, at minimum, a knowledge dimension that includes both software and hardware, and a sociological dimension that includes both social structure and behavioral norms.

In *The Origins of the Turbojet Revolution*, Constant (1980) discussed further the importance of community norms in engineering. He alleged that, at least in connection with complex systems, there are "fundamental social norms governing the behavior of technological practitioners which are very close in structure, spirit, and effect to the norms governing the behavior of scientists" (p. 21). Such norms guide the development of techniques and instruments and the reporting of data. Constant also argued for the existence of "counternorms" in engineering that are similar to those attributed to scientists by Meiksins and Smith (1993). Constant explains:

> Technological practitioners are required to be objective, emotionally neutral, rational, and honest. Yet technological practitioners often are—and protagonists of technological revolution usually are—passionate, determined, and irrationally recalcitrant in the face of unpleasant counter evidence bearing on their pet ideas. (p. 24)

Aerospace engineering might be thought of as a series of communities. Despite being a relatively new engineering discipline, aerospace engineering has diversified as it has grown. The aerospace engineering communities include a range of activities from basic science through very applied production engineering. The communities are held together because of a common use of aerospace-related knowledge. In Chapters 6 and 7, we provide data to demonstrate that aerospace engineers and scientists have varied duties and responsibilities and, consequently, differing information-seeking behaviors and information needs. These various behaviors and needs must be taken into account in the development of an effective system for diffusing aerospace knowledge.

CONCLUSIONS

We described the relationship between science and technology and the similarities and differences between engineers and scientists, and we demonstrated that there is a tendency for convergence in their duties and responsibilities. Self-identification as engineers and scientists remains a definitional problem, and the outputs from both groups are often quite similar. For example, engineers and scientists work together and do similar work in the national laboratories and large industrial research institutions. In aerospace in particular, there are likely to be fewer distinctions between scientists and engineers because most of the industry is focused on the development and exploitation of material artifacts to improve flight.

We find ourselves somewhat in agreement with the position that science and technology are not fully divergent. We argue, though, that the lack of a clear distinction further supports our contention that an engineering knowledge community exists. We believe that the factual and theoretical knowledge that engineers have and use is not clearly distinguishable from the knowledge used in science (Layton, 1992). Rather, the engineering knowledge community is based on a shared understanding among engineers on how to use factual and theoretical information. That is, engineering knowledge is based on engineers' understanding how to build into a design or a proposal ideas that demonstrate engineering. The engineering knowledge the community shares is the understanding of the engineering built into a design or proposal.

To give an example of the differences between science and engineering, we could assume that aerospace includes the following components: aerodynamics, structures, and propulsion. Scientists working in aerospace would base their work on the factual and theoretical foundations of one or more of the components. In contrast, engineers would examine one or more of the components but would also include one or more of the following—economics, sociology, psychology, market analysis, systems analysis, cultural analysis, and management issues—in their work. Because of the nature of the aerospace industry, many aerospace engineers and scientists could generate science as well as engineering in their everyday activities. We contend that the non-science part of engineers' work (the social part) distinguishes them from scientists. It is the inclusion of the social parts of their work that demonstrates engineering. The ability of engineers to understand the social part of their work is a distinguishing feature of the engineering knowledge community.

At the junction of (a) science and technology studies, (b) an awareness of an engineering knowledge community, and (c) an understanding of knowledge production, transfer, and use, we can best understand how innovation occurs in aerospace. Science and technology studies demonstrate how new ideas become integrated into the existing science or engineering communities, or in a few rare cases, transform the knowledge communities. An understanding of the engineering knowledge community demonstrates how change (technological innovation) happens in the daily workplace activities of engineers. Advances in aerospace knowledge cannot happen without effective knowledge production, transfer, and use.

IMPLICATIONS FOR DIFFUSING THE RESULTS OF FEDERALLY FUNDED AEROSPACE R&D

To maintain international competitiveness, U.S. aerospace must continually bring technological innovations into its product designs. Innovations are certainly not always the equivalent of radical technology, but an understanding of how new ideas become accepted into the engineering or science knowledge communities can help us comprehend how innovations can be brought into product designs. If U.S. aerospace engineers and scientists are to continue their leadership in design innovation in aerospace, they need effective tools not only to do the design but also to convince others that their innovations are "workable." That is, they need to demonstrate to others the value of their research and designs. They need to produce, transfer, and use knowledge to create innovation because innovation cannot happen without acceptance of the design by the engineering knowledge community.

There are multiple implications for diffusing the results of federally funded aerospace R&D based on the analysis of engineering and science and the distinctions between engineers and scientists presented in this chapter. *First*, the formal open literature should continue as the channel for diffusing the "science" resulting from federally funded aerospace R&D. Journal articles and conference and meeting papers are the information products of choice for diffusing the science part of aerospace knowledge. Federal knowledge diffusion policies should continue to support the release of this knowledge through these information products.

Second, the federal system that produces DoD and NASA technical reports that document the "engineering" resulting from federally funded aerospace R&D can be expected to continue as the channel for diffusing this engineering knowledge. However, federal policy may actually inhibit the documentation of "engineering" knowledge. Bikson, Quint, and Johnson (1984) pointed out the low level of support within the federal government and at the agency level for documenting and diffusing engineering knowledge. They concluded that documentation and diffusion of engineering knowledge are not viewed as important components of the R&D process. Further, studies show that a substantial portion of the knowledge produced under federal research grants and contracts remains undocumented because documentation in certain cases is not a requirement for receiving federal funding, and there is no penalty for failing to document research findings and deliver a report to the sponsoring organization in other cases. In the early 1980s, federal policy was changed to allow grantees and contractors to claim the results of federally funded R&D as proprietary information on the assumption that grantees and contractors could do a better job of commercializing new technology resulting from federally funded R&D than could the federal agencies that provided the funding. This policy restricts access to others who might be interested in or who could use the results of federally funded R&D. If engineering knowledge resulting from federally funded R&D goes undocumented, it cannot pass into the engineering knowledge community and is consequently lost.

Third, policy that concerns diffusing the results of federally funded aerospace R&D must recognize that the federal system must serve a variety of engineers and scientists in a variety of organizations performing a diversity of tasks. The system must meet the needs of the academic, government, and industry sectors. It must recognize that different sized organizations within each sector have differing needs and abilities to import and export information. The system must meet the unique information needs of aerospace engineers and scientists working in research, design and development, manufacturing and production, service and maintenance, and marketing and sales. It must support researchers on the cutting edge of technology as well as individuals who perform routine tasks.

CHAPTER REFERENCES

Allen, T.J. (1988). "Distinguishing Engineers From Scientists." In *Managing Professionals in Innovative Organizations*. R. Katz, ed. Cambridge, MA: Ballinger Publishing, 3-18.

Allen, T.J. (1977). *Managing the Flow of Technology: Technology Transfer and the Dissemination of Technological Information Within the R&D Organization*. Cambridge, MA: MIT Press.

Barber, B. (1962). *Science and the Social Order*. rev. ed. New York, NY: Collier Books.

Bikson, T.K.; B.E. Quint; and L.L. Johnson. (1984). *Scientific and Technical Information Transfer: Issues and Options*. Washington, DC: National Science Foundation. (Available NTIS; PB-85-150357; also available as Rand Note 2131.)

Blade, M.F. (1963). "Creativity in Engineering." In *Essays on Creativity in the Sciences*. M.A. Coler, ed. New York, NY: New York University Press, 110-122.

Cairns, R.W. and B.E. Compton. (1970). "The SATCOM Report and the Engineer's Information Problem." *Engineering Education* 60: 375-376.

Citro, C.F. and G. Kalton. (1989). *Surveying the Nation's Scientists and Engineers: A Data System for the 1990s*. Washington, DC: National Academy Press.

Constant, E.W. II. (1984). "Communities and Hierarchies: Structure in the Practice of Science and Technology." In *The Nature of Technological Knowledge: Are Models of Scientific Change Relevant?* R. Laudan, ed. Dordrecht, The Netherlands: D. Reidel, 22-46.

Constant, E.W. II. (1980). *The Origins of the Turbojet Revolution*. Baltimore, MD: Johns Hopkins University Press.

Danielson, L.E. (1960). *Characteristics of Engineers and Scientists: Significant for Their Motivation and Utilization*. Ann Arbor, MI: University of Michigan Press.

Dervin, B. and M. Nilan. (1986). "Information Needs and Use." Chapter 1 in the *Annual Review of Information Science and Technology 21*. M.E. Williams, ed. New York, NY: John Wiley, 3-33.

Donovan, A. (1986). "Thinking About Engineering." *Technology & Culture* 27: 674-677.

Doty, P.; A.P. Bishop; and C.R. McClure. (1991). "Scientific Norms and the Use of Electronic Research Networks." Paper presented at the 54th Annual Meeting of the American Society for Information Science. 28: 24-29.

Downey, G.L. and J.C. Lucena. (1995). "Engineering Studies." Chapter 8 in *Handbook of Science and Technology Studies*. S. Jasanoff; G.E. Markle; J.C. Petersen; and T. Pinch, eds. Newbury Park, CA: Sage Publications, 167-188.

Florman, S.C. (1987). *The Civilized Engineer*. New York, NY: St. Martin's Press.

Gaston, J. (1980). "Sociology of Science and Technology." In *A Guide to the Culture of Science, Technology, and Medicine*. P.T. Durbin, ed. New York, NY: Free Press, 465-526.

Gerstberger, P.G. and T.J. Allen. (1968). "Criteria Used by Research and Development Engineers in the Selection of an Information Source." *Journal of Applied Psychology* (August) 52(4): 272-279.

Grayson, L.P. (1993). *The Making of an Engineer: An Illustrated History of Engineering Education in the United States and Canada*. New York, NY: John Wiley.

Hagstrom, W.O. (1965). *The Scientific Community*. New York, NY: Basic Books.

Herner, S. (1954). "Information Gathering Habits of Workers in Pure and Applied Science." *Industrial Engineering and Chemistry* 46(1): 228-236.

Illinois Institute of Technology. (1968). *Technology in Retrospect and Critical Events in Science*. Washington, DC: National Science Foundation; AKA Project TRACES. (Available NTIS; PB-234767.)

Katz, D. and R.L. Kahn. (1966). *The Social Psychology of Organizations*. New York, NY: John Wiley.

Kaufman, H.G. (1983). *Factors Related to the Use of Technical Information in Engineering Problem Solving*. Brooklyn, NY: Polytechnic Institute of New York.

Kemper, J.D. (1990). *Engineers and Their Profession*. 4th ed. Philadelphia, PA: W.B. Saunders.

Kennedy, J.M.; T.E. Pinelli; and R.O. Barclay. (1995). "*A Comparison of the Information-Seeking Behaviors of Three Groups of U.S. Aerospace Engineers.*" Paper presented at the 33rd Aerospace Sciences Meeting & Exhibit of the American Institute of Aeronautics and Astronautics (AIAA), Reno, NV. AIAA 95-0706. (Available NTIS; 95N19127.)

King, D.W.; J. Casto; and H. Jones. (1994). *Communications by Engineers: A Literature Review of Engineers' Information Needs, Seeking Processes, and Use*. Washington, DC: Council on Library Resources.

Kintner, H.J. (1993). *Counting Engineers—A Latent Class Analysis of Self-Reported Occupation, Employer Administrative Records, and Educational Background*. GMR-8033. Warren, MI. General Motors Corporation, NAO Research and Development Center.

Kremer, J.M. (1980). *Information Flow Among Engineers in a Design Company*. Ph.D. Diss., University of Illinois at Urbana-Champaign. UMI 80-17965.

Kuhn, T. (1970). *The Structure of Scientific Revolutions*. 2nd ed. Chicago, IL: University of Chicago Press.

Langrish, J; M. Gibbons; W.G. Evans; and F.R. Jevons. (1972). *Wealth From Knowledge: A Study of Innovation in Industry*. New York, NY: John Wiley.

Latour, B. (1987). *Science in Action: How to Follow Scientists and Engineers Through Society*. Cambridge, MA: Harvard University Press.

Laudan, R. (ed.). (1984). "Introduction." In *The Nature of Technological Knowledge: Are Models of Scientific Change Relevant?* R. Laudan, ed. Boston, MA: Reidel, 1-26.

Law, J. (1987). "The Structure of Sociotechnical Engineering: A Review of the New Sociology of Technology." *Sociological Review* 35: 405-424.

Law, J. and M. Callon. (1988). "Engineering and Sociology in a Military Aircraft Project: A Network of Analysis of Technological Change." *Social Problems* 35: 115-142.

Layton, E.T. (1992). "Escape From the Jail of Shape: Dimensionality and Engineering Science." In *Technological Development and Science in the Industrial Age*. P. Kroes and M. Bakker, eds. Dordrecht, The Netherlands: Kluwer Academic Publishers, 69-98.

Layton, E.T. (1974). "Technology as Knowledge." *Technology & Culture* 15(1): 31-33.

Meiksins, P.F. and C. Smith. (1993). "Organizing Engineering Work: A Comparative Analysis." *Work and Occupations* 20(2): 123-146.

Menzel, H. (1964). "The Information Needs of Current Scientific Research." *Library Quarterly* (January) 34(1): 4-19.

Orr, R.H. (1970). "The Scientist as an Information Processor: A Conceptual Model Illustrated With Data on Variables Related to Library Utilization." In *Communication Among Scientists and Engineers*. C.E. Nelson and D.K. Pollock, eds. Lexington, MA: D.C. Heath, 143-189.

Pinelli, T.E. (1991). "The Information-Seeking Habits and Practices of Engineers." *Science & Technology Libraries* 11(3): 5-25.

Pinelli, T.E.; A.P. Bishop; R.O. Barclay; and J.M Kennedy. (1993). "The Information-Seeking Behavior of Engineers." In *Encyclopedia of Library and Information Science*. A. Kent and C.M. Hall, eds., 52:15 New York, NY: Marcel Dekker, 167-201.

Price, D.J. de Solla. (1965). "Is Technology Historically Independent of Science?" *Technology & Culture* (Summer) 6(3): 553-578.

Rip, A. (1992). "Science and Technology As Dancing Partners." In *Technological Development and Science in the Industrial Age*. P. Kroes and M. Bakker, eds. Dordrecht, The Netherlands: Kluwer Academic Publishers, 231-270.

Ritti, R.R. (1971). *The Engineer in the Industrial Corporation*. New York, NY: Columbia University Press.

Rogers, E.M. (1982). "Information Exchange and Technological Innovation." Chapter 5 in *The Transfer and Utilization of Technical Knowledge*. D. Sahal, ed. Lexington, MA: D.C. Heath, 105-123.

Rosenberg, V. (1967). "Factors Affecting the Preferences of Industrial Personnel for Information Gathering Methods." *Information Storage and Retrieval* (July) 3: 119-127.

Rosenbloom, R.S. and F.W. Wolek. (1970). *Technology and Information Transfer: A Survey of Practice in Industrial Organizations*. Boston, MA: Harvard Business School Press.

Rothstein, W.G. (1969). "Engineers and the Functionalist Model of Professions." In *The Engineers and the Social System*. R. Pericci and J.E. Gerstl, eds. New York, NY: John Wiley, 73-97.

Shapley, D. and R. Roy. (1985). *Lost at the Frontier: U.S. Science and Technology Policy Adrift*. Philadelphia, PA: ISI Press.

Shuchman, H.L. (1981). *Information Transfer in Engineering*. Glastonbury, CT: The Futures Group.

Storer, N.W. (1966). *The Social System of Science*. New York, NY: Holt, Rinehart and Winston.

Taylor, R.S. (1991). "Information User Environments." In *Progress in Communication Sciences 10*. Norwood, NJ: Ablex, 217-255.

Tushman, M.L. and D.A. Nadler. (1980). "Communication and Technical Roles in R&D Laboratories: An Information Processing Model." In *Management of Research and Innovation*. B.V. Dean and J.L. Goldhar, eds. New York, NY: North-Holland, 91-112.

U.S. Department of Defense, Office of the Director of Defense Research and Engineering. (1969). *Project Hindsight*. Washington, DC: U.S. Department of Defense. (Available NTIS: AD-495,905.)

Vincenti, W.G. (1990). *What Engineers Know and How They Know It: Analytical Studies From Aeronautical History*. Baltimore, MD: Johns Hopkins University Press.

Vincenti, W.G. (1992). "Engineering Knowledge, Type of Design, and Level of Hierarchy: Further Thoughts About *What Engineers Know* ...." In *Technological Development and Science in the Industrial Age*. P. Kroes and M. Bakker, eds. Dordrecht, The Netherlands: Kluwer Academic Publishers, 17-34.

Voight, M.J. (1960). *Scientists' Approaches to Information*. ACRL Monograph No. 24. Chicago, IL: American Library Association.

Weingart, P. (1984). "The Structure of Technological Change: Reflections on a Sociological Analysis of Technology." In *The Nature of Technological Knowledge: Are Models of Scientific Change Relevant?* R. Laudan, ed. Boston, MA: Reidel, 115-142.

Whalley, P. (1986). *The Social Production of Engineering Work*. Albany, NY: SUNY Press.

Young, J.F. and L.C. Harriott. (1979). "The Changing Technical Life of Engineers." *Mechanical Engineering* (January) 101(1): 20-24.

Ziman, J. (1984). *An Introduction to Science Studies: The Philosophical and Social Aspects of Science and Technology*. Cambridge, UK: Cambridge University Press.

Zussman, R. (1985). *Mechanics of the Middle Class*. Berkeley, CA: University of California Press.

Chapter 6

The NASA/DoD Aerospace Knowledge Diffusion Research Project—Overview, Study Populations, and Research Agenda

John M. Kennedy
Thomas E. Pinelli
Rebecca O. Barclay
Ann P. Bishop

SUMMARY

Chapter 6 provides an overview of the *NASA/DoD Aerospace Knowledge Diffusion Research Project*. The project was undertaken by researchers at the NASA Langley Research Center, the Indiana University Center for Survey Research, and Rensselaer Polytechnic Institute. The project began in 1987 as a 10-year effort to provide descriptive and analytical data regarding the diffusion of knowledge at the individual, organizational, national, and international levels and to examine both the channels and sources used to diffuse knowledge and the social system of the aerospace knowledge diffusion process. We present an organizing framework for the project, and discuss the limitations of the project's research design and methodology. Next we provide an overview and description of the project's four phases. We describe the study populations and selected demographic characteristics of the survey participants. We present the research agenda for the *NASA/DoD Aerospace Knowledge Diffusion Research Project* and discuss the variables that we believe are central to understanding aerospace knowledge diffusion. The Appendix to Chapter 6 lists the NASA technical reports that document the results of our work.

INTRODUCTION

Our research is concerned with the diffusion of knowledge resulting from federally funded aerospace research and development (R&D).

The *NASA/DoD Aerospace Knowledge Diffusion Research Project* has collected descriptive and analytical data about the diffusion of aerospace knowledge at the individual, organizational, national, and international levels and the information-seeking behaviors of aerospace engineers and scientists. This research has examined both the channels and sources used to diffuse knowledge and the social system within which aerospace knowledge diffuses. The people we studied are both producers and users of knowledge. They include aerospace engineers and scientists working in academia, government, and industry who represent the spectrum of aerospace R&D from research, design and development, manufacturing and production, flight testing, human factors, service and maintenance, and marketing and sales. We undertook this project with the conviction that an understanding of aerospace knowledge diffusion could be used to increase productivity, stimulate innovation, and improve and maintain the professional competence of U.S. aerospace engineers and scientists.

In addition to studying these professionals, we also studied aerospace engineering and science students and librarians working as information intermediaries in U.S. aerospace. Data were collected using mail (self-reported) and telephone surveys that elicited answers to questions about information use and production; a critical incident where information was used for a specific problem, project, or task; and the demographics of the respondents. The survey design is based primarily on Dillman's (1978) total design method (TDM). During the course of our research, we surveyed approximately 15 000 aerospace engineers and scientists, nearly 4000 engineering and science students, 182 industry libraries, and 75 academic libraries.

Our research is based on three assumptions: (a) knowledge production, transfer, and utilization are equally important components of aeronautical research, development, and production (RD&P); (b) the diffusion of the knowledge resulting from federally funded aeronautical R&D is indispensable in maintaining the international competitiveness and vitality of the U.S. aircraft industry; and (c) the U.S. government technical report plays an important, but as yet undefined, role in the aerospace knowledge diffusion process. The project is delimited by the following: (a) it focuses on technology rather than science; (b) it studies engineers rather than scientists as both producers and users of knowledge; and (c) it takes the position that knowledge resulting from federally funded aerospace R&D should be viewed more as an economic asset or resource than as a component of national security.

ORGANIZING FRAMEWORK FOR THE PROJECT

To establish a framework for the project, we focused on five specific areas: (a) the U.S. aerospace industry, (b) federal domestic and foreign policy, (c) federal technology policy and the U.S. aircraft industry, (d) the system used to disseminate the knowledge that results from federally funded R&D, and (e) science and technology as independent but interacting entities.

First, we looked at the aeronautics portion of the U.S. aerospace industry, an industry that occupies a unique position in the nation's industrial and economic structure. The U.S. aerospace industry is often described in superlatives. It is the largest positive industrial contributor to the U.S. balance of trade. It is the country's leading net exporter of manufactured goods, having sold $28 billion in 1995, and it produces the largest trade surplus of any U.S. manufacturing industry (about $21 billion in 1995). The work of Bluestone, Jordan, and Sullivan (1981); Bright (1978); Caves (1962); Golich (1992); Hayward (1990); and Jordan (1979) provides background information concerning the aeronautics portion of the U.S. aerospace industry. We also looked at the various studies that have investigated the competitive status of the large commercial aircraft portion of the U.S. aerospace industry. For a discussion of the competitive status of the large commercial aircraft sector, consult the following sources: Gellman Research Associates (1990), Lopez and Vadas (1993), U.S. Congress—Office of Technology Assessment (1991), and the U.S. International Trade Commission (1993).

Second, we looked at U.S. policies (i.e., domestic, foreign, and national security) because of their influence on commerce and the U.S. aerospace industry in particular. This industry can be considered a sheltered, rather than an exposed, economy because U.S. policies, in combination, influence the production and sale of commercial and military aircraft (Derian, 1990). Although they cut across political jurisdictions and are more or less subject to shifting political and economic alliances and imperatives, these policies ultimately affect the production and sale of aircraft and influence the strength and vitality of the industry. For a discussion of U.S. policy, see the work of Baldwin and Krugman (1988), Derian (1990), Golich (1989a, 1989b), Hall and Johnson (1970), and Hochmuth (1974).

Third, we focused on U.S. technology policy because of the role played by two federal agencies, the Department of Defense (DoD) and the National Aeronautics and Space Administration (NASA), as producers of federally funded aerospace R&D and the methods employed by these agencies to diffuse the results of their research to the U.S. aircraft industry. Since post World War II, U.S. technology policy has been dominated by "supply side" economics. Basic research, a major tenet of this policy, was assumed to be a federal responsibility because the benefits of basic research would accrue too far into the future to make industrial support (i.e., payback) feasible. Basic research would be conducted primarily, but not exclusively, in U.S universities. Applied research was to be a responsibility of government as well, but research programs would be fully developed and coordinated by the individual departments or agencies that would purchase or use the end product(s) of applied research. Department and agency support for applied R&D would take place in government laboratories and under contract to industry and universities. Commercialization would occur automatically as a by-product of federal support for basic research and department and agency support for applied R&D. For a discussion of U.S. technology policy, see the work of Dupree (1986), Hillis et al., (1992), Logsdon (1986), Mowery (1983), Smith (1990), and Tornatzky (1990). Mowery (1985) and Mowery and Rosenberg (1985, 1982) have conducted extensive studies of the impact of U.S. technology policy on the U.S. aircraft industry.

Fourth, we focused on the system used to transfer the knowledge that results from federally funded R&D. Three approaches have dominated knowledge transfer: appropriability, dissemination, and diffusion. The present system for transferring knowledge resulting from federally funded R&D is based on the dissemination model. The federal system is incompatible with diffusion-oriented technology policies. It is passive, for it does not take users into consideration except when they enter the system and request assistance. Furthermore, the dissemination model can encourage knowledge utilization only after the R&D results have been generated rather than during the idea development phase of the innovation process. Scholars point out that successful technological innovation rests more with the transfer and utilization of knowledge that results from federally funded R&D than with its production (Mowery, 1983). For an analysis of the federal system, consult the following: Bikson, Quint, and Johnson (1984); David (1986); Hernon and McClure (1984); Roberts and Frohman (1978); and Williams and Gibson (1990).

Fifth, we explored the relationship between science and technology and the differences between engineers and scientists. This relationship, which is hotly debated, is often presented as a continuous process or normal progression from basic research (science) through applied research (technology) to development (utilization). Technology is a process dominated by engineers rather than scientists, which leads to different philosophies and habits not only about contributions to the technical literature but also to use of the technical literature and other sources of information. Therefore, an understanding of the relationship between science and technology is critical to understanding the diffusion of knowledge resulting from federally funded aerospace R&D. For a discussion of the relationship between science and technology and the differences between engineers and scientists, consult the following sources: Adams (1991), Allen (1988 and 1977), Downey and Lucena (1995), Holmfeld (1970), Kemper (1990), Layton (1971), Price (1965), and Vincenti (1990).

LIMITATIONS

There are specific limitations, specific strengths, and specific weaknesses associated with each methodology used in research; survey research is no exception. All measurement involves error. In surveys, the two errors that researchers are most concerned about are bias and sample representation. We cannot be certain that each group we surveyed was fully representative of those in similar occupations and positions who were not surveyed. We are comfortable with the consistently high response rates to the surveys because they indicate the importance of the surveys to the respondents who recognized their professional responsibility to return the questionnaires. The fact that the demographic compositions were similar across surveys indicates less possibility of sample bias. Throughout the entire process we sought to maximize the strengths inherent to survey research, to compensate for its weaknesses, and to reduce measurement error as much as possible. To interpret our findings, the reader should be aware of the characteristics and the limitations of the methods used to collect data.

PROJECT OVERVIEW

The *NASA / DoD Aerospace Knowledge Diffusion Research Project* was conducted in four phases. Phase 1 examined the production and use of information by U.S. aerospace engineers and scientists. Phase 2 surveyed information intermediaries (principally librarians and technical information specialists) in the U.S. aerospace industry to explore the intermediary-user and the NASA-intermediary interfaces as part of the aerospace knowledge diffusion process. Phase 3 looked at aerospace engineering in U.S. academic settings, to include students, faculty, and information intermediaries (principally librarians). Phase 4 examined the national and international dimensions of aerospace knowledge diffusion. We examined the production and use of information by aerospace engineers, scientists, and students in western European nations, India, Israel, Japan, and the former Soviet Union. We used policy analysis to evaluate U.S. aeronautical and information policies.

Survey research, in the form of self-reported mail questionnaires and telephone surveys, was used to collect data about the use of, uses for, and users of the knowledge resulting from federally funded aerospace R&D. Survey research was selected because of the capability of this methodology to gather data on a population that is too large and geographically dispersed to observe directly. Questionnaires permitted large amounts of data to be collected and manipulated in a uniform manner, using statistical tests and methods.

Phase 1

The primary empirical research activities in the project were done as Phase 1 activities. During this phase, we examined the information-seeking behaviors of U.S. aerospace engineers and scientists. Survey data were collected from a variety of U.S. aerospace engineers and scientists over a 10-year period. Phase 1 investigated the information environment in which U.S. aerospace engineers and scientists work (i.e., the academic, government, and industrial sectors), their information-seeking behaviors, and the factors that influence their use of knowledge resulting from federally funded aerospace R&D. We collected data on technical communications in the workplace, such as the amount of time that U.S. aerospace engineers and scientists devoted to communicating and working with technical information, their use and production of specific technical information products, their present and anticipated use of electronic and information technologies, and their use of libraries and library services. Phase 1 also specifically examined the role of the U.S. government technical report in aerospace knowledge diffusion, the communications environment in U.S. aerospace, and the role of computer networks and communications in the U.S. aerospace workplace. To represent the full spectrum of aeronautical RD&P, survey participants were drawn from the American Institute of Aeronautics and Astronautics (AIAA), the American Society for Testing and Materials (ASTM), the Human Factors and Ergonomics Society (HF&ES), the Society of Automotive Engineers (SAE), the Society of Flight Test Engineers (SFTE), the Society for the Advancement of Material & Process Engineering (SAMPE), the Society of Manufacturing Engineers (SME), and NASA aerospace technologists (ASTs). To map the transition from student to entry-level professional, we also surveyed early career-stage aerospace engineers and scientists who belong to the AIAA.

Phase 2

Phase 2 explored the intermediary-user and the NASA-intermediary interfaces as part of the aerospace knowledge diffusion process. Phase 2 also investigated holdings of technical reports and the use and importance of print-based and electronic bibliographic tools. It assessed aerospace announcement and current awareness products and library outreach activities. NASA and the DoD rely on librarians and technical information specialists to complete the knowledge transfer process from producer to user. U.S. aerospace industry libraries and selected U.S. government libraries served as the population for the Phase 2 survey. This list was compiled from several sources, including the *Directory of Special Libraries and Information Centers* and the Special Libraries Association. To be eligible for participation, each library had to hold a substantial number of aerospace, aeronautical, or related collections.

Phase 3

Phase 3 investigated the information-seeking behaviors of U.S. aerospace engineering faculty and students and the role of academically affiliated information intermediaries in the aerospace knowledge diffusion process. The U.S. faculty sample was obtained primarily from four-year institutions that participated in the 1990 NASA/USRA (University Space Research Association) Capstone Design Program. Also included were faculty from other institutions that have aerospace programs accredited by the Accreditation Board for Engineering and Technology (ABET). The second faculty sample was obtained from the AIAA in 1996 and included the names of AIAA members who identified themselves as engineering educators.

One of the student samples included those students enrolled in a NASA/USRA-funded undergraduate capstone course in the spring of 1990. U.S. academic libraries in four-year colleges and universities with ABET-accredited engineering programs served as the information intermediary population for the Phase 3 survey. Four additional student surveys were conducted in the spring of 1993. One survey population was student members of the AIAA; another was composed of engineering and science students at the University of Illinois at Urbana-Champaign (UI-UC); a third was composed of aerospace engineering students at Texas A&M; and a fourth was composed of technology students at Bowling Green State University (BGSU). The data from these surveys helped us to understand information use in aerospace engineering programs from three perspectives: those of aerospace engineering faculty, students, and information intermediaries. In addition, the second wave of student surveys helped us to understand how aerospace engineering students are similar to and different from students in related majors and programs. Collectively, data obtained from the Phase 3 surveys would provide insight into knowledge diffusion (i.e., production, transfer, and use) within the academic portion of the aerospace community.

Phase 4

Phase 4 concentrated on describing and explaining the information environment in which non-U.S. aerospace engineers and scientists work. In this phase, we collected survey data from non-U.S. aerospace engineers and scientists and aerospace engineering students, similar to the data that were collected in Phase 1 from

U.S. aerospace engineers and scientists. Site-specific surveys of aerospace engineers were conducted in India, Israel, Japan, the Netherlands, and Russia. These surveys had small samples because they were conducted within specific organizations, for example, the Central Aero-Hydrodynamics Institute (TsAGI) in Russia. One large-scale survey with over 1000 respondents was conducted with the members of the Royal Aeronautical Society (RAeS) in the United Kingdom. Student surveys were conducted in India, Japan, Russia, and the United Kingdom at the following institutions: Indian Institute of Science, Indian Institute of Technology, University of Tokyo, Moscow Aviation Institute, Cranfield University, and the University of Southampton. The results of these surveys helped us to understand how aerospace engineers and scientists and aerospace engineering students are similar to and different from their U.S. counterparts in terms of their information-seeking behaviors. The results of Phase 4 also are useful for understanding how aerospace knowledge diffuses across international boundaries.

STUDY POPULATIONS

In this project we gathered data from (a) ten surveys of members of the AIAA; (b) four surveys based on a list of subscribers to *Aerospace Engineering* and one survey drawn from the membership of the Society of Automotive Engineers (SAE); (c) one survey based on a list of subscribers to *Manufacturing Engineering* whose Standard Industrial Classification (SIC) codes indicated that they were employed in the aerospace industry and one survey drawn from the membership list of the Society of Manufacturing Engineers (SME); (d) three surveys of NASA ASTs; (e) one survey that combined members of selected interest groups in the AIAA, members of the ASTM, and members of the SAMPE; (f) one survey of the members of the HF&ES; and (g) one survey of the members of the SFTE. Combined, these groups represent the types of engineers and scientists employed in the various sectors of U.S. aerospace. Surveying these groups lets us provide data from the full spectrum of the aircraft industry from research, design and development, manufacturing and production, service and maintenance, and marketing and sales.

Next, we present selected demographic characteristics of the U.S. aerospace engineers and scientists who participated in Phase 1 of this research. We also describe their duties and responsibilities, and we report how their work is funded. To understand the production, transfer, and use of knowledge in aerospace, it is necessary to understand the characteristics of those who create and use it. We use multiple demographics to analyze the roles played by U.S. aerospace engineers and scientists and to present various perspectives on the production, transfer, and use of aerospace knowledge. Using this information, we can begin to understand the diffusion of knowledge resulting from federally funded aerospace R&D at the individual, organizational, national, and international levels, and the social system within which aerospace knowledge diffuses.

AIAA Samples

Phase 1 included 10 surveys of AIAA members (Table 6.1). A pilot study for the project, conducted in 1988, focused on the technical communications practices of U.S. aerospace engineers and scientists. Three large-scale surveys were conducted in 1989 and 1990. These three studies focused on the use of various aerospace information products and services. The 1993 study of AIAA members focused on computer-mediated communication (CMC) and the communication of technical information in aerospace. In 1995, a

Table 6.1. Description of the AIAA Surveys

Sample	Respondents	Sample Size	Response Rate[a]	Completion Date	Survey Type
Pilot study	606	2000	30%	1988	Mail
Survey 1	2016	2894	70%	1989	Mail
Survey 2	975	1553	63%	1989	Mail
Survey 3	955	1462	65%	1990	Mail
Survey 4	1006	2000	55%	1993	Mail
Survey 5	341	750	53%	1995	Mail
Survey 6	264	700	49%	1995	Mail
Survey 7	312	700	51%	1995	Mail
Survey 8	134	300	51%	1996	Mail
Survey 9	142	300	57%	1996	Mail
Survey 10	84	200	56%	1996	Mail

[a]The response rate was adjusted for noneligible respondents, incorrect addresses, and other sample problems.

small survey gathered data on topics not covered in previous AIAA surveys to provide data comparable to those collected from other groups surveyed in Phase 1. Also in 1995, we conducted two surveys of early career-stage aerospace engineers and scientists who had transferred from student to professional memberships in the AIAA. In 1996 we conducted surveys among members of three technical interest groups to examine the production, transfer, and use of knowledge within specific technical areas.

The AIAA provided ten lists for us to use as sampling frames for the AIAA surveys. The pilot study was a random sample of all AIAA members in 1988. For Surveys 1, 2, and 3, the AIAA provided a list of approximately 6000 (a one-in-five sample) of its U.S. members in 1989. Students and retired persons were excluded from the frame. From the frame, we developed separate random samples for each survey. An analysis of the characteristics of the respondents who returned questionnaires early in Survey 1 indicated there would be too few academic respondents to provide statistically accurate estimates of their information-seeking activities, so we used the 1989 *AIAA Roster* to supplement each of the three samples with AIAA members who were identified as being located in academic institutions. The sample for Survey 4 was a random sample of 2000 members provided by AIAA in 1993.

The sample for Survey 5 was an all-industry sample drawn from the 1995 AIAA membership list. The demographic and locational characteristics of the engineers and scientists across the five AIAA samples were very similar. The sample for Surveys 6 and 7 came from the 1995 AIAA membership list. The sample contained individuals who had held "professional" membership for five years or less, having changed AIAA membership from "student" to "professional" status within that time frame. This sample represents aerospace professionals who are in the early stage of their careers. To help ensure that knowledge production, transfer, and use across the full spectrum of the U.S. aircraft industry was measured, three additional surveys (i.e., 8, 9, and 10) were conducted in 1996. Three all-industry samples were drawn from the following technical interest groups: propulsion and aircraft engines, aircraft design, service and maintenance, and marketing and sales. For descriptive purposes in this chapter, we present the results from AIAA Survey 1 and from AIAA Survey 10. Survey 1 was the largest and most comprehensive sample. Survey 10 represents a small but unique group of aerospace engineers—those whose interests are self-defined as service or maintenance and marketing or sales. While Survey 10 consisted of members of the AIAA, we analyze them separately in this chapter because of their unique status.

The large AIAA samples consisted primarily of U.S. aerospace engineers and scientists involved in aerospace engineering research (Pinelli, 1991). AIAA members surveyed in 1991 considered themselves research engineers, scientists, or technical managers of research. In 1991, about 25% identified themselves as design and development engineers. Many academic and government engineers and scientists belong to the AIAA, giving its membership an engineering research emphasis. We recognize the job diversity of AIAA membership but categorize this group as generally more oriented to the research end of aerospace engineering or, as Vincenti (1990) would classify their activities, to engineering science.

In an information-use context, we expected most AIAA members to behave more like scientists than engineers. For example, most academic and government engineers do not perform manufacturing and production work. Downey (1992) indicated that engineers themselves see differences among their colleagues who perform research, design and development, and manufacturing and production duties. The attitudes and behaviors that have shaped this distinction indicate that the duties and responsibilities of various types of aerospace engineers are different and noticeable. We expected that the AIAA members would behave differently from other aerospace engineers in their production, transfer, and use of knowledge. In particular, we expected the AIAA members to use more of the open literature (e.g., journals) and more formal channels and sources than other aerospace engineers use. (For a complete discussion of the results from the AIAA pilot study and AIAA surveys 1 through 10, refer to reports 1 through 6, 30, 33, 34, 37, 41, 42, and 46. See the Appendix to Chapter 6.)

SAE Samples

Phase 1 of the project also included four surveys of engineers who receive *Aerospace Engineering*, which is a trade magazine published by the Aerospace Division of the SAE. The fifth survey (not analyzed in this chapter) used a sample of SAE members whose technical interests included avionics and electrical components and subsystems. Both mail and telephone surveys were conducted over a five-year period (Table 6.2). The first telephone survey focused on engineering duties and the use of computers in aerospace. The second telephone survey focused on the production and use of knowledge, and was similar in content to AIAA surveys 1, 2, and 3. The third survey (mail) focused on technical communication in aerospace and the impact of uncertainty and complexity on engineering work. The fourth survey (mail) focused on the use of computers and computer networks in aerospace. The fifth survey (mail) focused on knowledge production, transfer, and use by SAE members.

The SAE provided a list of approximately 4000 *Aerospace Engineering* subscribers in 1991. Approximately 1000 names were used for each of the telephone surveys. The remaining 2000 names were used for the first SAE mail survey. For the telephone surveys, we selected from the subscribers who had provided either a work or a home telephone number. The remainder were assigned to the mail survey sample. Our analysis of the professional and demographic characteristics of the three samples indicates there were few differences among them. The SAE provided the fourth sample from the list of *Aerospace Engineering* subscribers in 1993. This sample was stratified by various job classifications to ensure that there would be a minimum number of respondents in selected engineering categories. We chose the SAE subscriber database because we thought it would represent the design and development portion of aerospace engineering. On the engineering spectrum, we assumed that the SAE engineers would be intermediate between the AIAA members and the SME members. We also expected they would use less of the formal open literature (e.g., journals) and more of the in-

Table 6.2. Description of the SAE Surveys

Sample	Respondents	Sample Size	Response Rate[a]	Completion Date	Survey Type
Survey 1	430	575	75%	1991	Phone
Survey 2	407	550	74%	1992	Phone
Survey 3	946	2000	67%	1991	Mail
Survey 4	950	2000	48%	1993	Mail
Survey 5	128	300	51%	1996	Mail

[a]The response rate was adjusted for noneligible respondents, incorrect addresses, and other sample problems.

formal channels and sources than the AIAA members would use, particularly technical reports, especially in-house technical reports.

The large SAE samples (Surveys 1-4) consisted of engineers involved primarily in design and development. About 60% of each sample identified themselves as design and development engineers. Very few identified themselves as scientists; about 16% identified themselves as managers. Both proportions are substantially less than the proportions of scientists and managers in the AIAA samples, except for the 1995 AIAA industry-only survey. For descriptive purposes in this chapter, we present the results from SAE Survey 3, the larger mail survey. This was the largest sample, and responses to many of the questions asked in that survey will be analyzed in chapters that follow. (For a complete discussion of the results of the SAE surveys, refer to reports 13, 14, 15, 24, and 39. See the Appendix to Chapter 6.)

SME Samples

During 1994, we conducted a mail survey of recipients of *Manufacturing Engineering*, a publication of the SME. Recipients whose SIC code indicated they were employed in aerospace firms were included in the sample. From a sample of 1500 subscribers, 463 respondents completed and returned the survey. The 43% response rate was relatively low because the sample did not efficiently target aerospace engineers and scientists and a substantial number of employment changes occurred during the survey period. We had expected to find engineers who were close to the manufacturing and production process in this list. A second survey conducted in 1996 used a sample of SME members who had identified themselves as working in aircraft manufacturing and production. From a sample of 500 SME members, we received 261 usable responses for an adjusted completion rate of 60%. The 1994 survey is analyzed in this chapter. The SME sample, together with the AIAA and the SAE samples, provided us with a sense that the range of engineers employed in the aerospace industry was included in our study. We expected the SME sample to make relatively little use of the formal literature (e.g., journals) but to make heavy use of informal channels or sources (e.g., in-house technical reports). (For a complete discussion of the results of the SME surveys, refer to reports 30 and 43. See the Appendix to Chapter 6.)

NASA AST Samples

We undertook three surveys of NASA ASTs as part of Phase 1 (Table 6.3). The first survey was conducted at five NASA field centers (Ames, Goddard, Langley, Lewis, and Marshall) and focused on knowledge production and use. All NASA ASTs, in all technical areas, were included in this survey. The second survey was conducted at the NASA Ames and Langley Research Centers and focused primarily on technical communication. The third survey was conducted at the NASA Langley Research Center. There are some differences in the characteristics of each NASA sample due to the roles and missions of the five field centers. Only ASTs involved in aerospace-related duties were included in NASA Survey 2. Because the second group is comparable to the AIAA samples, we describe the results of the NASA Survey 2 sample in this chapter.

NASA ASTs are more oriented toward research and engineering science than are the other groups included in the Phase 1 activities. Although about 70% identified themselves as "engineers," it appears that they are more oriented toward engineering science than to design and development or manufacturing and production engineering (Pinelli, 1991). NASA AST Survey 1 focused on the importance of knowledge production, which is more closely associated with the duties of scientists and researchers than with the duties of design and development or manufacturing and production engineers. Only 4% of the NASA ASTs considered themselves managers. (For a complete discussion of the results of the NASA AST surveys, refer to reports 12, 17, and 36. See the Appendix to Chapter 6.)

Table 6.3. Description of the NASA AST Surveys

Sample	Respondents	Sample Size	Response Rate[a]	Completion Date	Survey Type
NASA 1	550	1865	70%	1991	Phone
NASA 2	340	558	61%	1992	Mail
NASA 3	205	300	68%	1995	Mail

[a]The response rate was adjusted for noneligible respondents, incorrect addresses, and other sample problems.

Other Professional Society Samples

To ensure that we understood knowledge production, transfer, and use across the full spectrum of the U.S. aircraft industry, we surveyed professional societies whose members represented structures and materials, human factors and crew integration, and flight testing (Table 6.4). The sample for the structures and materials study (Mixed) came from a composite list taken from the following professional societies: the AIAA, ASTM, and SAMPE. The sample for the human factors and crew integration study came from the HF&ES. The SFTE provided the sample for the flight testing study.

Table 6.4. Description of Surveys from Other Professional Societies

Sample	Respondents	Sample Size	Response Rate[a]	Completion Date	Survey Type
Mixed	209	500	51%	1996	Mail
HF&ES	96	200	60%	1996	Mail
SFTE	68	200	51%	1996	Mail

[a]The response rate was adjusted for noneligible respondents, incorrect addresses, and other sample problems.

Group Selection and Representativeness

The groups selected for the Phase 1 surveys were assumed to include the range of aerospace engineers and scientists working in academia, government, and industry who represent the spectrum of aerospace research and development (R&D) from research, design and development, manufacturing and production, service and maintenance, and marketing and sales. Through group selection and sampling, we intended to produce a comprehensive scan of U.S. aerospace engineers and scientists within the context of the design and manufacture of an aircraft. The variety of groups we selected contains the broad range of engineers and scientists in aerospace, but we do not assume that the combination of all groups is a fully representative sample of U.S. aerospace engineers and scientists. We say this for multiple reasons. There is some overlap among the groups because aerospace engineers and scientists belong to multiple groups or receive multiple publications. Some respondents could have been included in more than one survey population.

However, in the 1996 studies, survey recipients were instructed not to respond if they had responded to any of our prior surveys. In addition, officials at the professional organizations that we surveyed believe they effectively target the types of engineers (and scientists) they serve.

Some engineers and scientists could not be included in any samples because many aerospace engineers and scientists do not belong to a professional society or do not receive either *Manufacturing Engineering* or *Aerospace Engineering*. Furthermore, each sample is not necessarily representative of other individuals who have similar job responsibilities but who were not included in the samples. For example, AIAA members who identify themselves as technical managers may be more oriented toward engineering science than technical managers who do not belong to the AIAA. We recognize that some types of engineers, such as those in service and maintenance and marketing and sales, are not very prevalent in our samples, but we assume that knowledge production, transfer, and use for these individuals is product-specific. Some relatively small groups of U.S. aerospace engineers and scientists may not have been included, but we do not think that their lack of inclusion affects the interpretation of the data. We suspect that many scientists working in aerospace belong to professional organizations other than the AIAA. Lack of membership in the AIAA is not a significant problem because if scientists belong to other professional organizations, they likely have needs for knowledge related to their scientific discipline rather than aerospace-specific knowledge.

We cannot estimate fully the amount of selection bias in the samples or the overall representativeness of the samples. In an earlier paper (Kennedy and Pinelli, 1990), we found some nonresponse bias when we examined the demographic characteristics of the AIAA respondents, but it was not clear from the analysis if the bias affected data quality. For example, using data furnished by the AIAA about the membership, we found that the survey respondents had higher levels of education than the nonrespondents had. Given the nature of the questionnaire—the use and production of aerospace knowledge and information channel or source selection—we expected that the response rate would be much higher among those who produced and used significant amounts of aerospace knowledge than among those did not produce and use significant amounts of aerospace knowledge. Those AIAA members not engaged in research produced and used less knowledge than those in research did, so the questionnaire had low salience for them and they were less likely to respond to it. We assume that the results from all surveys are skewed toward those who are current users of aerospace knowledge.

The results of our research could be used to improve access to and use of aerospace knowledge and to plan new aerospace knowledge (management) systems, so we feel this improves rather than harms data quality. That is, we are primarily interested in producers and users of aerospace knowledge. The somewhat lower response rate of the 1995 AIAA survey of U.S. industry aerospace engineers and scientists, further indicates that the higher the proportion of the sample engaged in information production and use, the higher the response rate. Including some minimal users and nonusers of aerospace knowledge helps us understand the diffusion of aerospace knowledge at the individual, organizational, national, and international levels and the information-seeking behaviors of aerospace engineers and scientists, but their inclusion is not critical to this research.

Engineers, Scientists, and Managers

This section presents a selected group of characteristics of the U.S. aerospace engineers and scientists that we surveyed. In the Phase 1 studies, we asked slightly different questions to measure various characteristics like occupational self-identification. Some questionnaires contained more than one question about occupational status and primary job activities. All responses to the various questions will not be reported; instead, we will use data selectively to illustrate the various characteristics.

Chapter 5 described the similarities and differences between engineers and scientists. This distinction is perhaps the most important determinant of aerospace knowledge production and use. We expected that the organizational responsibilities affecting the production and use of knowledge would differ for managers and nonmanagers.

We did not assume that current duties and professional identification are the sole determinants of knowledge production and use. We expected that some engineers and scientists had received training in disciplines different from their current occupation and that the training might affect their production and use of knowledge. Although in recent years the engineering curriculum has focused more on science, liberal arts, communications, and business topics, we thought that some basic differences between engineers and scientists in the patterns of knowledge production and use would result from differential training.

In each sample except the AIAA Survey 10, a majority of respondents identified themselves as engineers when asked such a question as "In your present job, do you consider yourself primarily an engineer, a scientist, or something else?" About 67% of the AIAA members, the NASA AST employees, and the SME sample self-identify as engineers (Table 6.5). About 90% of the respondents from the SAE database consider themselves engineers. Only 37% of the respondents in AIAA Survey 10 (i.e., AIAA 1996) classified themselves as engineers. A substantial portion of the AIAA members classify themselves as "other." An examination of the written explanations to that question indicates that most of the "other" responses were listed as "manager." Among the SME sample, those who wrote "other" tend to be engineering technologists. Engineers and scientists are expected to be heavy users of aerospace knowledge, but the amount of knowledge production and use among managers and technologists is not clear. We interpret the data in Table 6.5 to indicate that the variety of occupational positions in aerospace makes it unlikely that either a single or a simple model for knowledge diffusion will be effective.

The data in Table 6.5 also indicate that each of the groups had followed somewhat different career paths. The SAE and the SME samples exhibit the greatest consistency in training and duties. For example, about 90% of the SAE sample consider themselves engineers and were trained as engineers. The proportion of AIAA members and the NASA AST employees who were trained as engineers is substantially larger than the proportion who currently describe themselves as engineers. Among the AIAA 1996 respondents, 77% were trained as engineers but only 37% still identify themselves as engineers. There is an interesting difference between the AIAA members and the NASA ASTs that likely reflects the different tasks each does and the differences between industry and government career paths. Among the AIAA members, many of those trained as engineers now put themselves in the "other" category, which tends to be managers. Among the NASA employees, many of the engineers now consider themselves scientists.

Table 6.5. Training, Current Duties, and Current Responsibilities of Engineers and Scientists

	AIAA (1995)		SAE		SME		NASA		HF&ES		SFTE		AIAA (1996)	
	n	%	n	%	n	%	n	%	n	%	n	%	n	%
Training														
Engineer	1627	83	859	91	329	74	273	80	81	86	63	93	39	74
Scientist	235	12	68	7	11	3	58	17	7	8	2	3	5	9
Other	99	5	19	2	109	23	9	3	6	6	3	4	9	17
Current duties														
Engineer	1325	68	852	90	328	72	234	69	82	85	52	77	12	23
Scientist	168	8	24	3	5	1	92	27	4	4	1	2	1	2
Other	470	24	70	7	127	27	14	4	10	10	15	22	40	76
Responsibilities														
Research	328	17	66	7	12	3	279	82	7	7	0	0	0	0
Management	743	38	146	15	63	14	37	11	5	5	9	13	10	19
Design/development	556	28	569	60	74	15	21	6	53	55	10	15	2	4
Manufacturing/production	20	1	110	12	237	51	---	0	11	12	2	3	0	0
Quality assurance/control	---	---	---	---	---	---	---	---	11	12	---	---	---	---
Flight test	---	---	---	---	---	---	---	---	0	0	1	2	0	0
Marketing/sales	---	---	---	---	---	---	---	---	0	0	36	53	0	0
Teaching	202	10	3	1	---	---	---	0	---	---	2	3	38	72
Other	92	6	52	5	79	17	3	1	4	4	4	6	2	4

Note: --- indicates the response was not available in the question; 0 indicates less than 1% were in the category.

These differences in current self-identification, training, and transitions between training and current duties have implications for the production and use of aerospace knowledge. Survey respondents who were trained according to an engineering model probably use the communications and information use skills that they acquired in college. If their professional duties become those of scientists or managers, their communications and information use skills needs would probably change, although they might continue to use certain skills that they have acquired as engineers. For example, a new position requiring that an engineer do research and publish the findings is likely to change that engineer's perspective of the value and importance of knowledge as his or her need for knowledge expands. Our data suggest the complexity of knowledge needed by U.S. aerospace engineers and scientists and the difficulties of designing an effective system for diffusing aerospace knowledge.

Day-to-Day Responsibilities

We assumed that the primary day-to-day responsibilities of engineers and scientists would affect their technical communication practices, the amounts of information and knowledge that they produce and use, and their information-seeking behaviors. We were especially interested in the attitudes of managers in this regard because we believe that their attitudes would affect the production and use of information by others in their organizations. For example, if a manager does not use the library frequently and, in fact, does not value the library as a source of information, in all likelihood few of that manager's subordinates will use the library. This general attitude, one that devalues information, reduces the use of information throughout an organization.

As we observed with the current duties and training, we see substantial differences among the groups we surveyed in their identification of their primary day-to-day activities and duties (Table 6.5). Over 80% of the NASA ASTs consider themselves researchers. In fact, a substantial proportion of the NASA ASTs surveyed work alone on research projects, research being an activity more closely related to science than engineering. A substantial proportion of the AIAA members are involved in research, and many engineering faculty belong to the AIAA, a finding that implies involvement in research and other scholarly activities. The AIAA members are also most likely to be managers. The AIAA members' duties are varied but can best be described as engineering research and management.

The professional duties of the *Aerospace Engineering* subscribers (SAE) come closest to what we consider typical engineering work. Almost 75% of this group consider design and development and manufacturing and production their primary work activities. Fewer than 25% indicated that research and management are their primary activities. The SME sample, as expected, is composed primarily of manufacturing and production engineers. About 66% of both the AIAA (1996) and the SFTE samples do not identify their daily responsibilities in general engineering terms.

Table 6.5 indicates the diversity of professional day-to-day duties captured by the surveys conducted in Phase 1 of the project. These data confirm that we captured a broad range of aerospace occupations and duties. These data further demonstrate the complexity and diversity of the various aerospace engineering communities and disciplines. The continuing competitive success of U.S. aerospace depends on the successful performance of aerospace engineers and scientists across the spectrum of aerospace R&D from research through marketing and sales.

Education

We also expected that the level of education completed by aerospace engineers and scientists would affect their production and use of aerospace knowledge. Engineers with Ph.D.s are probably very similar to scientists in terms of their information needs and uses; we expected them to have different needs and uses for information than engineers with a bachelor's degree have. In general, we assumed that engineers and scientists with advanced degrees require more information to perform their daily activities than those who have completed less education require. Table 6.6 shows the distribution of the samples by educational levels.

A much higher proportion of NASA ASTs, AIAA (1995 and 1996) member samples than the other samples have completed a degree beyond the bachelor's. About 50% of the NASA ASTs and the AIAA (1996) sample have masters' degrees, and almost 33% of the AIAA (1995) members have completed a doctoral degree. About twice as many engineers with terminal bachelor's degrees appear in the SAE

Table 6.6. Education and Years of Employment of Engineers and Scientists

	AIAA (1995)		SAE		SME		NASA		HF&ES		SFTE		AIAA (1996)	
	n	%	n	%	n	%	n	%	n	%	n	%	n	%
Degree														
Bachelor's	543	27	498	53	213	45	91	27	52	54	38	56	22	42
Master's	774	39	255	27	90	21	158	46	32	33	26	38	26	49
Doctorate	618	31	48	5	6	1	91	27	8	8	3	4	1	2
Other	45	3	145	15	155	33	0	0	4	4	1	2	4	8
Years														
< 1-5	226	12	98	11	66	14	52	15	1	1	1	2	2	4
6-10	293	16	224	24	107	24	74	22	21	22	4	6	10	19
11-20	425	22	227	25	144	31	95	28	43	45	17	25	19	36
21-40	935	49	362	39	140	30	115	34	28	29	44	65	20	38
>40	137	1	18	1	5	1	4	1	3	3	2	3	2	4

sample as in the NASA and AIAA samples. Only about 25% of the NASA ASTs and the AIAA members have bachelor's degrees as the terminal degree. About 33% of the SME sample have not earned a bachelor's degree. For the most part, the SME respondents who answered "other" to the question about education indicated that they had two-year technology degrees.

Since World War II, the requirement for entry into professional engineering has been a bachelor's degree. Advanced degrees usually indicate either management responsibilities requiring an MBA or a master's in engineering management, or more engineering research responsibilities. We assume that engineers who completed their education with a bachelor's degree are doing different types of work than those with advanced degrees. We interpret the data in Table 6.5 to indicate that the AIAA members and the NASA ASTs are more likely to use knowledge that is external to their organizations to produce knowledge and to transfer the newly produced knowledge beyond their immediate organizations, probably because they are involved in engineering science. SAE and SME engineers, on the other hand, are probably more likely to use knowledge that resides within the organization, and they are probably less likely to produce aerospace knowledge and transfer it externally. We suspect the differences in the work responsibilities of the other three samples makes it less clear how much they use both internal and external information. Channel and source selection and use and the relative importance of channels and sources probably differ among the SAE and SME groups of engineers and the AIAA members and the NASA ASTs.

Years Employed in Aerospace

Overall, the U.S. aerospace engineers and scientists whom we studied are fairly evenly distributed by years of professional experience (Table 6.6), but there is considerable diversity across sample groups. Many aerospace engineers and scientists started their professional careers in the post-Sputnik period before 1970. The AIAA members and the SFTE appear to be the oldest groups. About 50% of the AIAA sample and 68% of the SFTE have more than 20 years of work experience in aerospace. The SME and HF&ES samples have the largest proportions with less than 20 years of work experience in aerospace. Except for the SME and HF&ES samples, at least 33% of the aerospace engineers and scientists we sampled are in their mid to late careers (over 20 years employed in aerospace). We expected that the production and use of aerospace knowledge would change over the course of engineers' and scientists' careers, with the greatest needs arising in the early years of a career.

Organization Type

Collectively, most of the U.S. aerospace engineers and scientists whom we sampled are employed in three types of organizations—academia, government, and industry. The Phase 1 AIAA participants were employed in all three types of organizations, making the AIAA sample the most diverse in terms of organizational affiliation. At the time of the first AIAA survey about 14% of the respondents worked in academia, about 25% worked in government, and about 57% worked in industry. Most of the remaining 4% worked in non-profit organizations (Pinelli, 1991). All of the NASA ASTs worked in the government sector. About 90% of the SAE sample and 100% of the SME sample were employed in industry. Participants in all of the 1996 Phase 1 studies were employed in industry.

We expected that the social system of the employing organization, with its values and norms, would influence the production, transfer, and use of aerospace knowledge. For example, we thought that the industry respondents would use knowledge internal to the organization most often and would value it more highly than external knowledge. Academics would use external aerospace knowledge, especially the scholarly and professional journals, to conduct research, prepare teaching materials, and transfer newly produced aerospace knowledge. Government-affiliated engineers and scientists would use a mixture of internal and external aerospace knowledge to produce new aerospace knowledge; that is, they would use scholarly and professional journals to support the engineering science portion of their work and technical reports to document the results of their applied engineering work.

We also assumed that knowledge production, transfer, and use would vary within organizations. In particular, we expected that the size of an organization (i.e., number of people employed) and its mission would affect aerospace knowledge production, transfer, and use. For example, aerospace engineers and scientists in research universities may have different responsibilities for research and teaching than do their counterparts in teaching colleges and universities. Differing responsibilities may affect the amounts and types of aerospace knowledge produced and used in academia by aerospace engineers and scientists. To contend with technical and marketplace uncertainty, large industry aerospace firms may have a greater need than small firms have for knowledge that resides external to the organization. Large aerospace industry firms would be more likely than small aerospace firms to have a knowledge support infrastructure (i.e., a library and information specialist). Further, large aerospace firms are more likely than small firms to have a diverse product line; hence, the aerospace engineers and scientists working in large firms may need more and different types of knowledge than their counterparts working in small firms may need. We also thought that knowledge production, transfer, and use would vary in government. In particular, we assumed that the amounts and types of knowledge produced and used would vary among aerospace engineers and scientists working for the DoD, the FAA, and NASA because of the unique missions assigned to each agency by the U.S. Congress.

When we designed the project, we also assumed that the numbers and types of information products produced and used by aerospace engineers and scientists in academia, government, and industry would differ. Aerospace engineers and scientists in academia produce new knowledge in conjunction with their duties as teachers and researchers. They may use a variety of information products, including technical reports, to produce new knowledge, most of which is diffused through the open literature as conference and meeting papers and journal articles. Government-affiliated aerospace engineers and scientists, directly or indirectly, are also engaged in producing new aerospace knowledge. This new knowledge is diffused as a mixture of information products including conference and meeting papers, journal articles, and technical reports. The goal of engineers and scientists in the aerospace industry is to produce such artifacts as airplanes and spacecraft. For these aerospace professionals, knowledge production is a by-product of their ultimate goal. They would be more likely to produce in-house technical reports and memoranda than conference or meeting papers and journal articles. Finally, we recognize that U.S. aerospace is diverse and that crossover probably occurs among and between the academic, government, and industry sectors in terms of knowledge production, transfer, and use. Hence, imposing a rigid classification scheme would severely limit our ability to study the production, transfer, and use of aerospace knowledge at the individual, organizational, national, and international levels.

Demographics

Demographic characteristics are important because they reflect specific attributes of the sample that can be used to analyze and interpret the study participants' responses in light of such parameters as gender, age, level of education, and years of professional experience. For example, we expected that aerospace engineers and scientists who hold advanced degrees and who work in research would be most likely both to produce and to use aerospace knowledge. We expected that those who identified themselves as scientists would produce and use more aerospace knowledge than those who identified themselves as engineers. We expected that aerospace engineers and scientists performing professional duties in research would produce and use more aerospace knowledge than their counterparts working in design and development, manufacturing and production, and marketing and service would produce and use.

RESEARCH AGENDA

Our research is concerned with the diffusion of knowledge resulting from federally funded aerospace R&D. In developing the research agenda for the *NASA/DoD Aerospace Knowledge Diffusion Research Project*, we identified a number of variables that we believe are central to understanding the diffusion of aerospace knowledge. Our focus for understanding knowledge diffusion is the large commercial aircraft (LCA). Chapter 1 presents an overview of the evolution of large commercial aircraft manufacturing in the U.S. from its beginnings as a collection of small entrepreneurial businesses to its present status as a strategic industry of global proportions characterized by a complex oligopolistic international production structure. U.S. policies, in combination, influence the production and sale of LCA. Chapter 2 explores how U.S. public policy has influenced technological innovation in LCA and the role played by public policy in shaping the innovation and transportation systems within which LCA are produced and operated. U.S. public policy and the production of federally funded aeronautical research and technology (R&T) are examined in Chapter 3. The influence of U.S. public policy on the dissemination of federally funded aeronautical R&T is investigated in Chapter 4. The focus of this research on technology, rather than science, prompted us to study engineers as both producers and users of knowledge. Chapter 5 builds the case for an engineering knowledge community by discussing the differences between science and technology and the similarities and differences between engineers and scientists. In Chapter 5 we also review previous research into the information-seeking behaviors of engineers.

The ability of aerospace engineers and scientists to identify, acquire, and utilize knowledge is of paramount importance to the efficiency of the R&D process. Studies reveal that engineers and scientists devote more time to communicating technical information than to any other professional activity (Mailloux, 1989). Research has also found strong relationships between the production, transfer, and use of knowledge and technical performance at both the individual and the group levels (Hall and Ritchie, 1975; Rubenstein, Barth, and Douds, 1971). A considerable portion of our research concentrates on investigating the communication practices and information-seeking behaviors of aerospace engineers and scientists. Chapter 6 describes the study populations and selected demographic characteristics of the survey participants. Chapter 7 examines the communication practices and the information-seeking behaviors of aerospace engineers and scientists from research to development and production through marketing and service as a critical element in understanding knowledge diffusion. (The complete background and methodology for these studies appear in reports 1 through 6, 13, 14, 20, 24, 31, 33, 34, 36, and 38 through 43. See the Appendix to Chapter 6.)

Tyson (1992) and Mowery (1985) state that the aerospace industry, in particular the commercial aviation sector, is characterized by a high degree of systemic complexity embodied in the design and development of its products. Industries such as aerospace must deal with technical and market uncertainty from outside the organization as well as work-related uncertainty concerning problem solving within the organization (Myers and Marquis, 1969). Miller (1971) states that organizations involved in technological innovation and R&D use internal and external information to reduce and moderate uncertainty. Chapter 8 looks at the relationship between technical uncertainty and information use by aerospace engineers and scientists. (The complete background and methodology for the technical uncertainty and information use studies appear in reports 15, 24, 25, 31, 33, 34, and 36. See the Appendix to Chapter 6.)

The effective communication of aerospace knowledge is critical to the professional success of aerospace engineers and scientists. Feedback from professional engineers and from engineers' supervisors about engineering competencies consistently ranks communication skills *high* in terms of their importance to engineering practice, but this same feedback ranks the communication skills of entry-level engineers *low* (Kimel and Monsees, 1979). Government and industry officials are generally satisfied with the technical knowledge preparation of new hires, but they find that entry-level engineers often lack effective communication skills. In Chapter 9 we examine the career goals, communication practices, and information-seeking behaviors of early-career stage U.S. aerospace engineers and scientists to gain their perspective. A deficiency in the communication skills of entry-level engineers is usually attributed to inadequate or inappropriate instruction at the undergraduate level. In Chapter 10 we investigate the career goals, communication practices, and information-seeking behaviors of U.S. aerospace engineering and science students to gather adequate and generalizable data about the instruction that they receive in communications and information use skills. (The complete background and methodology for the student and early-career stage studies appear in reports 8, 9, 23, 26, 27, 34, and 37. See the Appendix to Chapter 6.) In Chapter 11 we investigate the communication practices and information-seeking behaviors among aerospace engineering professionals and students across transnational boundaries. (The complete background and methodology for the foreign studies appear in reports 16, 17, 18, 25, 28, and 29. See the Appendix to Chapter 6.) In Chapter 12 we study the effects of years of academic work experience on the production and use of information by U.S. aerospace engineering faculty. (The complete background and methodology for this study appear in report 45. See the Appendix to Chapter 6.)

Communication, especially in the form of face-to-face oral communication, is recognized as an effective method of moderating uncertainty (Allen and Cohen, 1969). Therefore, a critical task for those involved in managing innovation and R&D is to develop organizational communications and information processing mechanisms that facilitate and promote the production, transfer, and use of knowledge among members of an organization. Chapter 13 focuses on the communication environment within U.S. aerospace and the influence of analyzability, equivocality, and uncertainty on communication within aerospace organizations. (The complete background and methodology for the communication environment study appear in report 30. See the Appendix to Chapter 6.)

Aerospace organizations, like other organizations involved in innovation and R&D, are making large investments in computers and computer networks to increase productivity, facilitate communication, and improve competitiveness. Federal policymakers are struggling to develop appropriate policies and network systems to ensure that potential users have access to the latest results of federally funded R&D. Little is known, however, about the extent of computer network use within and across aerospace organizations. Even less is known about the use of the results of federally funded R&D that are available in electronic form. Chapter 14 investigates the use of computer networks in aerospace. (The complete background and methodology for the computer network study appear in report 35. See the Appendix to Chapter 6.)

The model that we use to depict the transfer of the results of federally funded aerospace R&D is composed of two parts—an informal part that relies on collegial contacts and a formal part that relies heavily on libraries and librarians as information intermediaries to complete the "producer-to-user" knowledge transfer process. We know little, however, about the role that information intermediaries play in the knowledge diffusion process. A strong methodological basis for measuring or assessing their effectiveness is lacking. In addition, empirical findings on the effectiveness of information intermediaries and the role(s) that they play in knowledge transfer are sparse and inconclusive. Their impact is likely to be strongly conditional and limited to a specific institutional context. Chapter 15 examines the role of U.S. academic and industry libraries and librarians as information intermediaries. (The complete background and methodology for the academic and industry libraries and librarians study appear in reports 10, 21, and 22. See the Appendix to Chapter 6.)

Technical reports published by the DoD and NASA are a primary means by which the results of federally funded aerospace R&D are documented and distributed to U.S. aerospace. According to Auger (1975), the history of technical report literature in the U.S. coincides almost entirely with the development of aeronautics, the aviation industry, and the creation of the National Advisory Committee on Aeronautics (NACA), which issued its first technical report in 1917. However, as McClure (1988, p. 42) points out, "we know very little about the role, importance, and the impact of this literature vis-à-vis the transfer of federally funded R&D, technological innovation, and productivity." In Chapter 16 we report data that provide an empirical basis for understanding the role of the U.S. government technical report in the diffusion of knowledge resulting from federally funded aerospace R&D. (The complete background and methodology for the technical report studies appear in reports 6, 19, 20, and 47. See the Appendix to Chapter 6.)

Because of its high value for a technology based economy, the production of LCA enjoys a special niche in the national innovation systems and domestic and foreign policies of several countries. Chapter 17 examines knowledge diffusion in Western European and public policies affecting LCA. Chapter 18 focuses on Japanese technological innovation, public policies, and knowledge management as it pertains to LCA. Japan takes a diffusion-oriented approach to knowledge and values it as a *private* (rather than a public) economic resource and a key component of the R&D process. Redesigning the current system used to disseminate the results of federally funded aeronautical R&T requires a basic understanding of knowledge itself and its value and importance to the industry. Chapter 19, explores knowledge management as a strategy for bolstering the competitive status of the U.S. aircraft industry.

CONCLUSIONS

The Bureau of Labor Statistics estimated that in 1993 approximately 61 500 aerospace engineers were employed in the United States (Foley, 1995). About 34 000 worked in manufacturing, most in the commercial aviation sector. Of the 18 000 aerospace engineers not employed in manufacturing, most worked in engineering, research, and testing services. About 9000 worked in U.S. government agencies. These numbers do not include engineers whose primary duties were management, scientists employed in aerospace, or academic aerospace engineers and scientists. The National Science Foundation, using a broader definition based on both education and responsibilities, estimates that in 1994 there were about 160 000 aerospace engineers in the U.S. (National Science Foundation, 1995). For this research project, we focused on employed and practicing aerospace engineers and scientists, a number that we estimate is less than 100 000. (See Foley, 1995, for an explanation of why estimates of the numbers of aerospace engineers and scientists differ.)

In the *NASA/DoD Aerospace Knowledge Diffusion Research Project*, we sampled from multiple groups of professionals employed as aerospace engineers and scientists in the U.S.: members of the AIAA, SFTE, HF&SE, ASTM, SAMPE, aerospace engineers and scientists identified by the SAE and the SME, and NASA ASTs. When we drew our samples, the AIAA membership exceeded 32 000 and the SAE mailing list was about 60 000. This broad sampling strategy ensures representation across the spectrum of the U.S. aircraft industry, from research and development through technical services. We believe that combining all of these samples adequately covered the full range of practicing U.S. aerospace engineers and scientists.

The stratified sampling procedures and the multiple surveys make this project unique among studies of technical communication and information-seeking behavior because of the sheer numbers of individuals surveyed and the fact that they represented the spectrum of aerospace R&D from research, design and development, manufacturing and production, service and maintenance, and marketing and sales. The scope of the project, which included aerospace engineers and scientists in all phases of their careers from undergraduate students to early-career stage professionals through seasoned professionals, gives us a comprehensive understanding of their technical communication and information-seeking behaviors as they relate to the diffusion of aerospace knowledge resulting from federally funded R&T. Including the additional elements of our research agenda allows a comprehensive understanding of the social system within which aerospace knowledge diffuses.

Data from the project indicate substantial variety in the demographic characteristics of engineers and scientists employed in U.S. aerospace. The data further support our conclusions from Chapter 5 that there is a continuum of activities in aerospace that range from basic science, through engineering science, to design, development, production, and manufacturing. The data also indicate that the levels of education for aerospace engineers and scientists range from less than bachelor's degrees to Ph.D.s, and that the duties they perform range from those of bench engineers to those of managers.

The data lead us to agree with Derian (1990) that U.S. aerospace is a "sheltered" (as opposed to an exposed) culture. Federal funding is the most important source of research funds for U.S. aerospace engineers and scientists. Of the AIAA respondents who work in industry, 81% had received some level of federal funding. This proportion is substantially larger than the 67% of the academic sample of the AIAA that had received some level of federal funding. Among the AIAA sample, over 75% reported that the federal government had supplied the largest proportion of funds for their current work. About 73% of the respondents from the SAE sample and 41% of the respondents from the SME sample had received federal funding for their work. Most engineering and science professionals employed in aerospace depend on direct or indirect federal funding for their employment. This reliance on federal funding is evidenced by the current cutbacks in defense spending, which are primarily responsible for the recent substantial decrease in aerospace employment. The number of aerospace engineers and scientists employed in industry decreased from about 205 000 in 1991 to an estimated 132 000 in 1993 (Foley, 1995).

The dependence of U.S. aerospace on the federal government for funding creates market conditions for the aerospace industry that differ from those of other industries. In addition to product competition, there is a high degree of competition for research funds. In the aerospace industry, developing successful proposals to design and build new aircraft is often as important as selling the aircraft before it is produced. Engineers in U.S. aerospace reported that they spend substantial amounts of time working on proposals, and many engineers may spend more time writing proposals than doing design work. Proposals that win federal funds are critical to the success of an organization. Producing the next generation of commercially successful aircraft (i.e., a high-speed civil transport) may well determine the commercial viability of some U.S. aircraft companies and, overall, the ability of the large commercial aircraft portion of the U.S. aerospace industry to remain predominant in world markets.

The data also support the assumption that the diffusion of aerospace knowledge resulting from federally funded R&D is indispensible in maintaining the vitality and competitiveness of U.S. aerospace because nearly all of the Phase 1 AIAA survey respondents reported using some form of this knowledge in performing their professional duties. The use of this knowledge declined among the SAE and the SME survey respondents. These data suggest that the current system for disseminating the results of federally funded aerospace R&T does not address the information needs of aerospace engineers and scientists across the full spectrum of aerospace R&D.

IMPLICATIONS FOR DIFFUSING THE RESULTS OF FEDERALLY FUNDED AEROSPACE R&D

Three closely connected implications for diffusing the results of federally funded aerospace R&D emerge. *First*, the diversity of U.S. aerospace engineers and scientists requires that any system used to diffuse the results of federally funded aerospace R&D must be multifaceted. The system must include formal and informal channels and sources. Content-wise, the system must be capable of serving the full spectrum of the U.S. aircraft industry from research, design and development, manufacturing and production, service and maintenance, and marketing and sales. Differences among aerospace engineers and scientists lead us to believe that their production and use of knowledge cannot be assumed to be simple, linear, or consistent across time. In addition, it is not clear to us that it is appropriate to look at knowledge diffusion *only* as the movement of knowledge from the federal government to academia and industry.

The interlinking of academia, government, and industry makes it important to look carefully at the multiple relationships that exist among the three groups before developing *any* plan for aerospace knowledge diffusion. Perhaps the interlinking of the three is best illustrated by Smith (1993) in his discussion of the design of the Hubble Space Telescope:

> The Space Telescope's design and the manner in which it has been built have therefore resulted from international collaboration and astonishingly complex interactions among the scientific community, the government, and industry. (p. 3)

Further, Latour (1987) would argue that in technoscience (as he might describe aerospace), there is a continuum of engineering and science, and most technoscience is primarily engineering. An analysis of the funding of U.S. aerospace, and especially of U.S. aerospace research and technology, would support this claim. Latour (1987) would also argue that to understand technoscience, researchers must observe the activities of those involved in it. In the next chapter, we look further at those activities and see that the diversity reflected in Chapter 7 results in different information use behaviors among aerospace engineers and scientists across their various occupational and demographic categories. Our analysis will show that multiple knowledge communities exist in aerospace. To be successful, an aerospace knowledge diffusion system must meet the differing information needs of the aerospace engineers and scientists residing in multiple knowledge communities.

Second, given that U.S. aerospace is a "sheltered" (as opposed to an exposed) culture, the system should be viewed as a national trust. Considering the close intermingling of academia, government, and industry; the nature of the federal support for aerospace R&D; and the influence of the federal government as a major buyer of aircraft, there is ample justification for the federal government to assume the oversight responsibility for the system in partnership with U.S. aerospace. *Third*, any system used to diffuse aerospace knowledge *must* be designed and developed to accommodate the needs of many different types of users. Consequently, the way in which aerospace engineers and scientists produce and use information becomes critical. An understanding of the demands that they would place on the system and the satisfaction that they derive from their efforts are necessary items for the wise planning and developing of aerospace information policy and systems.

Christopher Hill (1993) of the National Academy of Engineering describes the importance of knowledge diffusion:

> In the old era, only a few large wealthy firms had to be cognizant of the latest in fundamental science and engineering developments to be able to exploit breakthroughs as they occurred. In the new era, however, a greater number of smaller and a wider variety of firms— many of them small and medium sized—need access to a base of rapidly evolving new knowledge in science and technology. (p. 283)

The ability of engineers and scientists in U.S. aerospace to access the results of federally funded aerospace R&D and the ability of the knowledge diffusion system to meet their needs for knowledge will have a tremendous impact on the competitiveness of U.S. aerospace in this new era.

CHAPTER REFERENCES

Adams, J.L. (1991). *Flying Buttresses, Entropy, and O-Rings: The World of an Engineer*. Cambridge, MA: Harvard University Press.

Allen, T.J. (1988). "Distinguishing Engineers From Scientists." In *Managing Professionals in Innovative Organizations*. R. Katz, ed. Cambridge, MA: Ballinger Publishing, 3-18.

Allen, T.J. (1977). *Managing the Flow of Technology: Technology Transfer and the Dissemination of Technological Information Within the R&D Organization*. Cambridge, MA: MIT Press.

Allen, T.J. and S. Cohen. (1969). "Information Flow in R&D Labs." *Administrative Science Quarterly* 14: 12-19.

Auger, C.P. (1975). *Use of Technical Reports Literature*. Hamden, CT: Archon Books.

Baldwin, R.E. and P. Krugman. (1988). "Industrial Policy and International Competition in Wide-Bodied Jet Aircraft." Chapter 3 in *Trade Policy Issues and Empirical Analysis*. R.E. Baldwin, ed. Chicago, IL: University of Chicago Press, 43-77.

Bikson, T.K.; B.E. Quint; and L.L. Johnson. (1984). *Scientific and Technical Information Transfer: Issues and Options*. Washington, DC: National Science Foundation. (Available NTIS; PB-85-150357; also available as Rand Note 2131.)

Bluestone, B.; P. Jordan; and M. Sullivan. (1981). *Aircraft Industry Dynamics: An Analysis of Competition, Capital, and Labor*. Boston, MA: Auburn House.

Bright, C.D. (1978). *The Jet Makers: The Aerospace Industry From 1945 to 1972*. Lawrence, KS: The Regents Press of Kansas.

Caves, R.E. (1962). *Air Transport and Its Regulators: An Industry Study*. Cambridge, MA: Harvard University Press.

David, P.A. (1986). "Technology Diffusion, Public Policy, and Industrial Competitiveness." In *The Positive Sum Strategy: Harnessing Technology for Economic Growth*. R. Landau and N. Rosenberg, eds. Washington, DC: National Academy Press, 373-391.

Derian, J.C. (1990). *America's Struggle For Leadership in Technology*. Cambridge, MA: MIT Press.

Dillman, D.A. (1978). *Mail and Telephone Surveys: The Total Design Method*. New York, NY: John Wiley.

Downey, G.L. (1992). "CAD/CAM Saves the Nation? Toward an Anthropology of Technology." *Knowledge & Society* 9: 143-168.

Downey, G.L. and J.C. Lucena. (1995). "Engineering Studies." Chapter 8 in *Handbook of Science and Technology Studies*. S. Jasanoff; G.E. Markle; J.C. Petersen; and T. Pinch, eds. Newbury Park, CA: Sage Publications, 167-188.

Dupree, A.H. (1986). *Science in the Federal Government: A History of Policies and Activities*. Baltimore, MD: Johns Hopkins University Press.

Foley, T.M. (1995). "Finding a Job in Aerospace." *Aerospace America*. (March) 33(3): 35-41.

Gellman Research Associates. (1990). *An Economic and Financial Review of Airbus Industrie*. Washington, DC: U.S. Department of Commerce, International Trade Administration. (Available NTIS; PB90-243817.)

Golich, V.L. (1992). "From Competition to Collaboration: The Challenge of Commercial-Class Aircraft Manufacturing." *International Organization* (Autumn) 46(4): 900-934.

Golich, V.L. (1989a). *The Political Economy of International Air Safety: Design for Disaster?* New York, NY: St. Martin's Press.

Golich, V.L. (1989b). *The Political Economy of International Commercial Aviation*. London: Macmillan Press.

Hall, G.R. and R.E. Johnson. (1970). "Transfers of United States Aerospace Technology to Japan." In *The Technology Factor in International Trade*. R. Vernon, ed. New York, NY: Columbia University Press, 305-358.

Hall, K.R. and E. Ritchie. (1975). "A Study of Communication Behavior in an R&D Laboratory." *R&D Management* 5: 243-245.

Hayward, K. (1994). *The World Aerospace Industry: Competition and Collaboration*. London, UK: Royal United Services Institute Whitehall Papers/Duckworth.

Hernon, P. and C.R. McClure. (1984). *Public Access to Government Information: Issues, Trends, and Strategies*. Norwood, NJ: Ablex Press.

Hill, C.T. (1993). "New Manufacturing Paradigms: New Manufacturing Policies." In *Science and Technology Policy Yearbook*. A. Teich, S. Nelson, and C. McEnaney, eds. Washington, DC: American Association for the Advancement of Science, 267-284.

Hillis, W.D.; D.F. Burton; R.B. Costello; R.M. White; M. Weidenbaum; L. Georghiou; U. Columbo; L. Schnieder; T.H. Lee; and J.F. Gorte. (1992). "Technology Policy: Is America on the Right Track?" *Harvard Business Review* (May/June) 70(3): 140-157.

Hochmuth, M.S. (1974). "Aerospace." Chapter 8 in *Big Business and the State*. R. Vernon, ed. Cambridge, MA: Harvard University Press, 145-169.

Holmfeld, J.D. (1970). *Communication Behavior of Scientists and Engineers*. Ph.D. Diss., Case Western Reserve University. UMI 70-25874.

Jordan, W.A. (1979). *Airline Regulation in America: Effects and Imperfections*. Westport, CT: Greenwood Press.

Kemper, J.D. (1990). *Engineers and Their Profession*. 4th ed. Philadelphia, PA: Saunders College Publishing.

Kennedy, J.M. and T.E. Pinelli. (1990). "The Impact of a Sponsor Letter on Mail Survey Response Rates." Paper presented at the Annual Meeting of the American Association for Public Opinion Research, Lancaster, PA. (Available NTIS: 92N28112.)

Kimel, W.R. and M.E. Monsees. (1979). "Engineering Graduates: How Good Are They?" *Engineering Education* (November) 70(2): 210-212.

Latour, B. (1987). *Science in Action: How to Follow Scientists and Engineers Through Society*. Cambridge, MA: Harvard University Press.

Layton, E.T. (1971). "Mirror Image Twins: The Communities of Science and Technology in 19th Century America." *Technology & Culture* 12: 562-580.

Logsdon, J.M. (1986). "Federal Policies Towards Civilian Research and Development: A Historical Overview." In *Technological Innovation: Strategies for a New Partnership*, D.O. Gray; T. Solomon; and W. Hetzner, eds. New York, NY: North-Holland, 9-26.

Lopez, V.C. and D.H. Vadas. (1993). *The U.S. Civil Aircraft Industry. Can it Retain Leadership?* Washington, DC: Aerospace Industries Association of America, Inc.

Mailloux, E.N. (1989). "Engineering Information Systems." Chapter 6 in *Annual Review of Information Science and Technology*. 25 M.E. Williams, ed. Amsterdam, The Netherlands: Elsevier Science Publishers, 239-266.

McClure, C.R. (1988). "The Federal Technical Report Literature: Research Needs and Issues." *Government Information Quarterly* 5(1): 27-44.

Miller, R.E. (1971). *Innovations, Organizations, and Environment: A Study of Sixteen American and West European Steel Firms*. Sherbrooke, PQ, Canada: University of Sherbrooke Press.

Mowery, D.C. (1985). "Federal Funding of R&D in Transportation: The Case of Aviation." Paper commissioned for a workshop on *The Federal Role in Research and Development* sponsored by the National Academy of Sciences, National Academy of Engineering, and Institute of Medicine in Washington, DC, 21-22 November.

Mowery, D.C. (1983). "Economic Theory and Government Technology Policy." *Policy Sciences* 16: 27-43.

Mowery, D.C. and N. Rosenberg. (1985). *The Japanese Commercial Aircraft Industry Since 1945: Government Policy, Technical Development, and Industrial Structure*. An Occasional Paper of the Northeast Asia: United States Forum on International Policy. Stanford, CA: Stanford University.

Mowery, D.C. and N. Rosenberg. (1982). "The Commercial Aircraft Industry." Chapter 3 in *Government and Technical Progress: A Cross-Industry Analysis*. R.R. Nelson, ed. New York, NY: Pergamon Press, 101-161.

Myers, S. and D.G. Marquis. (1969). *Successful Industrial Innovation: A Study of Factors Underlying Innovation in Selected Firms*. NSF 69-17. Washington, DC: National Science Foundation.

National Science Foundation. (1995). *Nonacademic Scientists and Engineers: Trends from the 1980 and 1990 Censuses.* NSF95-306. Washington, DC: Government Printing Office.

Pinelli, T.E. (1991). *The Relationship Between the Use of U.S. Government Technical Reports by U.S. Aerospace Engineers and Scientists and Selected Institutional and Sociometric Variables.* NASA TM-102774. Washington, DC: National Aeronautics and Space Administration. (Available NTIS: 91N18898.)

Price, D.J. de Solla. (1965). "Is Technology Historically Independent of Science?" *Technology & Culture* (Summer) 6(3): 553-578.

Roberts, E.B. and A.L. Frohman. (1978). "Strategies for Improving Research Utilization." *Technology Review* (March/April) 80(5): 32-39.

Rubenstein, A.H.; R.T. Barth; and C.F. Douds. (1971). "Ways to Improve Communications Between R&D Groups." (November) 14(6): 49-59.

Smith, B.L.R. (1990). *American Science Policy Since World War II.* Washington, DC: The Brookings Institute.

Smith, R.W. (1993). *The Space Telescope: A Study of NASA, Science, Technology, and Politics.* New York, NY: Cambridge University Press.

Tornatzky, L.G. (1990). "Issues of Science and Technology Policy." Chapter 10 in *The Processes of Technological Innovation.* L.G. Tornatzky and M. Fleischer, eds. Lexington, MA: Lexington Books, 235-258.

Tyson, L. (1992). *Who's Bashing Whom? Trade Conflict in High-Technology Industries.* Washington, DC: Institute for International Economics.

U.S. Congress, Office of Technology Assessment. (1991). "Government Support of the Large Commercial Aircraft Industries of Japan, Europe, and the United States." Chapter 8 in *Competing Economies: America, Europe, and the Pacific Rim.* OTA-ITE-498. Washington, DC: Government Printing Office. (Available NTIS: PB92-115575.)

U.S. International Trade Commission. (1993). *Global Competitiveness of U.S. Advanced-Technology Manufacturing Industries: Large Civil Aircraft.* Publication 2667. Washington, DC: U.S. International Trade Commission.

Vincenti, W.G. (1990). *What Engineers Know and How They Know It: Analytical Studies From Aeronautical History.* Baltimore, MD: Johns Hopkins University Press.

Williams, F. and D.V. Gibson. (1990). *Technology Transfer: A Communication Perspective.* Newbury Park, CA: Sage Publications.

Appendix. NASA/DoD Aerospace Knowledge Diffusion Research Project:
Report Publications List

No.	Report Title

1A Pinelli, T.E.; M. Glassman; W.E. Oliu; and R.O. Barclay. (1989). *Technical Communications in Aerospace: Results of Phase 1 Pilot Study.* Washington, DC: National Aeronautics and Space Administration. NASA TM-101534. 106 p. (Available NTIS: 89N26772.)

1B Pinelli, T.E.; M. Glassman; W.E. Oliu; and R.O. Barclay. (1989). *Technical Communications in Aerospace: Results of a Phase 1 Pilot Study.* Washington, DC: National Aeronautics and Space Administration. NASA TM-101534. 83 p. (Available NTIS: 89N26773.)

2 Pinelli, T.E.; M. Glassman; W.E. Oliu; and R.O. Barclay. (1989). *Technical Communications in Aerospace: Results of Phase 1 Pilot Study: An Analysis of Managers' and Nonmanagers' Responses.* Washington, DC: National Aeronautics and Space Administration. NASA TM-101625. 58 p. (Available NTIS: 90N11647.)

3 Pinelli, T.E.; M. Glassman; W.E. Oliu; and R.O. Barclay. (1989). *Technical Communications in Aerospace: Results of Phase 1 Pilot Study: An Analysis of Profit Managers' and Nonprofit Managers' Responses.* Washington, DC: National Aeronautics and Space Administration. NASA TM-101626. 71 p. (Available NTIS: 90N15848.)

4 Pinelli, T.E.; J.M. Kennedy; and T.F. White. (1991). *Summary Report to Phase 1 Respondents.* Washington, DC: National Aeronautics and Space Administration. NASA TM-102772. 8 p. (Available NTIS: 91N17835.)

5 Pinelli, T.E.; J.M. Kennedy; and T.F. White. (1991). *Summary Report to Phase 1 Respondents Including Frequency Distributions.* Washington, DC: National Aeronautics and Space Administration. NASA TM-102773. 53 p. (Available NTIS: 91N20988.)

6 Pinelli, T.E. (1991). *The Relationship Between the Use of U.S. Government Technical Reports by U.S. Aerospace Engineers and Scientists and Selected Institutional and Sociometric Variables.* Washington, DC: National Aeronautics and Space Administration. NASA TM-102774. 350 p. (Available NTIS: 91N18898.)

(continued)

Appendix (Continued)

No.	Report Title

7   Pinelli, T.E.; J.M. Kennedy; and T.F. White. (1991). *Summary Report to Phase 2 Respondents Including Frequency Distributions*. Washington, DC: National Aeronautics and Space Administration. NASA TM-104063. 42 p. (Available NTIS; 91N22931.)

8   Pinelli, T.E.; J.M. Kennedy; and T.F. White. (1991). *Summary Report to Phase 3 Faculty and Student Respondents*. Washington, DC: National Aeronautics and Space Administration. NASA TM-104085. 8 p. (Available NTIS; 91N24943.)

9   Pinelli, T.E.; J.M. Kennedy; and T.F. White. (1991). *Summary Report to Phase 3 Faculty and Student Respondents Including Frequency Distributions*. Washington, DC: National Aeronautics and Space Administration. NASA TM-104086. 42 p. (Available NTIS; 91N25950.)

10 Pinelli, T.E.; J.M. Kennedy; and T.F. White. (1991). *Summary Report to Phase 3 Academic Library Respondents Including Frequency Distributions*. Washington, DC: National Aeronautics and Space Administration. NASA TM-104095. 42 p. (Available NTIS; 91N33013.)

11 Pinelli, T.E.; M. Henderson; A.P. Bishop; and P. Doty. (1992). *Chronology of Selected Literature, Reports, Policy Instruments, and Significant Events Affecting Federal Scientific and Technical Information (STI) in the United States: 1945-1990*. Washington, DC: National Aeronautics and Space Administration. NASA TM-101662. 130 p. (Available NTIS; 92N17001.)

12 Glassman, N.A. and T.E. Pinelli. (1992). *An Initial Investigation Into the Production and Use of Scientific and Technical Information (STI) at Five NASA Centers: Results of a Telephone Survey*. Washington, DC: National Aeronautics and Space Administration. NASA TM-104173. 80 p. (Available NTIS; 92N27170.)

13 Pinelli, T.E. and N.A. Glassman. (1992). *Source Selection and Information Use by U.S. Aerospace Engineers and Scientists: Results of a Telephone Survey*. Washington, DC: National Aeronautics and Space Administration. NASA TM-107658. 27 p. (Available NTIS; 92N33299.)

(continued)

Appendix (Continued)

No.	Report Title

14 Pinelli, T.E.; J.M. Kennedy; and T.F. White. (1992). *Engineering Work and Information Use in Aerospace: Results of a Telephone Survey.* Washington, DC: National Aeronautics and Space Administration. NASA TM-107673. 25 p. (Available NTIS; 92N34233.)

15 Pinelli, T.E.; N.A. Glassman; L.O. Affelder; L.M. Hecht and J.M. Kennedy; and R.O. Barclay. (1993.) *Technical Uncertainty and Project Complexity as Correlates of Information Use by U.S. Industry-Affiliated Aerospace Engineers and Scientists: Results of an Exploratory Investigation.* Washington, DC: National Aeronautics and Space Administration. NASA TM-107693. 68 p. (Available NTIS; 94N17291.)

16 Pinelli, T.E.; J.M. Kennedy; and R.O. Barclay. (1993). *A Comparison of the Technical Communications Practices of Russian and U.S. Aerospace Engineers and Scientists.* Washington, DC: National Aeronautics and Space Administration. NASA TM-107714. 56 p. (Available NTIS; 93N18160.)

17 Barclay, R.O.; T.E. Pinelli; and J.M. Kennedy. (1993). *A Comparison of the Technical Communication Practices of Dutch and U.S. Aerospace Engineers and Scientists.* Washington, DC: National Aeronautics and Space Administration. NASA TM-108987. 69 p. (Available NTIS; 94N11352.)

18 Pinelli, T.E.; R.O. Barclay; and J.M. Kennedy. (1993). *A Comparison of the Technical Communication Practices of Aerospace Engineers and Scientists in India and the United States.* Washington, DC: National Aeronautics and Space Administration. NASA TM-109006. 68 p. (Available NTIS; 94N13057.)

19 Pinelli, T.E.; R.O. Barclay; and J.M. Kennedy. (1994). *The U.S. Government Technical Report and the Transfer of Federally Funded Aerospace R&D: An Analysis of Five Studies.* Washington, DC: National Aeronautics and Space Administration. NASA TM-109061. 114 p. (Available NTIS; 94N24660.)

(continued)

Appendix (Continued)

No.	Report Title

20 Pinelli, T.E.; R.O. Barclay; and J.M. Kennedy. (1994). *The Use of Selected Information Products and Services by U.S. Aerospace Engineers and Scientists: Results of Two Surveys.* Washington, DC: National Aeronautics and Space Administration. NASA TM-109022. 61 p. (Available NTIS; 94N24649.)

21 Pinelli, T.E.; R.O. Barclay; and J.M. Kennedy. (1994). *U.S. Aerospace Industry Librarians and Technical Information Specialists as Information Intermediaries: Results of the Phase 2 Survey.* Washington, DC: National Aeronautics and Space Administration. NASA TM-109064. 65 p. (Available NTIS; 94N24709.)

22 Pinelli, T.E.; R.O. Barclay; and J.M. Kennedy. (1994). *U.S. Academic Librarians and Technical Information Specialists as Information Intermediaries: Results of the Phase 3 Survey.* Washington, DC: National Aeronautics and Space Administration. NASA TM-109067. 61 p. (Available from NTIS; 94N30150.)

23 Pinelli, T.E.; R.O. Barclay; and J.M. Kennedy. (1994). *The Communications Practices of U.S. Aerospace Engineering Faculty and Students: Results of the Phase 3 Survey.* Washington, DC: National Aeronautics and Space Administration. NASA TM-109085. 55 p. (Available from NTIS; 94N30149.)

24 Pinelli, T.E.; R.O. Barclay; and J.M. Kennedy. (1994). *The Technical Communications Practices of U.S. Aerospace Engineers and Scientists: Results of the Phase 1 SAE Mail Survey.* Washington, DC: National Aeronautics and Space Administration. NASA TM-109088. 52 p. (Available from NTIS; 94N32837.)

25 Pinelli, T.E.; R.O. Barclay; and J.M. Kennedy. (1994). *The Technical Communications Practices of British Aerospace Engineers and Scientists: Results of the Phase 4 RAeS Mail Survey.* Washington, DC: National Aeronautics and Space Administration. NASA TM-109098. 45 p. (Available from NTIS; 94N32836.)

(continued)

Appendix (Continued)

No.	Report Title

26 Pinelli, T.E.; L.M. Hecht; R.O. Barclay; and J.M. Kennedy. (1994). *The Technical Communication Practices of Aerospace Engineering Students: Results of the Phase 3 AIAA National Student Survey.* Washington, DC: National Aeronautics and Space Administration. NASA TM-109121. 90 p. (Available NTIS; 95N18950.)

27 Pinelli, T.E.; R.O. Barclay; L.M. Hecht; and J.M. Kennedy. (1994). *The Technical Communication Practices of Engineering and Science Students: Results of the Phase 3 Survey Conducted at the University of Illinois.* Washington, DC: National Aeronautics and Space Administration. NASA TM-109122. 89 p. (Available NTIS; 9521015.)

28 Pinelli, T.E.; L.M. Hecht; R.O. Barclay; and J.M. Kennedy. (1994). *The Technical Communication Practices of Aerospace Engineering and Science Students: Results of the Phase 4 Cross-National Surveys.* Washington, DC: National Aeronautics and Space Administration. NASA TM-109123. 78 p. (Available NTIS; 95N20162.)

29 Pinelli, T.E.; R.O. Barclay; and J.M. Kennedy. (1994). *The Technical Communications Practices of Japanese and U.S. Aerospace Engineers and Scientists.* Washington, DC: National Aeronautics and Space Administration. NASA TM-109164. 53 p. (Available NTIS; 95N18953.)

30 Murphy, D.J. (1994). *Computer-Mediated Communication (CMC) and the Communication of Technical Information in Aerospace.* Washington, DC: National Aeronautics and Space Administration. NASA CR-194973. 284 p. (Available NTIS; 95N15585.)

31 Pinelli, T.E.; R.O. Barclay; and J.M. Kennedy. (1994). *The Technical Communication Practices of U.S. Aerospace Engineers and Scientists: Results of the Phase 1 SME Mail Survey.* Washington, DC: National Aeronautics and Space Administration. NASA TM-109169. 51 p. (Available NTIS; 95N20163.)

(continued)

Appendix (Continued)

No.	Report Title

32 Pinelli, T.E.; R.O. Barclay; and J.M. Kennedy. (1994). *Descriptive Findings and Their Implications for Diffusing the Results of NASA Aeronautical research and Development.* Washington, DC: National Aeronautics and Space Administration. NASA TM-109167. (Not available from NTIS.)

33 Pinelli, T.E.; R.O. Barclay; and J.M. Kennedy. (1995). *The Technical Communication Practices of U.S. Aerospace Engineers and Scientists: Results of the Phase 1 AIAA Mail Survey.* Washington, DC: National Aeronautics and Space Administration. NASA TM-110180. 51 p. (Available NTIS 95N34217.)

34 Pinelli, T.E.; R.O. Barclay; and J.M. Kennedy. (1995). *Technical Communications in Aerospace: How Early Career-Stage Engineers and Scientists Obtain Information.* Washington, DC: National Aeronautics and Space Administration. NASA TM-110181. 49 p. (Available NTIS 95N10999.)

35 Bishop, A.P. (1995). *The Use of Computer Networks in Aerospace Engineering.* Washington, DC: National Aeronautics and Space Administration. NASA CR-198170. 402 p. (Available NTIS; 95N42860.)

36 Pinelli, T.E.; R.O. Barclay; and J.M. Kennedy. (1995). *The Technical Communications Practices of U.S. Aerospace Engineers and Scientists: Results of the Phase 1 NASA Langley Research Center Mail Survey.* Washington, DC: National Aeronautics and Space Administration. NASA TM-110208. (Available NTIS; 95N18066.)

37 Morrison, E.W. and T.E. Pinelli. (1996). *Factors Motivating and Impeding Information-Seeking By Early Career-Stage U.S. Aerospace Engineers and Scientists—Results of an Initial Investigation.* Washington, DC: National Aeronautics and Space Administration. NASA TM-110213. 37 p. (Available NTIS 96N33991.)

38 Pinelli, T.E.; R.O. Barclay; and J.M. Kennedy. (1996). *The Technical Communication Practices of U.S. Aerospace Engineers and Scientists: Results of the Phase 1 Mail Survey—Flight Test Engineers Perspective.* Washington, DC: National Aeronautics and Space Administration. NASA TM-110231. 51 p. (Available NTIS; 97N10135.)

(continued)

Appendix (Continued)

No.	Report Title

39 Pinelli, T.E.; R.O. Barclay; and J.M. Kennedy. (1996). *The Technical Communication Practices of U.S. Aerospace Engineers and Scientists: Results of the Phase 1 Mail Survey—Avionics and Electrical Components and Subsystems Perspective.* Washington, DC: National Aeronautics and Space Administration. NASA TM-110232. 51 p. (Available NTIS; 97N10134.)

40 Pinelli, T.E.; R.O. Barclay; and J.M. Kennedy. (1996). *The Technical Communication Practices of U.S. Aerospace Engineers and Scientists: Results of the Phase 1 Mail Survey—Human Factors and Crew Integration Perspective.* Washington, DC: National Aeronautics and Space Administration. NASA TM-110233. 51 p. (Available NTIS; 97N10133.)

41 Pinelli, T.E.; R.O. Barclay; and J.M. Kennedy. (1996). *The Technical Communication Practices of U.S. Aerospace Engineers and Scientists: Results of the Phase 1 Mail Survey—Propulsion and Aircraft Engine Perspective.* Washington, DC: National Aeronautics and Space Administration. NASA TM-110234. 51 p. (Available NTIS; 97N10130.)

42 Pinelli, T.E.; R.O. Barclay; and J.M. Kennedy. (1996). *The Technical Communication Practices of U.S. Aerospace Engineers and Scientists: Results of the Phase 1 Mail Survey—Aircraft Design Perspective.* Washington, DC: National Aeronautics and Space Administration. NASA TM-110235. 51 p. (Available NTIS; 97N10132.)

43 Pinelli, T.E.; R.O. Barclay; and J.M. Kennedy. (1996). *The Technical Communication Practices of U.S. Aerospace Engineers and Scientists: Results of the Phase 1 Mail Survey—Manufacturing and Production Perspective.* Washington, DC: National Aeronautics and Space Administration. NASA TM-110236. 51 p. (Available NTIS; 97N10138.)

44 Pinelli, T.E.; R.O. Barclay; and J.M. Kennedy. (1996). *The Technical Communication Practices of U.S. Aerospace Engineers and Scientists: Results of the Phase 1 Mail Survey—Structures and Materials Perspective.* Washington, DC: National Aeronautics and Space Administration. NASA TM-110237. 51 p. (Available NTIS; 97N10131.)

(continued)

Appendix (Concluded)

No.	Report Title

45 Pinelli, T.E.; R.O. Barclay; and J.M. Kennedy. (1996). *The Technical Communication Practices of U.S. Aerospace Engineers and Scientists: Results of the Phase 3 U.S. Aerospace Engineering Educators Survey.* Washington, DC: National Aeronautics and Space Administration. NASA TM-110238. 50 p. (Available NTIS; 96N31050.)

46 Pinelli, T.E.; R.O. Barclay; and J.M. Kennedy. (1996). *The Technical Communication Practices of U.S. Aerospace Engineers and Scientists: Results of the Phase 1 Mail Survey—Service/Maintenance and Marketing/Sales Perspective.* Washington, DC: National Aeronautics and Space Administration. NASA TM-110268. 63 p. (Available NTIS; 9710145N.)

47 Pinelli, T.E.; R.O. Barclay; and J.M. Kennedy. (1996). *Survey of Reader Preferences Concerning the Format of NASA Langley-Authored Technical Reports.* Washington, DC: National Aeronautics and Space Administration. NASA TM-110253. 47 p. (Available NTIS; 96N33259.)

No.	Dissertation Title

1 Bishop, A.P. (1995). *The Use of Computer Networks in Aerospace Engineering.* Ph.D. Diss. Syracuse University. UMI 95-44896.

2 Manuel, D.M. (1993). *Technology Transfer: Techniques and Approaches in the Scientific Community.* Ph.D. Diss. The Claremont Graduate School. UMI 93-30355.

3 Murphy, D.J. (1994). *Computer-Mediated Communication and the Communication of Technical Information in Aerospace.* Ph.D. Diss. Rensselaer Polytechnic Institute. UMI 95-11160.

4 Pinelli, T.E. (1990). *The Relationship Between the Use of U.S. Government Technical Reports by U.S. Aerospace Engineers and Scientists and Selected Institutional and Sociometric Variables.* Ph.D. Diss. Indiana University. UMI 91-19750.

5 Sato, Y. (1998). *Culture, Social Networks, and Information Sharing: An Exploratory Study of Japanese Aerospace Engineers' Information-Seeking Processes and Habits.* Ph.D. Diss. University of Pittsburgh. UMI Pending.

Chapter 7

The Production and Use of Information by U.S. Aerospace Engineers and Scientists—From Research through Production to Technical Services

John M. Kennedy
Thomas E. Pinelli
Rebecca O. Barclay

SUMMARY

Chapter 7 describes the communication practices and information-related activities of U.S. aerospace engineers and scientists who represent the spectrum of aerospace from research through development and production to technical services. Data were collected from 10 mail (self-reported) surveys conducted as a Phase 1 activity of the *NASA/DoD Aerospace Knowledge Diffusion Research Project* in 1995 and 1996. Survey participants included members of seven professional societies and NASA aerospace technologists. Questions covered such topics as the number of hours spent communicating information, the kinds and amounts of information produced and used, the use of libraries and computer networks, the strategies employed in seeking information, the reasons for using specific information products, and the ratings of specific information products. We also investigated aerospace engineers' and scientists' use of knowledge resulting from federally funded research and development (R&D). An analysis of the results indicates a diversity of communication practices and information-seeking behaviors among survey respondents and across the spectrum of U.S. aerospace. In a separate study, two surveys were used to collect data about aerospace engineers' and scientists' use of the bibliographic products designed to provide awareness of and linkages to the results of federally funded R&D. The results indicate that respondents seldom use these bibliographic products. The chapter closes by presenting implications for diffusing the results of federally funded (U.S.) aerospace R&D.

INTRODUCTION

The research, development, and production (RD&P) of (subsonic) large commercial aircraft (LCA) is a complex process spanning several years that requires significant capital investment and the efforts of thousands of engineers, skilled workers, and technical specialists. The RD&P of LCA encompasses the integration of numerous systems and subsystems and literally thousands of components. In many respects RD&P is a compromise between what is feasible (technology and innovation) and what is required or desired by the marketplace (user requirements). According to Sabbagh (1996), the Boeing 777 represents a paradigm shift away from a technology-driven, innovation-based strategy to one that focuses on market and user requirements. Knowledge is critical to the RD&P of LCA. In the *historic* model, technology and innovation (i.e., new knowledge) were used to design LCA whose technical performance, in terms of speed, range, and fuel efficiency, significantly exceeded that of its predecessors. The price of the LCA was dictated by the cost of the technology embodied in its RD&P.

According to Rubbert (1994), the *new* model is "market or customer" driven. Knowledge takes on a different focus in the new model; technology and innovation are used to improve the reliability, dependability, safety, and overall economic performance of LCA, the parameters of which are determined by the marketplace. From a "balanced scorecard" approach, technological advances that reduce fuel consumption but add to the cost of design, fabrication, installation, and in-service maintenance are unacceptable to the user. In the new model, LCA manufacturers face the challenge of applying knowledge to design and build aircraft that deliver the best in economic performance (i.e., the ability of LCA to do their job at less overall cost to the airline and to the traveling public) and to deliver them when the marketplace wants and needs them.

The new model for LCA RD&P requires fundamental changes in organization and management. Chief among the changes is the use of knowledge as "intellectual capital." Knowledge is central to innovation and crucial to the technical performance of U.S. aerospace engineers and scientists. Managed wisely and responsibly, knowledge can provide the U.S. aircraft industry with a competitive advantage. Yet little is known about the process of knowledge diffusion, how organizational and workplace culture influence knowledge diffusion, and how the results of federally funded R&D diffuse throughout U.S. aerospace. At the most basic level, we know very little about the communication practices and information-related activities of U.S. aerospace engineers and scientists that play a key role in the production, transfer, and use of knowledge. Such an understanding, we assert, is essential for the wise planning of aerospace information systems and policy. In this chapter, we report on the communication practices and information-related activities of U.S. aerospace engineers and scientists who represent the spectrum of aerospace from research through development and production to technical services. We also report data about their use of the bibliographic products designed to provide awareness of and linkages to the results of federally funded aerospace R&D.

BACKGROUND

Our research assumes that the diffusion of knowledge resulting from federally funded aerospace R&D is indispensable in maintaining the vitality and international competitiveness of U.S. aerospace. Scholars often cite federal involvement in aerospace as a model for government intervention in civilian R&D and precommercial research cooperation between industry and government. Key to the success of the U.S. government in aerospace is the application of a policy framework that recognizes the relationship between knowledge production, transfer, and use as equally important components of innovation and includes a system that effectively transmits and encourages the utilization of complex research results (Mowery, 1983).

Implications for Diffusing the Results of Federally Funded R&D

An examination of federal involvement in aerospace suggests several points that are pertinent to the larger issue of government involvement in precommercial research. *First*, any attempt at intervention and stimulation of civilian R&D must take into account the unique characteristics of the industry in question, its previous experience with the federal government, and its abilities and limitations to exploit the results of extramural research. The market driven nature of aerospace exerts substantial pressure to innovate as a means of remaining competitive in a global economy. Although aerospace companies devote sizable resources to experimenting with, screening, and adapting new technology to their own specific needs, few can afford to build and operate the facilities and to invest in long term, high risk R&D. Thus, they are ideal clients for a federally funded R&D program that produces new knowledge in the form of technology and innovation, works on specific discipline-related problems, and makes the results available to the companies. *Second*, the structure of the industry, which is the presumed beneficiary of the government's largess, must lend itself to taking advantage of the research results, the leaders of the industry must be interested in and support the research programs, and the government-industry relationship must be based on mutual trust. The aerospace industry and the U.S. government have an established relationship that dates back more than 70 years. *Third*, careful attention needs to be given to maintaining a balance between user (industry) needs and the government's capabilities and interests. The availability of research results is not sufficient to ensure their utilization and application. In aerospace, advisory committees and panels composed of industry and academic leaders are used to help ensure that the research undertaken is relevant and desirable and has high potential payoff. *Fourth*, there must be a system for coupling the producers and users with the research results. Aerospace has had a formal program for disseminating the results of federally funded R&D since the establishment of the National Advisory Committee for Aeronautics (NACA). Within aerospace, the U.S. government technical report is used as a primary means of transferring the results of this research to the user community.

The Existing Structure, Organization, and Management of Knowledge Resulting from Federally Funded R&D

The existing structure, organization, and management of knowledge resulting from federally funded R&D may account for its underutilization. We contend that the *existing* system does not take into account the characteristics of individuals most likely to use the results of federally funded R&D. The majority of these individuals are engineers, not scientists. R&D and technological innovation are knowledge-intensive activities that are embedded within a larger problem-solving process (Vincenti, 1990). The fact that this knowledge may be physically or hardware encoded does not alter the fact that R&D and technological innovation are fundamentally and foremost information-processing activities (Allen, 1977). Engineers, unlike scientists, work within time constraints; they are seldom as interested in theory, source data, and guides to the literature (a fact supported later in this chapter) as much as they are in reliable answers to specific questions. What engineers usually want is a specific answer, in terms and format that are intelligible to them, not a collection of documents that they have to sift, evaluate, and translate before they can apply them. Their search for information seems to be based more on the need for a specific problem solution than around a search for general opportunity (Pinelli, 1991).

In the existing system, knowledge is seldom organized and structured according to problem relevance (i.e., user context) because it is packaged and transferred by producers. From the user's perspective, putting knowledge to work frequently requires repackaging it in a context or format that is quite different from how it was produced or originally packaged. The fact that knowledge is frequently organized along traditional disciplinary lines, as are subject matter indexes, abstracts, and key words, often frustrates potential users. This organizational scheme makes multidisciplinary utilization of knowledge difficult for users and intermediaries alike. Contemporary storage and retrieval systems may exacerbate problems of the existing structure, organization, and management of knowledge. In fact, these systems may now be contributing to the very problems they were designed to solve. Lancaster (1991) states that although technological advances have undoubtedly increased physical access to sources of information, it is doubtful that intellectual access has increased significantly, if at all. Although advances in computer and information technology may provide greater access to available knowledge, they do not provide an effective means of filtering it in terms of quality or problem relevance. Continued growth in the amount of knowledge that can be stored and retrieved is not the only problem confronting users of storage and retrieval systems, however. Contemporary storage and retrieval systems, according to Parsaye (1989), "are primitive and prevent the full utilization of the information stored therein." These systems do not inform (i.e., change the knowledge of) users on the subject of their inquiry. They merely inform users on the existence (or nonexistence) and whereabouts of information packages related to their request. For the most part, these systems do not retrieve information; rather, they retrieve citations. Bibliographic citations seldom reflect the rich network of interrelationships that exist in R&D and technological innovation.

The Need for a Diffusion-Oriented System

The *new* system, in which the structure, organization, and management of knowledge change dramatically, represents a "striking break with the past" (Dougherty, 1990). Criticism of the existing system notwithstanding, a new system for organizing, managing, and transferring the knowledge resulting from federally funded R&D is needed to meet the changed demands of Rubbert's (1994) new model of the RD&P process for LCA. In the new model, knowledge is managed as a strategic resource to achieve competitive (economic) advantage. The new system should be "user focused" and grounded in the belief that the knowledge resulting from federally funded aerospace R&D is intellectual capital that the U.S. aerospace industry can use to enhance its economic competitiveness. Initiatives that utilize the full capability of computers and information technology (i.e., intelligent, knowledge-based systems), the proficiency and talents of information professionals to design and operate these systems, and the human and monetary resources needed to develop, test, and implement them will be required. The "supply-side," one-way, source-to-user, passive transmission procedures that characterize the existing system are seldom responsive in a user context and should be replaced with interactive, two-way communication procedures.

The new system is based on the assumption that knowledge production, transfer, and use should be equally important components of federal R&D. In the existing system, transfer and use receive such a low level of support in comparison to knowledge production that many critics of the existing system consider transfer and use to be afterthoughts, rather than components of the federal R&D process. The cost of knowledge transfer is an organizational overhead. Pressure to reduce overhead results in passively structured operations that are understaffed and inadequately funded. In such a passive environment, information is arrayed for the taking, and users must request or search out the information that they may need. The new system replaces passivity with proactivity. It recognizes knowledge as intellectual capital; funds mechanisms that foster knowledge transfer and use; and structures, organizes, and manages knowledge in contexts and formats based on user-defined needs or requirements. Learning to organize and structure knowledge according to problem relevance (i.e., user context) begins with an understanding of the communication practices and information-related activities of engineers and scientists working across the spectrum of U.S. aerospace.

Focus on the User

Unlike the existing system in which users' (the majority of whom are engineers) requirements and the characteristics of the task (i.e., R&D and technological innovation) are seldom known or considered, users and the users' environments become the focus of the new system. As Menzel (1966) states,

> The way in which [aerospace] engineers and scientists make use of information, the demands they make on information systems, and the satisfaction achieved by their efforts are among the items of knowledge which are necessary. (p. 42)

A description of aircraft design from Vincenti (1990) further illustrates this concept:

> A pilot and aircraft, taken together, form a single dynamic system, with feedback loops to the pilot via the feel of the cockpit controls plus cues from instruments and from vehicle orientation and acceleration. The pilot is a dynamic part of this closed loop system ... and senses him- or herself as such. ... Engineers of the 1930's ... tended to see the airplane as an open looped system ... with the pilot as an external agent who supplied whatever more or less quasi-static actions were required. (p. 76)

We contend that the current system used to disseminate the results of federally funded aerospace R&D resembles aircraft design in the early 1930s in that users' (i.e., the pilot) input is not an integral part of the design and development process. To develop a closed loop system in aircraft design, aerospace engineers included information from human factors studies of pilots. Changing the current dissemination based system for transferring the results of federally funded aerospace R&D requires moving to a closed loop system that incorporates an understanding of the communication practices and information-related activities of U.S. aerospace engineers and scientists.

To be effective, the new system would recognize that aerospace engineers and scientists typically seek information in the context of finding a solution to a specific problem rather than in the context of general opportunity. In addition, the need for information that provides a solution is frequently time sensitive, if not time critical. Furthermore, aerospace knowledge may be *scientific* or *technical* in nature, it may be *explicit* or *tacit*, it may be *product* or *process* oriented, and it may exist at the *managerial* or *systems* integration level. The tacit nature of much aerospace knowledge makes it particularly challenging to articulate and specify a problem and identify potentially useful and reliable sources of information. Therefore, informal oral communication networks, which play a key role in diffusing tacit knowledge, must be accommodated in the new system, and the knowledge itself must be modeled to address users' needs.

CONCEPTUAL FRAMEWORK

Developing and implementing the new model requires incorporating what is known about U.S. aerospace and the communication practices and the information-related activities of U.S. aerospace engineers and scientists. Incorporating this knowledge would result in a system that is responsive in a user context and seamless in terms of its day-to-day use by U.S. aerospace engineers and scientists.

Conceptualizing U.S. Aerospace

We use Figure 7.1 to establish a conceptual framework for U.S. aerospace. In doing so, we acknowledge that U.S. aerospace, in particular the LCA sector, is heavily influenced by various national (public) policies. These policies and the corresponding federal agencies appear in the upper portion of Figure 7.1. As Tornatzky and Fleischer (1990) point out, industries such as aerospace, that are engaged in R&D and technological innovation, are affected by these policies and agency programs "that cut across political jurisdictions and the idiosyncratic missions and mandates of single agencies which are more or less responsive to a series of shifting political alliances and imperatives" (p. 241). Public policies influence the production, transfer, and utilization of knowledge, which is a building block of R&D and technological innovation. In the lower portion of Figure 7.1, we have included the various kinds of knowledge associated with R&D and technological innovation.

The center portion of Figure 7.1 represents the RD&P of LCA. Admittedly, we have simplified a very complex process. Historically, the RD&P of LCA has relied on a strategy of innovation and technol-

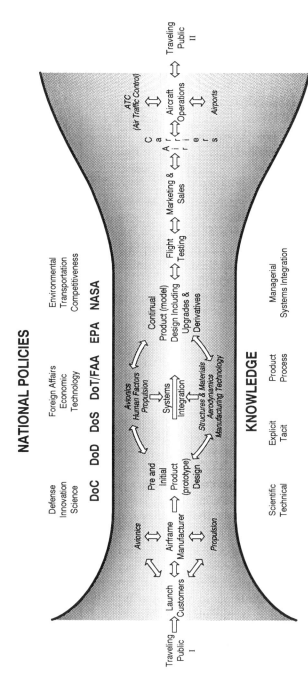

Figure 7.1. The Research, Development, and Production (RD&P) of a (Subsonic) Large Commercial Aircraft.

ogy, accompanied by the required processes for design, fabrication, testing, and certification. This strategy, according to Rubbert (1994), enabled the Wright brothers to reduce to practice the dream of powered flight, and it led to the development of modern LCA like the Boeing 747. The new model, which produced the Boeing 777 and included customers in "design-build teams," has replaced the historical "technology-driven" model with a strategy that is "market and user" driven. This change in strategy does not imply that advances in technology and innovation are no longer worthy of pursuit. To the contrary. Advances in technology and innovation are still valued. What the change in strategy means is that the worth or value of advances in technology is now determined by the user. Figure 7.1 is based on the new model in which the value of new technology and innovation is determined by the user (i.e., the traveling public and the launch customers).

The RD&P of LCA is presented as an intricate, value-added process that begins with the user. As mentioned in Chapter 1, the traditional pyramid, with an airframe manufacturer (i.e., prime contractor) atop the pyramid, illustrates the process in the U.S. In simple terms, the airframe manufacturer performs the aerodynamics, manufactures most of the main structure, and performs final assembly, systems integration, and testing. Subcontractors, both within and outside the U.S., add value by providing various subsystems and components. Literally thousands of vendors and suppliers (i.e., fabricators and machine shops) add value by furnishing a variety of specialized components and devices that make up the subsystems and subsystem assemblies. Components from the smallest fasteners to the largest components are built and tested according to Federal Aviation Administration (FAA) specifications. Further, all LCA must successfully complete a battery of ground and flight tests (e.g., brakes, decompression, engines, heating and air conditioning, and emergency evacuation) before FAA certification is granted and the airline is licensed to operate it within the nation's air transportation system.

The launch customer and the airframe manufacturer negotiate such features as the engines (e.g., General Electric, Pratt & Whitney, and Rolls-Royce) to be used, the seating configuration, and color schemes. In this sense, each LCA becomes a customized product and is manufactured on completion of a general purchase agreement. LCA are designed as a "family of aircraft" that incorporate various distance (i.e., range in nautical miles), weight, and

passenger capacities. For example, the Boeing 777 can be purchased in three configurations with two additional configurations
planned for the future. These different airplanes have been
designed to deliver the value of family commonality. The pilots,
flight attendants, and maintenance personnel trained on one B-777
will, in effect, be trained on all B-777s. All models draw from the
same body of spares, reducing airline investment and operation.

The RD&P of LCA embodies much of what is state-of-the-art
knowledge. Some is new—used for the first time; some is evolutionary and represents what David (1986) refers to as "advances in systemic knowledge of the useful arts" (p. 376). Throughout the RD&P
process, knowledge is constantly being diffused (i.e., produced,
transferred, and used) internally between and among individuals,
units, and departments in an attempt to add value to the product,
the LCA. A substantial amount of knowledge used in the RD&P
process is imported from sources external to the organization, including subcontractors and vendors, colleges and universities, government research laboratories, and airlines. For example, the
knowledge used to design the flight deck for the B-777 came from
several scientific and technical disciplines; engineers and scientists
working in the fields of human factors, ergonomics, and cognitive
and computer sciences; and airline pilots and engineers around the
world.

The need for more frequent and more effective use of knowledge
characterizes the strategic vision of today's competitive marketplace. The marketplace is characterized by a growing number of
stakeholders that are constantly changing. This implies that a
broader array of knowledge that can be interpreted and analyzed is
required for decisionmaking. The need for interpretation and analysis is critical because there is less time for decisionmaking and the
half-life of knowledge is getting shorter. Finally, increasing U.S. collaboration with foreign producers is resulting in a more international manufacturing environment. Such alliances will result in a more
rapid diffusion of knowledge, increasing pressure on U.S. aerospace
to maximize the use of knowledge for product improvement.

Conceptualizing Information-Seeking in Aerospace

An attribute common to all engineers is their use of information.
Mailloux (1989) estimates that 20% of an engineer's time is spent

in such intellectual activities of engineering as conceiving, sketching, and calculating. The remaining 80% of an engineer's time is spent on activities associated with "creating, accessing, receiving, manipulating, or transferring information" (p. 239). To relate this claim to aerospace, we undertook a telephone survey using a sample drawn from the Society of Automotive Engineers' subscriber database. We asked the participants to estimate the proportion of the previous work week that was spent doing activities that they considered to be engineering (Pinelli, Kennedy, and White, 1992). Overall, the respondents estimated that about 70% of their time was spent doing engineering-related activities (Table 7.1). Respondents were also asked to describe the engineering and non-engineering activities they had done in the previous work week. A number of activities were classified in both categories, and some activities were considered engineering by some engineers and non-engineering by other engineers. The respondents indicated that engineering and non-engineering activities involved writing reports, proposals, or researching new designs. We interpret these findings to mean that a substantial part of an engineer's time is spent in communication and information-related activities.

While acknowledging that other models exist (see, for example Kuhlthau, 1993 and Dervin and Nilan, 1986), we have chosen to view the information-seeking behaviors of aerospace engineers within a conceptual framework of the engineer as an information processor. Our conceptual framework, shown in Figure 7.2, is based on the work of Paisley (1968); Allen (1977); and Mick, Lindsey, Callahan, and Spielberg (1979) and represents an extension of Orr's (1970) scheme of the engineer-scientist as an information processor. The aerospace engineer occupies the center of the framework. According to Sayer (1965):

> Engineering is a production system in which data, information, and knowledge comprise new materials. Whatever the purpose of the engineering effort, the engineer is an information processor who is constantly faced with the problem of effectively acquiring, using, producing, and transferring data, information, and knowledge. (p. 25)

We assume that, individual differences notwithstanding, an internal, consistent logic governs the information-seeking behaviors of U.S. aerospace engineers and scientists.

Table 7.1. Engineering and Non-Engineering Activities Performed by U.S. Aerospace Engineers and Scientists

Activities	n	%[a]
Engineering		
Part and product design	94	31.6
System/product analysis	87	29.3
Writing, including technical proposals and specifications	56	18.9
Management, including cost estimates and contract negotiations	38	12.8
Testing design	37	12.5
Troubleshooting	25	8.4
Research/information gathering	19	6.4
Review designs	18	6.1
Computer software design/programming/ applications	17	5.7
Non-Engineering		
Paperwork/clerical/word processing	31	28.2
Bookkeeping/cost estimates	25	22.7
Personnel management/performance reviews/training	23	20.9
Writing/communicating with others	21	19.1
Meetings	16	14.5
Management	11	10.0
Sales/marketing/customer relations	6	5.5
Scheduling	8	7.3
Research/data search	3	2.7

[a]Percentages do not total 100 because respondents could provide more than one response.

Engineering is often defined as the application of scientific knowledge to the creation or improvement of technology for human use (Kemper, 1990). The term "technology" as used in the context of describing engineering work encompasses products, systems, structures, and processes. Engineering work can be described as a process that originates with an idea for a new or improved technology. Engineering work can also be described in terms of projects, tasks, or problems that engineers attend to on a day-to-day basis (Pinelli, Bishop, Barclay, and Kennedy, 1993). The need for

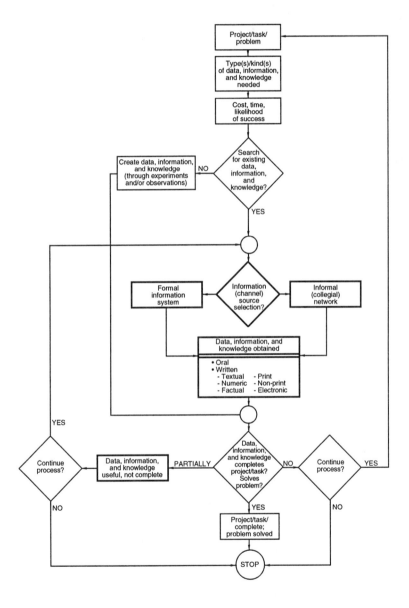

Figure 7.2. The U.S. Aerospace Engineer–Scientist as an Information Processor.

information is central to our conceptual framework. This need for information may, in turn, be *internally* or *externally* induced and is referred to by Orr (1970) as *inputs* or *outputs*, respectively.

Inputs originate within the mind of the individual engineer-scientist and include information needed to keep up with advances in one's profession, to perform one's professional duties, to interact with peers, colleagues, and coworkers, and to obtain stimulation and feedback from others (Hagstrom, 1965; Menzel, 1964; Storer, 1966; Voight, 1961). Outputs frequently, but not exclusively, result from an external stimulus or impetus. Outputs serve a variety of functions, including responding to a request for information from a supervisor, coworker, peer, or colleague; reporting progress; providing advice; reacting to inquiries; defending; advocating; and proposing. Inputs and outputs require the use of specific kinds and types of information and information products.

Our conceptual framework assumes that, in response to a project, task, or problem, specific kinds and types of information and information products are needed by the engineer-scientist. In response to this scenario, U.S. aerospace engineers and scientists are confronted with two basic alternatives: they can create the information through experimentation or observation or they can search the existing information. Existing information can be either *explicit* or *tacit* and may reside as codified knowledge in writing or as experience and understanding in the mind(s) of peers, colleagues, and coworkers. If the engineer–scientist acts rationally, the decision to "make or buy" the information will depend on a subjective perception of the relative likelihood of success in acquiring the desired information by these two alternatives within an acceptable time, and on the perception of the relative cost (i.e., money or effort) of these alternatives.

If a decision is made to search the existing information, the engineer–scientist must choose between two information channels and sources. One is *informal* and *collegial*; it is characterized by interpersonal (oral) communication with peers, coworkers, colleagues, gatekeepers, vendors, consultants, key personnel, and supervisors and by personal collections of information. The other is *formal* and includes libraries, librarians, information products and services, and electronic tools and bibliographic databases. Two variables—*accessibility* and *technical quality*—are assumed to influence the choice of a particular information channel or source.

Buckland (1983) states that "it is known that accessibility is a dominant factor in information behavior" (p. 114;173). He cites the early work of Rosenberg (1967), Gerstberger and Allen (1968), and Harris (1966) to support this position. Orr (1970) states that technical quality, not accessibility, is a prime consideration in channel and source selection. Such accessible sources as peers and colleagues may also offer information that is both relevant and high in technical quality. In a telephone survey of U.S. aerospace engineers and scientists, respondents overwhelmingly reported that the primary reason for using coworkers to obtain technical information stemmed from the fact that the information they had was relevant and of high technical quality (Glassman and Pinelli, 1992).

Information resulting from the search is subjectively evaluated. The engineer-scientist, as an information processor, is faced with three possible courses of action. *First*, if the acquired-obtained information completes the project or task or solves the problem, the search process is successfully terminated. *Second*, if the acquired-obtained information is useful but only partially completes the project or task or partially solves the problem, a decision is made either to continue the process by reevaluating the information channel or source selection or to terminate the process. *Third*, if the acquired-obtained information is not applicable or does not complete the project or task or solve the problem, a decision is made either to continue the process by redefining or rethinking the project, task, or problem or to terminate the process.

The *NASA/DoD Aerospace Knowledge Diffusion Research Project* is concerned with the diffusion of knowledge resulting from federally funded aerospace R&D. The information-seeking behaviors of U.S. aerospace engineers and scientists is an important part of that research to the extent that it helps us understand knowledge diffusion (i.e., production, transfer, and use) within U.S. aerospace. For these reasons, the conceptual framework is investigated but not validated.

METHODOLOGY

The conceptual design for U.S. aerospace shown in Figure 7.1 was used to establish the ten disciplines associated with the RD&P of LCA. The disciplines include research, design, avionics, structures, propulsion, human factors, production, flight test, sales, and service. Next, we identified an organization or professional society to represent each of the disciplines and to establish a sample frame for each survey (see Table 7.2). Ten mail (self-reported) surveys were used to collect the reported data. For reporting purposes, responses from the sales and service surveys were combined. Data were collected for research in 1995 and for all other disciplines in 1996 as a Phase 1 activity of the *NASA/DoD Aerospace Knowledge Diffusion Research Project*.

The research sample was composed of aerospace technologists (ASTs) employed at the NASA Langley Research Center in Hampton, VA, who worked in the Research and Technology Group. Samples for the remaining disciplines were engineers and scientists who were members of the American Institute of Aeronautics and Astronautics (AIAA), the American Society for Test and Materials (ASTM), the Human Factors and Ergonomics Society (HF&ES), the Society of Automotive Engineers (SAE), the Society for the Advancement of Material & Process Engineering (SAMPE), the Society of Flight Test Engineers (SFTE), and the Society of Manufacturing Engineers (SME). Within each professional society, sample frames were drawn from members whose interest area or professional duties were aligned with the discipline. We also attempted to limit the sample frame to include engineers and scientists working in industry. Sample lists were compared and duplicate entries removed. In the cover letter, addressees were instructed to inform us (and to return the survey) if they had previously participated in any *NASA/DoD Aerospace Knowledge Diffusion Research Project* surveys.

DESCRIPTIVE FINDINGS

We present data about the communication practices and information-related activities of U.S. aerospace engineers and scientists who represent the spectrum of aerospace from research through development and production to technical services. Data are presented according to the following topics: sample demographics, technical

Table 7.2. Survey Samples and Related Information

Discipline	Society/ Organization	Sample Size	Response Rate (%)	Useable Surveys	Reference[a]
Research	NASA Langley	300	75	222	1995
Design	AIAA	300	57	142	1996a
Avionics	SAE	200	51	128	1996b
Structures	AIAA, ASTM, SAMPE	500	51	209	1996h
Propulsion	AIAA	300	51	134	1996f
Human factors	HF&ES	200	60	96	1996d
Production	SME	500	60	261	1996e
Flight test	SFTE	200	51	68	1996c
Sales/service	AIAA	200	56	87	1996g

[a]Pinelli, Barclay, and Kennedy, 1995, 1996a-g

communication practices, collaborative writing practices, the kinds and amounts of information produced and used, the use and importance of libraries and computer networks, the strategies employed in seeking information, reasons for using specific information products and ratings of specific information products, and the use of knowledge resulting from federally funded R&D. Data were also collected about the use of the bibliographic products designed to provide awareness of and linkages to the results of federally funded aerospace R&D. Survey respondents are U.S. aerospace engineers and scientists employed in government and industry.

Survey Demographics

Demographics for the 1347 engineers and scientists who participated in the ten surveys appear in Table 7.3. Based on an analysis of previous surveys, we assumed that these respondents would be male, would hold at least a bachelor's degree, had been educated as engineers and would be performing engineering duties. A substantial portion of the respondents in each survey were male. The disciplines with the highest percentages of males were propulsion and production (97%). The disciplines with the highest percentages of females were avionics (24%) and research (15%). Among respondents, the levels of education varied. With the exception of human factors, production, and flight test, the majority of respondents in each group held a graduate degree. About 54% of the respondents in avionics held a doctorate.

Table 7.3. Survey Demographics [N=222; 142; 128; 209; 134; 96; 261; 68; 87]

Demographics	Research	Design	Avionics	Struc-tures	Propul-sion	Human Factors	Produc-tion	Flight Test	Sales/Service
	%	%	%	%	%	%	%	%	%
Gender									
Male	85.3	97.2	76.4	90.9	97.8	94.8	97.3	95.6	97.7
Female	14.7	2.8	23.6	9.1	2.2	5.2	2.7	4.4	2.3
Degree									
Bachelor's	14.5	38.7	8.6	31.6	27.6	54.2	44.1	55.9	42.5
Master's	49.1	50.7	37.5	40.7	53.7	33.3	23.4	38.2	50.6
Doctorate	35.9	7.7	53.9	24.4	17.2	8.3	3.4	4.4	2.3
Educational preparation as:									
Engineer	78.2	94.4	22.8	81.3	97.0	86.2	69.2	92.6	77.0
Scientist	20.0	4.9	56.7	15.8	2.2	7.4	6.9	2.9	8.0
Other	1.8	0.7	20.5	2.9	0.7	6.4	23.8	4.4	14.9
Primary duties									
Engineer	68.2	89.4	42.5	82.3	86.6	85.4	67.3	76.5	36.8
Scientist	26.4	2.1	40.2	7.7	0.7	4.2	1.5	1.5	1.1
Other	5.5	8.5	17.3	10.0	12.7	10.4	31.2	22.1	62.1
Current work federally funded?									
Yes	98.6	59.2	76.6	73.2	66.4	43.8	41.0	52.9	44.8
No	0.0	38.0	21.1	25.4	32.1	53.1	49.4	47.1	52.9
Don't know	1.4	2.8	2.3	1.4	1.5	3.1	9.6	---	2.3
	$\bar{x}$	$\bar{x}$	$\bar{x}$	$\bar{x}$	$\bar{x}$	$\bar{x}$	$\bar{x}$	$\bar{x}$	$\bar{x}$
Aerospace work experience; years	19.9	22.4	17.6	21.4	21.0	19.0	17.9	25.7	22.2

The majority of respondents in each discipline were educated as engineers. Disciplines with the highest percentages of respondents educated as engineers were propulsion (97%), design (94%), and flight test (93%). The highest percentages of scientists were found in avionics (57%), research (20%), and production (16%). Similarly, the majority of respondents in each discipline also performed engineering duties. About 57% of the respondents in avionics and 20% of the respondents in research performed primary duties as scientists. About 62% of the respondents in sales and service performed duties as neither engineers nor scientists. The majority of respondents in all but three disciplines (human factors, production, and flight test) indicated that the funding for their current work was provided by the federal (U.S.) government. The mean number of years of aerospace work experience was highest for respondents in flight test ($\bar{x}$ = 25.7) and lowest for respondents in production ($\bar{x}$ = 17.9) and avionics ($\bar{x}$ =17.6).

Communication Practices

Surveys of engineers, including both supervisors and practitioners, indicate that the ability to communicate effectively is critical to professional success (Pinelli, Barclay, Keene, Kennedy, and Hecht, 1995). We asked a series of questions designed to determine the significance of certain communication factors in the careers of U.S. aerospace engineering and scientists. Table 7.4 shows mean responses to survey questions about the importance of communicating information effectively in the performance of the respondents' duties. Importance was measured on a 5-point scale. The responses indicate that specific disciplines notwithstanding, U.S. aerospace engineers and scientists consider the ability to communicate effectively to be important. The amount of time that survey respondents spend communicating information is also an important indicator of its significance. Overall, respondents in all disciplines spend a considerable number of hours each week communicating information to others and working with information received from others. The average number of hours spent each week communicating ranged from a low of 27.8 for research to a high of 46.0 for service and sales. Overall, respondents spent more time each week communicating information to others than they spent working with information received from others. Respondents in all disciplines except for design and avionics spent more time communicating information to others *orally* than they did in writing. The majority of

Table 7.4. Technical Communication Practices of U.S. Aerospace Engineers and Scientists [N=222; 142; 128; 209; 134; 96; 261; 68; 87]

	Research	Design	Avionics	Struc-tures	Propul-sion	Human Factors	Produc-tion	Flight Test	Sales/ Service
Technical communication factors	$\bar{x}$	$\bar{x}$	$\bar{x}$	$\bar{x}$	$\bar{x}$	$\bar{x}$	$\bar{x}$	$\bar{x}$	$\bar{x}$
Importance of communicating effectively	4.9	4.7	5.0	4.9	4.8	4.7	4.6	4.7	4.8
Time per week communicating —									
Written, hours	7.7	11.7	14.0	12.3	9.5	12.1	11.0	11.1	11.1
Orally, hours	8.1	13.0	10.6	11.7	10.5	12.2	12.0	11.4	15.6
Time per week working with —									
Written information	7.2	10.3	10.7	9.1	8.2	11.0	9.9	10.7	10.5
Oral information	5.7	9.0	7.3	8.0	7.3	8.6	7.6	7.8	9.8
Changes over time	%	%	%	%	%	%	%	%	%
Compared with 5 years ago, time spent communicating technical information has:									
Increased	58.1	71.1	54.8	65.6	60.4	70.8	71.5	57.4	67.8
Stayed the same	29.7	24.6	37.3	28.2	32.1	24.0	23.1	29.4	27.6
Decreased	12.2	4.2	7.9	6.2	7.5	5.2	5.4	13.2	4.6
As you have advanced professionally, time spent communicating technical information has:									
Increased	57.2	66.9	52.0	60.1	64.7	66.7	69.6	63.2	67.4
Stayed the same	32.9	28.2	37.8	31.7	24.1	29.2	22.3	19.1	19.8
Decreased	9.9	4.9	10.2	8.2	11.3	4.2	8.1	17.6	12.8

respondents in all disciplines indicated that the amount of time they spent communicating technical information had increased in the last five years. Furthermore, respondents in all of the disciplines reported that as they have advanced professionally, the amount of time they spend communicating technical information has actually increased.

Collaborative Writing

Engineering in general, and R&D in particular, is a collaborative process. More and more companies are abolishing rigid hierarchical organizations and replacing them with teams and groups that cut across functions. Lunsford and Ede (1990), in their survey of working professionals, reported that 87% of those individuals surveyed indicated that they wrote as part of a team or group. Three questions were used to determine the role of collaborative writing in the professional lives of U.S. aerospace engineers and scientists. Table 7.5 shows that the majority of respondents in all disciplines reported that they had done some of their writing with others in the past six months. The percentages of respondents writing collaboratively ranged from a high of 100% for respondents in production to a low of about 67% for respondents in flight test. Respondents in all disciplines except structures indicated that writing in groups was more productive than writing alone. Of those who wrote with others, respondents in all disciplines, except avionics and flight test, indicated that most of their collaborative writing was done with the same group. The numbers of individuals in a group ranged in size from 3.5 (research) to 7.1 (design) and 7.2 (human factors). The number of groups with whom respondents wrote collaboratively ranged from 2.9 (research) to 6.4 (flight test). Finally, the number of individuals in each group averaged 3.7 for research and increased to 7.6 for flight test.

Information Products Produced and Used

Table 7.6 contains the mean numbers of information products that the U.S. aerospace engineers and scientists in the various disciplines produced and used. For discussion purposes, we divided the information products into informal (i.e., letters) and formal (i.e., conference and meeting papers, journal articles, and in-house technical reports). Regardless of their disciplines, the aerospace engineers and scientists whom we surveyed produced more informal than formal information products. Survey respondents produced

Table 7.5. Collaborative Writing Practices of U.S. Aerospace Engineers and Scientists
[N=222; 142; 128; 209; 134; 96; 261; 68; 87]

	Research	Design	Avionics	Struc-tures	Propul-sion	Human Factors	Produc-tion	Flight Test	Sales/Service
Writing practices	%	%	%	%	%	%	%	%	%
Write alone	24.7	22.5	11.7	23.4	29.9	29.2	0.0	32.4	28.7
Write with others	75.3	77.5	88.3	76.6	70.1	70.8	100.0	67.6	71.4
Group productivity	%	%	%	%	%	%	%	%	%
More	39.9	48.6	49.1	38.2	45.7	42.6	47.0	43.2	45.6
Same	37.5	18.3	17.6	21.0	25.5	19.1	17.9	22.7	8.8
Less	22.0	33.0	33.0	40.8	28.7	35.3	33.9	31.8	43.9
Difficult to judge	0.6	---	---	---	---	2.9	1.2	2.3	1.8
Number of groups	%	%	%	%	%	%	%	%	%
Worked with same group:									
Yes	55.8	58.7	47.3	50.3	57.6	61.2	60.0	64.4	48.3
No	44.2	41.3	52.7	49.7	42.4	38.8	40.0	35.6	51.7
Size of group	$\bar{x}$	$\bar{x}$	$\bar{x}$	$\bar{x}$	$\bar{x}$	$\bar{x}$	$\bar{x}$	$\bar{x}$	$\bar{x}$
Number in group	3.5	7.1	3.9	5.8	4.7	7.2	3.9	6.0	6.1
Number of groups	2.9	4.2	3.4	5.7	3.8	3.9	4.1	6.4	4.5
Number in each group	3.7	6.6	5.0	4.3	5.1	4.9	4.7	7.6	4.8

more memoranda and letters than any other kind of information product. U.S. aerospace engineers and scientists also produced a substantial number of drawings and specifications (respondents in production preparing the most) and technical talks and presentations. Considering the three formal information products, respondents in research produced more conference and meeting papers. Respondents in all other disciplines produced more in-house technical reports—which are likely to be informal products in most aerospace organizations—than they prepared conference and meeting papers and journal articles. The relatively low production of formal information products is not surprising. As Allen (1977) notes, most engineers, unlike scientists, are neither expected nor hired to make "original" contributions to the literature. To confirm this finding, in our telephone survey of SAE members, only about 35% of those participants said that it is important in their job to publish new ideas or make original contributions to the literature.

Table 7.6 also lists the average number of information products that the U.S. aerospace engineers and scientists had used in the six months prior to the survey. The list contained 10 informal (e.g., letters) and 3 formal information products (e.g., journal articles). In general, U.S. aerospace engineers and scientists used considerably more information products than they produced. They also used more informal than formal information products. Use of two informal information products—letters and memoranda—is notable. The mean number of letters and memoranda used is high for industry-affiliated survey respondents in all disciplines. The number of letters and memoranda used was lowest among respondents in research ($\bar{x}$ = 7.1, 6.9) and highest among respondents in design ($\bar{x}$ = 17.5, 42.0) and sales and service ($\bar{x}$ = 31.4, 48.1). The use of drawings and specifications was also high among industry-affiliated survey respondents. The number of drawings and specifications used was lowest among respondents in avionics ($\bar{x}$ = 8.5) and highest among survey respondents in human factors ($\bar{x}$ = 48.3) and production ($\bar{x}$ = 52.0). Use of two formal information products—conference and meeting papers and journal articles—was highest among respondents in research, avionics, and human factors. Except in research ($\bar{x}$ = 1.8), use of in-house technical reports was fairly consistent among industry-affiliated respondents. The overall use of technical talks and presentations tends to support claims in the literature regarding the importance of oral communication in R&D.

Table 7.6. Mean Number of Technical Information Products Produced and Used in Past Six Months by U.S. Aerospace Engineers and Scientists [N=222; 142; 128; 209; 134; 96; 261; 68; 87]

Produced	Research	Design	Avionics	Structures	Propulsion	Human Factors	Production	Flight Test	Sales/Service
	X̄	X̄	X̄	X̄	X̄	X̄	X̄	X̄	X̄
Abstracts	1.2	0.7	1.6	2.0	0.3	0.6	0.6	0.2	0.4
Journal articles	0.6	0.2	0.8	0.4	1.8	0.1	0.1	0.1	0.1
Conference/meeting papers	2.4	1.2	1.7	1.4	0.6	3.8	2.6	1.9	1.5
Drawings/specifications	2.1	6.0	2.5	6.6	2.7	13.1	18.4	2.8	2.5
Letters	5.4	11.7	17.1	17.0	16.7	12.1	14.9	13.7	34.0
Memoranda	5.0	22.5	16.0	25.6	20.1	21.0	19.4	23.4	37.4
Technical proposals	0.6	1.6	2.3	1.9	1.7	3.6	2.4	3.1	7.9
Computer program/documentation	0.5	0.7	0.3	0.9	0.7	4.2	8.7	0.6	0.8
In-house technical reports	0.7	1.9	2.7	5.5	4.0	7.3	4.8	4.6	3.4
Technical talks/presentations	2.9	5.5	3.9	5.6	6.3	2.3	3.4	4.0	6.0

Used	Research	Design	Avionics	Structures	Propulsion	Human Factors	Production	Flight Test	Sales/Service
	X̄	X̄	X̄	X̄	X̄	X̄	X̄	X̄	X̄
Abstracts	20.2	4.5	14.2	7.7	2.3	12.8	1.6	0.9	3.1
Journal articles	14.8	5.1	19.4	9.4	4.9	15.2	4.2	4.0	8.1
Conference/meeting papers	9.6	3.2	12.3	8.5	4.7	8.0	3.9	3.9	3.0
Drawings/specifications	3.4	22.4	8.5	26.3	22.4	48.3	52.0	13.7	16.8
Letters	7.1	17.5	22.1	18.7	15.9	10.1	16.6	31.4	31.4
Memoranda	6.9	42.0	19.8	23.8	39.3	17.8	22.7	22.6	48.1
Technical proposals	2.3	3.0	3.3	3.9	3.0	3.3	3.2	2.5	8.9
Computer program/documentation	4.2	5.1	5.8	3.8	5.4	8.2	13.6	5.4	3.3
In-house technical reports	1.8	8.6	5.6	7.9	8.8	9.6	6.2	9.6	8.7
Technical talks/presentations	6.6	11.5	13.0	9.4	6.9	7.6	3.8	5.1	8.9

Use, Importance, and Frequency of Use of Selected Information Products

Survey respondents were asked to report their use of, the frequency of use, and the importance they assign to five technical information products: conference and meeting papers, journal articles, in-house technical reports, DoD technical reports, and NASA technical reports (Table 7.7). Conference and meeting papers and journal articles constitute part of what is often called the "open or published literature"; it is available to anyone with access to a library. In-house technical reports often form the basis of an (aerospace) organization's internal (technical) knowledge base. DoD and NASA technical reports constitute a significant portion of government's internal (scientific and technical) knowledge base and serve as a primary means of disseminating the results of federally funded aerospace R&D to U.S. aerospace.

The use of conference and meeting papers and journal articles was consistently high among all disciplines but highest among survey respondents in research (97.3%, 96.3%) and avionics (96%, 95.3%). The same pattern of use emerged for in-house technical report use. The use of DoD technical reports was highest among respondents in research (67.6%), avionics (79.5%), and structures (74.4%). NASA technical report use was highest among respondents in research (94.4%), avionics (75.2%), structures (73.6%), and propulsion (68%). Survey respondents rated the importance of the five technical information products in terms of performing their present professional duties. Considering the five products as a group, the highest importance ratings were assigned by respondents in research, avionics, and structures. These same three disciplines reported the highest individual importance rating for conference and meeting papers and journal articles. Consistently high importance ratings were assigned to in-house technical reports by respondents in all disciplines. The importance ratings for DoD and NASA technical reports varied, with the highest ratings for DoD reports assigned by respondents in avionics ($\bar{x}$ = 3.5), structures ($\bar{x}$ = 3.3), and research ($\bar{x}$ = 3.1). The highest importance ratings for NASA technical reports were recorded for respondents in research ($\bar{x}$ = 3.9), avionics ($\bar{x}$ = 3.3), and structures ($\bar{x}$ = 3.1).

Survey respondents were asked to indicate the number of times they had used each of the five technical information products in a

Table 7.7. Use, Importance, and Frequency of Use of Selected Technical Information Products by U.S. Aerospace Engineers and Scientists [N=222; 142; 128; 209; 134; 96; 261; 68; 87]

	Research	Design	Avionics	Structures	Propulsion	Human Factors	Production	Flight Test	Sales/Service
Use	%	%	%	%	%	%	%	%	%
Conference/meeting papers	97.3	66.2	96.0	85.2	80.2	60.6	67.1	73.1	78.6
Journal articles	96.3	62.5	95.3	89.9	67.4	68.5	70.9	51.5	71.4
In-house technical reports	93.0	88.6	82.1	96.0	96.3	85.1	85.9	86.8	91.7
DoD technical reports	67.6	48.5	79.5	74.4	55.9	45.1	40.8	41.9	53.8
NASA technical reports	94.4	54.8	75.2	73.6	68.0	42.2	32.5	30.2	49.4
Importance	$\bar{x}$	$\bar{x}$	$\bar{x}$	$\bar{x}$	$\bar{x}$	$\bar{x}$	$\bar{x}$	$\bar{x}$	$\bar{x}$
Conference/meeting papers	4.2	2.9	3.9	3.4	3.2	3.0	3.0	3.2	3.3
Journal articles	4.2	2.8	3.9	3.2	2.9	3.1	2.9	2.7	2.9
In-house technical reports	3.8	3.9	3.9	4.2	4.3	4.0	3.8	4.0	3.8
DoD technical reports	3.1	2.8	3.5	3.3	2.7	2.7	2.5	2.4	2.8
NASA technical reports	3.9	2.8	3.3	3.1	2.9	2.6	2.3	2.3	2.5
Frequency of use	$\bar{x}$	$\bar{x}$	$\bar{x}$	$\bar{x}$	$\bar{x}$	$\bar{x}$	$\bar{x}$	$\bar{x}$	$\bar{x}$
Conference/meeting papers	9.6	3.2	12.3	8.5	4.7	8.0	3.9	3.9	3.0
Journal articles	14.8	5.1	19.4	9.4	4.9	15.2	4.2	4.0	8.1
In-house technical reports	1.9	8.6	5.6	7.9	8.8	9.6	6.2	9.6	8.7
DoD technical reports	0.9	3.6	8.7	3.4	1.6	5.6	1.2	1.2	1.7
NASA technical reports	3.5	3.0	3.5	3.5	2.0	1.8	0.9	0.5	0.6

6-month period in performing their professional duties. Survey respondents were told to use the six months prior to completing the survey as their frame of reference for answering this question. Overall frequency of use was highest at the beginning of the RD&P process (i.e., research), tapering off toward the end of the process (i.e., sales and service). Of the "open literature" products, journal articles were used more frequently than were conference and meeting papers. Frequency of use for journal articles was highest among respondents in avionics ($\bar{x}$ = 19.4) and research ($\bar{x}$ = 14.8). Frequency of use for conference and meeting papers was highest among respondents in avionics ($\bar{x}$ = 12.3), research ($\bar{x}$ = 9.6), and structures ($\bar{x}$ = 8.5). Frequency of use for in-house technical reports was consistently high among all disciplines except research. Frequency of use for DoD technical reports was highest among respondents in avionics ($\bar{x}$ = 8.7), human factors ($\bar{x}$ = 5.6), and design ($\bar{x}$ = 3.6). Frequency of use for NASA technical reports was highest among respondents in research ($\bar{x}$ = 3.5), avionics ($\bar{x}$ = 3.5), and structures ($\bar{x}$ = 3.5).

Factors Affecting the Use of Selected Information Products

Survey respondents were asked to indicate how important each of eight factors would be in their decision to use a particular information product (Tables 7.8 and 7.9). Two of the factors are concerned with *accessibility*: "easy to obtain physically" and "can be obtained at a nearby location or source." Three factors focus on *cost* (i.e., monetary or effort): "are inexpensive," "are easy to read or use," and "had good prior experience using them." Three factors are *content* attributes: "are relevant to my work," "have comprehensive data," and "have good technical quality." Importance was measured on a 5-point scale. Data are presented for the four technical information products most frequently used by the survey respondents: conference and meeting papers, journal articles, in-house technical reports, and NASA technical reports.

These factors were previously evaluated by Gerstberger and Allen (1968), who concluded that accessibility was more important than technical quality when engineers decided on an information source. Later work by Kaufman (1983) indicates that technical quality was more important than accessibility. Based on our findings regarding the behavior patterns used to select information needed to solve a technical problem, we expected that accessibility

Table 7.8. Factors Affecting Use of Conference/Meeting Papers and Journal Articles by U.S. Aerospace Engineers and Scientists [N=222; 142; 128; 209; 134; 96; 261; 68; 87]

	Research	Design	Avionics	Structures	Propulsion	Human Factors	Production	Flight Test	Sales/Service
	$\bar{x}$	$\bar{x}$	$\bar{x}$	$\bar{x}$	$\bar{x}$	$\bar{x}$	$\bar{x}$	$\bar{x}$	$\bar{x}$
Conference/meeting papers									
Easy to obtain physically	4.0	4.1	4.2	4.2	4.0	4.0	4.0	4.1	3.9
Easy to use or read	4.0	4.1	3.8	4.1	3.9	4.0	4.1	4.2	4.0
Inexpensive	2.8	3.5	3.4	3.4	3.1	3.5	3.6	3.5	3.6
Good technical quality	4.5	4.5	4.6	4.6	4.5	4.5	4.4	4.5	4.4
Comprehensive data and information	4.2	4.3	4.2	4.4	4.4	4.3	4.3	4.2	4.2
Relevant to my work	4.8	4.7	4.9	4.8	4.7	4.7	4.5	4.5	4.5
Obtainable at nearby location or source	3.6	3.6	3.6	3.7	3.6	3.6	3.7	3.5	3.5
Good prior experiences with use	3.4	3.5	3.5	3.5	3.5	3.3	3.3	3.4	3.4
Journal articles									
Easy to obtain physically	3.9	4.0	4.0	4.1	3.9	3.9	3.9	3.9	3.8
Easy to use or read	4.0	4.1	3.8	4.0	3.9	4.0	4.0	4.1	3.9
Inexpensive	2.8	3.5	3.3	3.3	3.1	3.5	3.6	3.4	3.6
Good technical quality	4.6	4.5	4.7	4.6	4.5	4.5	4.4	4.5	4.2
Comprehensive data and information	4.3	4.2	4.4	4.4	4.4	4.3	4.2	4.2	4.1
Relevant to my work	4.8	4.7	4.8	4.7	4.6	4.6	4.4	4.5	4.4
Obtainable at nearby location or source	3.5	3.7	3.5	3.6	3.4	3.5	3.4	3.5	3.5
Good prior experiences with use	3.4	3.4	3.6	3.5	3.4	3.3	3.3	3.4	3.4

Table 7.9. Factors Affecting Use of In-house and NASA Technical Reports by U.S. Aerospace Engineers and Scientists [N=222; 142; 128; 209; 134; 96; 261; 68; 87]

	Research	Design	Avionics	Structures	Propulsion	Human Factors	Production	Flight Test	Sales/ Service
	X̄	X̄	X̄	X̄	X̄	X̄	X̄	X̄	X̄
In-house reports									
Easy to obtain physically	3.9	4.2	3.9	4.1	3.9	3.9	4.0	4.2	3.9
Easy to use or read	4.0	4.1	3.7	4.0	3.9	4.1	4.1	4.1	4.0
Inexpensive	2.7	3.0	2.9	2.9	2.7	3.1	3.3	3.0	3.2
Good technical quality	4.5	4.4	4.4	4.6	4.4	4.6	4.4	4.5	4.4
Comprehensive data and information	4.3	4.3	4.2	4.4	4.4	4.4	4.3	4.4	4.3
Relevant to my work	4.7	4.7	4.7	4.7	4.7	4.6	4.4	4.5	4.5
Obtainable at nearby location or source	3.5	3.8	3.5	3.6	3.4	3.6	3.6	3.8	3.7
Good prior experiences with use	3.3	3.5	3.4	3.5	3.4	3.5	3.5	3.5	3.5
	X̄	X̄	X̄	X̄	X̄	X̄	X̄	X̄	X̄
NASA reports									
Easy to obtain physically	3.9	4.0	4.0	4.1	3.9	3.8	3.9	3.9	3.6
Easy to use or read	4.0	4.0	3.7	4.0	3.8	4.0	4.0	4.0	3.9
Inexpensive	2.7	3.4	3.3	3.3	3.0	3.4	3.5	3.3	3.5
Good technical quality	4.5	4.4	4.5	4.6	4.5	4.4	4.3	4.3	4.2
Comprehensive data and information	4.4	4.3	4.3	4.4	4.3	4.2	4.1	4.2	4.0
Relevant to my work	4.7	4.7	4.8	4.6	4.5	4.4	4.3	4.5	4.3
Obtainable at nearby location or source	3.5	3.6	3.4	3.5	3.4	3.4	3.5	3.5	3.4
Good prior experiences with use	3.4	3.4	3.5	3.4	3.4	3.2	3.3	3.2	3.3

would be the most important factor that influenced the use of the various products. Journal articles and in-house technical reports are likely to be most accessible to engineers and scientists. Journals should be accessible to all sample groups either because of their membership in a professional or technical society or because aerospace libraries subscribe to them. We also expected that technical quality would have a significant influence on the choice of information product.

Overall, relevance, technical quality, and comprehensiveness of data and information appear to have the greatest influence on the use of conference and meeting papers followed by the ease with which they can be obtained and used or read. Overall, five factors—relevance, technical quality, comprehensiveness of data and information, the ease with which they can be obtained, and used or read—appear to exert the greatest influence on the use of journal articles and conference and meeting papers. Similarly, the same factors influence the use of in-house and NASA technical reports. Across all disciplines involved in the RD&P of a LCA, it appears that the same three factors—relevance, good technical quality, and comprehensiveness of data and information—are the primary influences on use. Of the factors considered, relevance appears to be the most important factor in the decision to use a product. By itself, relevance is obviously the most important factor and can be used to benchmark the other reasons for product use. Because engineers work under time constraints and deadlines, it is unlikely that few, if any, of them in our samples would take the time to identify and obtain hard-to-find information unless they considered it to be extremely relevant. Consequently, these engineers might be forced to "satisfice," choosing information that might be accessible but less relevant. Based on the ratings, it is clear that product accessibility is important to the members of our samples. However, we believe it is secondary to relevance, technical quality, and comprehensiveness of data and information in the decision to use an information source or product.

Finally, it is interesting to note that the patterns of factors influencing use are the same across all groups and for all products. There are few differences among the disciplines in the relative ratings of the reasons that influence the use of products. There are also few differences among the disciplines in the ratings of the products themselves as to the various factors that influence product use. These findings confirm what we expected about the decision-making processes of U.S. aerospace engineers and scientists concerning information use. They are not as concerned about the source of information as they are about its relevance, technical quality, comprehensiveness, and accessibility. Within each discipline, NASA technical reports were rated about equal to the other information products. This indicates that U.S. aerospace engineers and scientists have no special reasons to choose or not to choose NASA technical reports when they are looking for information.

Ratings of Selected Information Products

Survey respondents were further asked to rate the same four information products—conference and meeting papers, journal articles, in-house technical reports, and NASA technical reports—according to the same eight factors previously used to determine choice of product. The respondents' ratings of the four information products appear in Tables 7.10 and 7.11. The data in Tables 7.10 and 7.11 differ from the data in Tables 7.8 and 7.9. Data in Tables 7.10 and 7.11 indicate the overall rating of the product on each factor. Tables 7.8 and 7.9 indicate the influence of the factor on the decision to use the product. To illustrate the difference, if being easy to obtain influenced the use of a particular information product, respondents were now being asked to rate that product as to how easy or difficult *it is* to obtain the information product physically. A 5-point scale was used to measure respondents' ratings.

The data in Tables 7.10 and 7.11 indicate substantial differences in the overall ratings assigned to the factors influencing use and how the respondents actually rate the four information products on each of the eight factors. For example, survey respondents in design indicated that relevance ($\bar{x}$ = 4.7) exerted the greatest influence on their use of conference and meeting papers. However, when they were asked to rate the relevance of conference and meeting papers to their work, the score was substantially less ($\bar{x}$ = 2.9). An analysis of the similarities and differences in engineers' and scientists' opinions of the four products provides a clearer picture of how they decide to use a particular information products. Considering conference and meeting papers, survey respondents in all disciplines assigned the highest ratings to two factors: relevance to work and good prior experience with use. For journal articles, respondents ascribed the highest ratings to two factors: good techni-

Table 7.10. Opinions of U.S. Aerospace Engineers and Scientists Concerning Conference/Meeting Papers and Journal Articles [N=222; 142; 128; 209; 134; 96; 261; 68; 87]

	Research	Design	Avionics	Structures	Propulsion	Human Factors	Production	Flight Test	Sales/Service
	$\bar{x}$	$\bar{x}$	$\bar{x}$	$\bar{x}$	$\bar{x}$	$\bar{x}$	$\bar{x}$	$\bar{x}$	$\bar{x}$
Conference/meeting papers									
Physically obtainable	3.6	2.9	2.7	2.8	2.9	2.7	2.8	2.8	3.0
Usable or readable	3.5	3.0	3.4	3.1	3.1	2.9	3.0	3.0	3.0
Inexpensive cost	3.6	2.9	2.9	2.9	3.1	3.0	2.9	2.9	2.9
Good technical quality	3.1	3.2	3.0	3.0	3.2	3.0	3.1	3.1	3.1
Comprehensive data and information	2.8	2.9	2.7	2.7	2.9	2.9	3.0	2.9	3.1
Relevant to work	3.8	2.9	3.4	3.2	3.2	3.1	3.1	3.1	3.3
Obtainable from nearby location/source	3.8	2.7	2.5	2.6	2.7	2.7	2.7	2.6	2.7
Experience with use	3.8	3.0	3.3	3.2	3.3	3.1	3.1	3.2	3.1
Journal articles									
Physically obtainable	4.0	3.2	3.4	3.4	3.3	3.1	3.2	3.2	3.4
Usable or readable	3.6	3.0	3.3	3.2	3.1	3.2	3.2	3.1	3.2
Inexpensive cost	3.5	3.0	2.9	3.2	3.0	3.2	3.2	3.2	3.1
Good technical quality	3.9	3.4	3.6	3.4	3.5	3.2	3.3	3.2	3.4
Comprehensive data and information	3.5	3.1	3.4	3.1	3.0	3.1	3.2	3.0	3.2
Relevant to work	3.9	2.9	3.4	3.2	3.1	3.2	3.1	2.8	3.2
Obtainable from nearby location/source	4.0	3.0	3.0	3.1	3.1	3.0	3.1	2.9	3.2
Experience with use	4.0	3.1	3.6	3.3	3.3	3.3	3.2	3.1	3.3

Table 7.11. Opinions of U.S. Aerospace Engineers and Scientists Concerning In-house and NASA Technical Reports [N=222; 142; 128; 209; 134; 96; 261; 68; 87]

	Research	Design	Avionics	Structures	Propulsion	Human Factors	Production	Flight Test	Sales/Service
	$\bar{x}$	$\bar{x}$	$\bar{x}$	$\bar{x}$	$\bar{x}$	$\bar{x}$	$\bar{x}$	$\bar{x}$	$\bar{x}$
In-house reports									
Physically obtainable	4.2	4.0	4.1	3.9	4.1	3.7	3.8	3.7	4.1
Usable or readable	3.9	3.4	3.7	3.4	3.6	3.3	3.4	3.3	3.6
Inexpensive cost	4.3	4.4	4.3	4.4	4.5	4.1	4.0	4.3	4.3
Good technical quality	3.8	3.5	3.3	3.3	3.6	3.3	3.4	3.2	3.5
Comprehensive data and information	3.8	3.4	3.3	3.3	3.7	3.3	3.3	2.9	3.5
Relevant to work	3.7	3.8	3.7	3.8	4.2	3.6	3.7	3.7	4.0
Obtainable from nearby location/source	4.4	4.0	4.0	4.1	4.3	3.8	3.8	3.8	4.0
Experience with use	3.8	3.7	3.6	3.6	3.9	3.4	3.5	3.5	3.8
NASA reports									
Physically obtainable	4.3	3.0	2.7	2.9	3.1	2.8	2.7	2.8	2.7
Usable or readable	3.9	3.2	3.3	3.2	3.3	3.2	3.0	3.2	3.2
Inexpensive cost	4.2	3.1	3.2	3.4	3.5	3.3	3.0	3.0	3.4
Good technical quality	3.9	3.6	3.5	3.5	3.6	3.2	3.5	3.3	3.5
Comprehensive data and information	3.8	3.5	3.4	3.2	3.5	3.1	3.4	3.1	3.4
Relevant to work	3.7	3.0	3.4	3.2	3.3	2.9	3.0	2.8	3.1
Obtainable from nearby location/source	4.3	2.9	2.5	2.8	2.8	2.7	2.7	2.6	2.7
Experience with use	3.9	3.4	3.5	3.3	3.4	3.2	3.1	3.1	3.2

cal quality and good prior experience with use (with relevant to work running a close third). For in-house technical reports, respondents assigned the highest ratings to three factors: easy to obtain physically, good prior experience using them, and can be obtained at a nearby location or source. Considering NASA technical reports, respondents assigned the highest ratings to three factors: good technical quality, comprehensive data and information, and good prior experience using them. A comparison of the data reveals that, with minor exceptions, U.S. aerospace engineers and scientists recorded higher scores for the factors affecting use of a particular information product and lower opinion scores for the same factors for each of the four information products. For example, respondents in avionics identified relevance ($\bar{x}$ = 4.8), good technical quality ($\bar{x}$ = 4.5), and comprehensiveness of data and information ($\bar{x}$ = 4.3) as the factors affecting their use of NASA technical reports. However, when asked their opinions of NASA technical reports, avionics survey respondents evaluated these same three factors accordingly: relevance ($\bar{x}$ = 3.4), good technical quality ($\bar{x}$ = 3.5), and comprehensiveness of data and information ($\bar{x}$ = 3.4). These data suggest that the relevance, technical quality, and the comprehensiveness of data and information of NASA technical reports may be problematic.

Use and Importance of Electronic (Computer) Networks

Approximately equal percentages of respondents in all disciplines reported using electronic (computer) networks in performing their present professional duties (Table 7.12). Network use varied from a high of about 92% for respondents in design to a low of about 76% for respondents in sales and service. Percentages of respondents reporting "no access" as a reason for not using computer networks ranged from 14% (sales and service) to 2% (human factors). Respondents were asked to rate the importance of computer networks in performing their present professional duties. Importance was measured on a 5-point scale. U.S. aerospace engineers and scientists indicate that computer networks are important to performing their present professional duties. Importance scores ranged from a mean high of 4.4 for respondents in research and design to a low of 4.0 for respondents in flight test. Respondents in all disciplines reported high use rates for computer networks with scores ranging from means of 15.3 hours per week for respondents in human factors to 10.7 hours per week for those in research.

Table 7.12. Electronic (Computer) Networks and U.S. Aerospace Engineers and Scientists
[N=222; 142; 128; 209; 134; 96; 261; 68; 87]

	Research	Design	Avionics	Structures	Propulsion	Human Factors	Production	Flight Test	Sales/Service
Use networks	%	%	%	%	%	%	%	%	%
Yes	90.9	91.5	86.5	84.5	84.3	91.7	76.9	89.4	75.6
No	8.2	4.2	5.6	7.7	9.7	6.3	11.5	7.6	10.5
No access	0.9	4.2	7.9	7.7	6.0	2.1	11.5	3.0	14.0
Purpose of use	%	%	%	%	%	%	%	%	%
Connect to geographically distant sites	78.4	71.7	86.1	74.7	72.9	66.3	60.1	71.9	85.2
Electronic mail	99.5	89.8	99.1	97.7	96.5	95.4	86.7	96.6	96.8
Electronic bulletin boards or conferences	51.1	42.3	64.7	46.9	38.9	57.3	47.0	43.6	62.7
Access/search library's catalog	73.0	47.5	66.7	46.4	44.9	61.9	43.8	37.5	33.9
Order documents from library	55.4	22.7	31.6	25.0	26.7	39.8	28.5	14.8	10.9
Search electronic (bibliographic) databases	64.0	36.4	72.5	49.1	42.9	52.4	45.5	37.0	37.9
Prepare scientific and technical papers with colleagues at geographically distant sites	46.7	31.0	60.8	31.4	24.3	29.6	22.4	21.8	24.6
Information search and data retrieval									
FTP	85.2	42.2	58.5	40.3	45.6	35.5	29.2	27.3	25.0
Gopher	34.7	25.5	38.5	17.9	15.0	26.0	17.0	10.9	11.5
WAIS	16.5	7.6	14.8	6.6	10.9	8.8	5.8	3.7	4.0
World Wide Web (WWW)	85.1	59.7	84.5	70.5	53.8	52.5	51.7	30.9	58.9
	$\bar{x}$	$\bar{x}$	$\bar{x}$	$\bar{x}$	$\bar{x}$	$\bar{x}$	$\bar{x}$	$\bar{x}$	$\bar{x}$
Importance of networks in performing duties	4.4	4.4	4.2	4.1	4.1	4.3	4.3	4.0	4.3
Hours networks used past week in performing duties	10.7	15.1	11.7	12.0	12.0	15.3	14.7	12.9	12.4

U.S. aerospace engineers and scientists use computer networks for various purposes. Overall, about equal percentages of respondents used computer networks for electronic mail and to connect to geographically distant sites. Overall, respondents used computer networks least often to order documents from the library and to prepare scientific and technical papers with colleagues at geographically distant sites. Respondents were asked about their use of the Internet for information search and retrieval. With the exception of respondents in flight test (30.9%), a simple majority of respondents in the other disciplines used the World Wide Web (WWW) for information search and retrieval.

Use and Importance of Libraries

U.S. aerospace engineers and scientists were asked a series of questions about their use of and the importance of a library in the context of performing their present professional duties (Table 7.13). With minor exceptions, about equal percentages of respondents in all disciplines reported that their organization has a library. (About 20% of respondents in sales and service and production and about 18% of the respondents in flight test indicated that their organization does not have a library.) Survey respondents were also asked to indicate the number of times that they had used a library during the past six months. Respondents in human factors, avionics, production, and sales and service reported the highest library use rates during the past six months prior to the survey.

U.S. aerospace engineers and scientists were asked to rate the importance of a library in performing their present professional duties. Importance was measured on a 5-point scale. Respondents in research assigned the highest rating ($\bar{x}$ = 4.3) to the importance of a library in performing their present professional duties. They were also asked to indicate the extent to which the proximity of their workplace (i.e., office) to a library was an important consideration in terms of their use of a library. With the exception of those respondents in research, the data indicate that workplace proximity to a library appears to affect its use moderately. Finally, we asked those respondents who had not used a library in the past six months their reasons for not doing so. The most frequently reported reason was "information needs were more easily met some other way," followed by "I had no information needs," and "have my own library and do not need another library."

Table 7.13. Libraries and U.S. Aerospace Engineers and Scientists [N=222; 142; 128; 209; 134; 96; 261; 68; 87]

	Research	Design	Avionics	Structures	Propulsion	Human Factors	Production	Flight Test	Sales/Service
Institution has library	%	%	%	%	%	%	%	%	%
Yes	96.8	94.4	84.9	94.3	94.8	88.5	80.0	82.1	79.1
No	3.2	5.7	15.1	5.8	5.3	11.5	19.9	17.9	20.9
Importance, proximity, and use	$\bar{x}$	$\bar{x}$	$\bar{x}$	$\bar{x}$	$\bar{x}$	$\bar{x}$	$\bar{x}$	$\bar{x}$	$\bar{x}$
Importance of library for performing your present duties	4.3	3.4	3.7	3.6	3.3	3.8	3.7	3.4	3.5
Extent to which proximity of workplace to library affects use	2.8	3.2	3.2	3.3	3.2	3.4	3.0	3.2	2.9
Times used in past 6 months	8.7	8.5	15.8	8.4	8.8	15.9	14.3	8.2	13.2
Reasons for nonuse	%	%	%	%	%	%	%	%	%
Had no information needs	40.7	29.2	10.0	24.2	37.5	10.0	43.8	81.3	37.5
Information needs were more easily met some other way	92.3	92.6	90.9	90.9	93.8	90.0	94.7	92.9	92.3
Tried the library before; couldn't find the information needed	0.0	4.5	0.0	9.4	14.3	20.0	13.8	0.0	12.5
Library staff is not cooperative or helpful	0.0	0.0	10.0	3.1	7.1	10.0	3.4	0.0	0.0
Library staff does not understand information needs	0.0	0.0	11.1	6.3	7.1	0.0	7.4	0.0	0.0
Library did not have the information needed	22.7	18.2	30.0	18.8	41.4	30.0	48.3	7.7	25.0
Have my own library and do not need another library	20.8	21.7	55.5	25.0	17.2	30.0	38.7	25.0	25.0
Library is too slow in getting information needed	9.1	21.7	20.0	28.1	10.7	0.0	26.7	7.1	50.0
Have to pay to use library	0.0	0.0	11.1	3.1	0.0	0.0	7.1	7.7	0.0
Are discouraged from using library	0.0	8.7	0.0	0.0	0.0	10.0	3.6	8.3	0.0

Information Seeking and Source Selection

Survey respondents were given a list of the following information sources: (a) used personal stores of technical information, (b) spoke with coworkers inside the organization, (c) spoke with colleagues outside the organization, (d) spoke with a librarian or technical information specialist, (e) used literature resources in the organization's library, and (f) searched (or had someone search for me) an electronic (bibliographic) database. U.S. aerospace engineers and scientists were asked to identify the steps they had followed to obtain the information they had used in completing or solving their most important project, task, or problem in the past six months by sequencing these items (i.e., #1, #2, #3, #4, and #5) (Table 7.14). They were instructed to place an "X" beside any information source not used during the process of obtaining the information.

The process of seeking information might not be unidimensional (i.e., individual sources might be consulted more than once during the search process). Furthermore, by identifying the order in which the survey respondents consulted the various sources, we hoped to determine the general or typical search strategy that U.S. aerospace engineers and scientists use when seeking information. In general, the steps mentioned most often were the first steps they used to obtain information. The quest for information begins with an informal search. Most respondents first used their personal stores of technical information; then, they asked coworkers within the organization and colleagues outside the organization. If the need for information has not been met to this point, the search for information begins a more formalized process.

Disciplines notwithstanding, the strategies of U.S. aerospace engineers and scientists we surveyed were to (a) search personal stores of technical information; (b) discuss problems with coworkers or key persons within the organization; (c) discuss problems with colleagues outside the organization; (d) use literature resources in a library; (e) search (or have someone search) an electronic (bibliographic) database; and (f) ask a librarian. Among all disciplines, the first step—using personal stores of technical information—does not specify whether respondents are using a formal or an informal information source. We did not ask the types of information contained in the personal stores, but if Table 7.14 is a guide, it is likely that many more informal than formal sources are used. The U.S.

Table 7.14. Information Sources Used To Solve Project, Task, or Problem by U.S. Aerospace Engineers and Scientists [N=222; 142; 128; 209; 134; 96; 261; 68; 87]

Source	Research %	Design %	Avionics %	Struc-tures %	Propul-sion %	Human Factors %	Produc-tion %	Flight Test %	Sales/Service %
Personal store of technical information	93.2	95.8	96.9	95.7	92.5	97.9	92.0	92.6	87.4
Spoke with coworkers inside organization	94.6	93.7	89.8	94.7	91.8	96.9	90.8	89.7	90.8
Spoke with colleagues outside organization	89.2	85.2	85.9	87.6	82.1	80.2	73.6	76.5	75.9
Used literature resources in organization's library	74.3	63.4	67.2	69.9	65.7	62.5	55.9	48.5	52.9
Spoke with librarian/technical information specialist	36.9	45.1	39.1	44.5	26.1	41.7	30.7	16.2	31.0
Searched (or had someone search) electronic (bibliographic) database	62.6	56.3	68.0	56.9	38.1	51.0	38.7	29.4	42.5

aerospace engineers and scientists we surveyed next use other informal sources most often. They then discuss the problems with coworkers inside and colleagues outside their organizations and with their managers. They may receive recommendations or leads to formal information sources in these discussions, but our sense is that does not happen often. Only after informal channels and sources have been utilized do they enter the formal information structure—using a library or a librarian or searching a database for information.

Formal information channels and sources are likely to be the last used. Information is gathered through the most accessible means until an answer is found. Formal channels and sources of information may play only a supporting role (verifying informal information) or be consulted when the problem is especially challenging. These findings support the results of Allen (1977) but allow us to explore information-seeking behavior more fully. Engineers are reputed to be "social" researchers (i.e., they gather information from coworkers and colleagues). Although the assumption appears valid, we must remember that they first go to the most accessible sources (i.e., their personal stores of information). Within this collection, they make their own subjective decisions about the quality, reliability, and relevance of the information contained therein. According to Allen, there is little evidence that engineers will search further if they find an answer in their personal stores. Consequently, if the information in their personal stores is not complete, not of high quality, and not current, then they are not using the best information in problem solving.

Allen (1977) also reported that there is a known but somewhat vague limit on the amount of help that engineers can receive from coworkers and colleagues. Whereas interpersonal communication, group projects, and teams may be the norm in aerospace, an engineer who asks for "too much" help would not be considered fully competent. In those cases where engineers feel they are passing the threshold of asking for too much information too often, they will then rely on other sources. With the advent of electronic communications and the Internet, the possibility of acquiring other sources of information increases. However, although Internet access may increase the availability of information, it may also increase the likelihood of obtaining information that is unreliable, outdated, or simply incorrect.

One final comment regarding engineers and their use of informal information sources. The data show that the U.S. aerospace engineers and scientists we surveyed consider coworkers inside the organization to be an important source of information. The data in Table 7.14 show that respondents consulted with their coworkers almost as often as they used their personal stores of information. Obviously coworkers are accessible, but there is no indication that they can or do provide the best information. The responses to a question in an SAE telephone survey (see Table 7.15) about why engineers use coworkers as information sources demonstrates why coworkers are valued as information sources (Pinelli, Kennedy, and White, 1992). The primary reason is that a coworker will provide information that is relevant to the work being done. A much smaller proportion report that the technical quality of the information is the reason for asking a coworker. These data further indicate the potential for U.S. aerospace engineers and scientists not having access to or using information that is technically accurate, comprehensive, and state-of-the art.

Table 7.15. Reasons U.S. Aerospace Engineers and Scientists Use Coworkers as Sources of Information

Reasons	n	%
When you perform your job, coworkers in your place of employment are more important sources of information than are outside sources of information	236	78.9
Your preferred method of obtaining technical information is to communicate with coworkers in your place of employment	242	80.6
Your primary reason for using coworkers to obtain technical information:		
They are accessible	35	13.3
The information they have is relevant to your job	131	49.8
The information they have is of high technical quality	45	17.1
A combination of the above	52	19.8

Use of Federally Funded R&D

A majority of respondents in each discipline reported using the results of federally funded aerospace R&D in performing their present professional duties. Exploring the use of federally funded research in the critical tasks and projects that U.S. aerospace engineers and scientists undertake might be more instructive. Survey respondents were asked a series of questions designed to learn about their use of federally funded aerospace R&D in completing or solving the most important project, task, or problem they had worked on in the past six months. Survey respondents were asked to categorize their most important project, task, or problem (Table 7.16). Most respondents categorized it as design and development, research, or management. Survey respondents were also asked to categorize the nature of the duties they performed in completing or solving the most important project, task, or problem they had worked on in the past six months. A majority of respondents categorized their duties as engineering, followed by management. A majority of respondents indicated that they had worked with others in completing or solving their most important project, task, or problem. The number of groups ranged from a high of 4.9 and 4.8 for production and design, respectively, to a low of 2.3 and 2.4, respectively, for research and avionics. The number of people per group ranged from a high of 10.4 for design and production to a low of 5.2 and 5.6, respectively, for research and avionics.

In addition, the survey respondents rated the overall complexity and uncertainty of their most important project, task, or problem (Table 7.16). Mean complexity scores were highest for respondents in research ($\bar{x}$ = 4.3 out of a possible 5.00) and lowest for respondents in sales and service ($\bar{x}$ = 3.6). Mean uncertainty scores were somewhat lower than the complexity scores and ranged from a high of $\bar{x}$ = 3.9 for research to a low of $\bar{x}$ = 3.0 for sales and service. Correlation coefficients (Pearson's r) were calculated to compare overall level of project, task, or problem complexity and technical uncertainty and the level of project, task, or problem by category and technical uncertainty. The correlation coefficients for most of the disciplines were .4 and .5, which indicate a fairly predictable relationship. As complexity increases, so does uncertainty.

The data in Table 7.17 show that those U.S. aerospace engineers and scientists who used the results of federally funded aerospace R&D in their most important project actively seek it from multiple sources. Respondents in all of the disciplines used coworkers inside and colleagues outside their organizations and NASA and DoD technical reports at or about the same rate to learn about or to ob-

Table 7.16. Correlation of Project Complexity and Technical Uncertainty by Categorization of Most Important Project, Task, or Problem Completed by U.S. Aerospace Engineers and Scientists in Past Six Months [N=222; 142; 128; 209; 134; 96; 261; 68; 87]

	Research	Design	Avionics	Structures	Propulsion	Human Factors	Production	Flight Test	Sales/Service
	r	r	r	r	r	r	r	r	r
Complexity Uncertainty Correlations									
Overall	0.5**	0.4**	0.4**	0.4**	0.5**	0.3**	0.2**	0.4**	0.4**
Research	0.5**	0.9**	0.4**	0.5**	0.6**	0.8*	0.4	0.8	0.3
Design/development	0.2	0.3	0.5**	0.3*	0.5**	0.3*	0.3**	0.5*	0.3
Manufacturing/production	---	0.4	---	0.5*	1.0	0.1	0.1	0.9	-1.0**
Quality assurance/control	---	---	---	0.7	---	0.3	---	0.5	---
Computer applications	0.3	1.0	---	---	0.9	0.0	-0.1	---	---
Management	0.7**	0.5*	0.7	0.1	0.5**	0.8	0.4*	0.5*	0.5**
Other	0.5	0.5	-0.2	0.6	0.3	-0.1	0.7*	---	0.2
	$\overline{x}$	$\overline{x}$	$\overline{x}$	$\overline{x}$	$\overline{x}$	$\overline{x}$	$\overline{x}$	$\overline{x}$	$\overline{x}$
Overall score									
Complexity	4.3	4.2	4.1	4.1	4.1	4.0	3.9	3.7	3.6
Uncertainty	3.9	3.7	3.8	3.7	3.6	3.4	3.3	3.2	3.0

*r values are statistically significant to $p \leq 0.05$.
**r values are statistically significant to $p \leq 0.01$.

Table 7.17. Sources Used to Learn About and Problems Encountered Using the Results of Federally Funded Aerospace R&D by U.S. Aerospace Engineers and Scientists [N=222; 142; 128; 209; 134; 96; 261; 68; 87]

Source	Research	Design	Avionics	Structures	Propulsion	Human Factors	Production	Flight Test	Sales/Service
	%	%	%	%	%	%	%	%	%
Coworkers inside organization	95.5	90.3	72.7	83.6	89.6	93.1	81.0	92.0	83.3
Colleagues outside organization	81.1	71.4	87.4	80.3	70.5	69.0	61.1	69.2	87.5
NASA and DoD contacts	83.1	62.3	79.8	73.5	59.7	40.0	52.9	56.0	82.6
Publications such as NASA STAR	21.0	19.0	7.5	12.5	12.3	20.7	27.5	18.2	22.7
NASA and DoD sponsored and co-sponsored conferences and workshops	58.8	31.7	44.7	60.5	44.4	33.3	41.2	33.3	63.6
NASA and DoD technical reports	86.2	75.7	77.8	81.5	66.2	56.7	72.2	64.0	72.7
Professional and society journals	76.9	72.1	87.4	73.9	57.9	83.9	72.7	65.4	62.5
Librarians inside organization	33.5	43.1	27.7	33.9	25.7	41.4	32.1	36.4	34.8
Trade journals	32.0	56.1	47.7	54.0	36.0	83.3	63.5	40.9	58.3
Searches of computerized databases	61.8	53.0	70.1	53.6	26.4	56.7	58.5	28.6	40.9
Professional and society meetings	57.0	41.5	78.4	64.3	44.7	48.4	50.0	50.0	70.8
Visits to NASA and DoD facilities	48.8	44.8	42.9	46.6	46.7	26.7	25.0	31.8	73.9
Problem	%	%	%	%	%	%	%	%	%
Time and effort to locate results	38.4	66.7	64.9	52.8	45.7	69.7	55.6	60.7	58.6
Time and effort to obtain results	37.8	64.1	64.9	57.5	46.9	63.6	52.4	67.9	55.2
Accuracy, precision, and reliability of results	21.1	28.2	23.4	26.8	28.4	36.4	25.4	21.4	37.9
Distribution limitations or security restrictions of results	15.7	39.7	28.7	27.6	27.2	24.2	25.4	32.1	24.1
Organization or format of results	17.8	17.9	13.8	15.0	24.7	21.2	15.9	21.4	20.7
Legibility or readability of results	20.5	21.8	13.8	15.7	25.9	9.1	15.9	17.9	13.8

tain the results of federally funded R&D. Use of NASA and DoD contacts was highest among respondents in sales and service, research, and avionics. Attendance at NASA and DoD sponsored workshops, visits to NASA and DoD facilities, and publications such as *NASA STAR* were used to a lesser extent to learn about or to obtain the results of federally funded R&D.

The results in Table 7.17 clearly indicate that the federal initiatives designed to create awareness and disseminate the results of federally funded aerospace R&D are limited in terms of their effectiveness, especially among manufacturing and production engineers. This is not surprising given the nature of NASA aeronautical research and technology (R&T), the focus of which is not on production. But, it also points out that NASA-funded work that results in new production techniques and processes or that leads to improvement in existing production techniques and processes may not be getting to manufacturing and production engineers directly. These results may get filtered through colleagues and trade journals, but this does not ensure that the most appropriate or best information is reaching manufacturing and production engineers. Further, the data in Table 7.17 suggest that the overall use of such formal sources as librarians and publications like NASA *STAR* is very small. The moderate use of computerized (bibliographic) databases also indicates that few engineers in industry use the formal channels that are designed to disseminate the results from federally funded R&D to them.

Table 7.17 describes some of the problems that U.S. aerospace engineers and scientists faced when using the results of federally funded R&D. To confirm the problems with the formal system, the respondents in all disciplines reported that the time and effort it took to *locate* the results and the time and effort it took to *obtain* the results were the primary difficulties they had with accessing the results of federally funded R&D. Distribution limitations and security restrictions do not appear to be much of an obstacle to using the results, but the engineers might not be aware of the existence of the most secure or restricted information. When they obtain the results of federally funded R&D, they report few problems with it in terms of accuracy, reliability, readability, or organization. This finding is consistent among respondents in all disciplines.

Finally, data indicate that the use of federally funded R&D is not uniform among respondents representing the spectrum of aerospace from research through development and production to technical services. However, those who use it appear to be satisfied with its quality. The data also indicate that although the use of DoD and NASA technical reports varies among respondents across the spectrum, the reports are considered to be relevant, to have good technical quality, and to contain comprehensive data and information. Furthermore, DoD and NASA technical reports appear to serve as a permanent link between DoD and NASA and the U.S. aerospace industry. The problems associated with awareness and widespread use have little to do with the reports themselves but rather with the dissemination system, its design, and implementation.

USE OF THE BIBLIOGRAPHIC PRODUCTS THAT PROVIDE AWARENESS OF AND LINKAGES TO THE RESULTS OF FEDERALLY FUNDED AEROSPACE R&D

In a separate study, two surveys were used to collect data about aerospace engineers' and scientists' use of the bibliographic products designed to provide awareness of and linkages to the results of federally funded R&D (Pinelli, Barclay, and Kennedy, 1994). AIAA members in academia, government, and industry were surveyed regarding the use of current awareness and bibliographic products. We asked those U.S. aerospace engineers and scientists a variety of questions concerning these products that covered use, familiarity with, frequency of use, reasons for nonuse, and the factors affecting use. In addition, we asked a series of questions regarding respondents' use of, frequency of use, reasons for nonuse, and problems encountered using federally funded aerospace R&D. Because the initial results of the study indicated little use of these products by aerospace engineers and scientists, we did not include the questions in our surveys of other groups. The AIAA members, we believe, would be most likely to use these products because among all the samples, the AIAA sample contained the greatest proportion of scientists and research engineers.

Background

The process of disseminating the results of federally funded aerospace R&D includes three functions: (a) production, (b) distribution, and (c) awareness. The production process appears to create a quality product. The initial or primary distribution of DoD and NASA technical reports is made to libraries and technical information centers in academia, government, and industry, with copies of unclassified and unlimited distribution reports going to the National Technical Information Service (NTIS)—the U.S. government's repository for government technical reports. A limited number of reports are provided to authors for "scientist to scientist" exchanges, but most are sent to surrogates for secondary and subsequent distribution. For those who do not receive the reports directly, both end users (i.e., engineers and scientists) and information intermediaries (i.e., librarians) are expected to facilitate delivery by using the various bibliographic products.

Bibliographic Products

Surrogates serve as technical report repositories for the producers of technical reports. These include the Defense Technical Information Center (DTIC), the NASA Center for Aero Space Information (CASI), and NTIS. These surrogates created such technical report announcement print journals as *CAB* (Current Awareness Bibliographies), *STAR* (Scientific and Technical Aerospace Reports), *NASA SP-7037*, and *GRA&I* (Government Reports Announcement and Index) and such computerized retrieval systems as *DROLS* (Defense RDT&E Online System), *RECON* (Research Connection), and *NTIS On-line* that permit online access to databases containing U.S. government technical reports. The federal system that disseminates the results of federally funded R&D assumes that both end users and information intermediaries will use these products. In reality, they are most often used by intermediaries.

RECON, *DROLS*, and *NTIS On-line* are computerized databases. *RECON* and *DROLS* are not available commercially and have "different access levels." *NTIS On-line* is available commercially through Dialog. The *AIAA Aerospace Database* includes citations to technical reports in the NASA database and open literature and is available on CR-ROM (Compact Disc-Read Only Memory) and online commercially through Dialog. The availability of and access to *RECON* and *DROLS* may be restricted or limited. Further, the assistance of a librarian or a working knowledge of NASA or DoD systems is required to formulate the search process. For end users who do not know these systems, the start is even more difficult.

The on-line version of *DROLS* is an unlikely candidate for end-user searching because physical access is limited, and its search structure is difficult at best for even the most highly trained intermediary. However, the CD-ROM version has a "user friendly" interface. *RECON* can be searched by end-users, however, a reasonable level of training is necessary to gain proficiency. Use of Dialog to search *NTIS On-line* and the *AIAA Aerospace Database* can be costly and, for that reason, is perhaps best left to trained searchers. Both the *AIAA Aerospace Database* and *NTIS On-line* are available to librarians as CD-ROMs and some end user searching is encouraged.

For the end user, four print journals are often available in technical libraries, which, in general, would make them more accessible to end users. *STAR* and *CAB* (*CAB* is no longer produced) are biweekly journals that list recently released government technical reports. *NASA SP-7037* (no longer produced) is a monthly journal that includes journal articles and other documents in addition to technical reports. *GRA&I* contains a wider range of government reports. For the end user, the primary cost of searching is the time required to find the information that is sought.

Use, Frequency of Use, and Familiarity with Bibliographic Products

We asked the AIAA members about their use of the four print and three electronic bibliographic and current awareness products. The responses indicate that, overall, they made little use of these products (Table 7.18). NASA *STAR* was used most frequently but by fewer than 25% of the AIAA members. Fewer than 10% used NASA *SP-7037*, DoD *CAB*, and NTIS *GRA&I*. Among the users, NASA *STAR* was used "sometimes" or "frequently" by about 16% of the U.S. aerospace engineers and scientists. Most users of the three products reported that they seldom used them. The AIAA members who did not use the four print products were asked if they were familiar with them. With the exception of NASA *STAR* (25% of the nonusers indicated familiarity), most were not familiar with the four print products.

There was even less use of and familiarity with the electronic bibliographic and current awareness products: NASA *RECON*, DoD *DROLS*, and *NTIS On-line*. The sample members made very little

Table 7.18.   Use, Frequency of Use, and Familiarity with Selected Announcement, Current Awareness, and Bibliographic Products by U.S. Aerospace Engineers and Scientists

Products	No %	Yes % Frequently	Sometimes	Seldom	If No, Familiar With % No	Yes
Print:						
STAR	77.5	3.8	12.0	6.7	74.1	25.9
NASA SP-7037 [a]	93.6	0.8	3.5	2.1	90.2	9.8
CAB [a]	98.3	0.3	0.6	0.8	96.2	3.8
GRA&I	96.3	0.6	1.5	1.6	96.6	3.4
Electronic:						
RECON	88.2	2.3	5.0	4.5	93.8	6.2
DROLS	96.7	0.4	1.9	1.0	98.1	1.9
NTIS File	82.7	3.1	8.7	5.5	86.1	13.9

[a]Ceased publication.

use of these products. *NTIS On-line* was used most often but by only 17% of the sample. NASA *RECON* was used by 11%. Among the nonusers, there is little awareness of these electronic products. About 14% of the nonusers knew about *NTIS File*, but only 2% were aware of *DROLS*. Because these products are designed for use by those who need access to information, the relatively low use and awareness of these products indicates a substantial gap between the systems designers and the end users.

Reasons for NonUse

We asked those who were aware of the bibliographic products but did not use them to tell us why not (Tables 7.19 and 7.20). The reasons for nonuse varied by product. Among the print products, availability and accessibility and relying on others to search for information were the reasons given most often. Among the electronic sources, some noted that they could get the information more easily from other sources. Very few said that it was lack of skill with computers or that the sources were not user friendly. The federal dissemination system is intermediary based (see Chapter 4), but there are also supposed to be direct links from the clearinghouses to the end users and from end users to producers. Because there were such a small number of respondents in each study who

Table 7.19.   Reasons   for   Nonuse   of   Selected   (Print) Announcement, Current Awareness, and Bibliographic Products by U.S. Aerospace Engineers and Scientists

Reasons	NASA STAR		NASA SP-7037 [a]		DoD CAB [a]		NTIS GRA&I	
	n	%[b]	n	%[b]	n	%[b]	n	%[b]
Not easily available/accessible	74	36.1	32	31.4	15	24.6	13	23.6
Not relevant	55	26.8	22	21.6	10	16.4	9	16.4
Don't use technical reports	12	5.9	4	3.9	3	4.9	5	9.1
Get same information more easily from another source	36	17.6	16	15.7	8	13.1	7	12.7
Rely on others to search for needed information	79	38.5	38	37.3	15	24.6	12	21.8
Difficult to physically obtain what is in there	11	5.4	4	3.9	2	3.3	2	3.6
Other	16	7.8	7	6.9	3	4.9	3	5.5

[a]Ceased publication.
[b]Percentages do not total 100 because respondents could provide more than one response.

Table 7.20.   Reasons   for   Nonuse   of   Selected   (Electronic) Announcement, Current Awareness, and Bibliographic Products by U.S. Aerospace Engineers and Scientists

Reasons	RECON		DROLS		NTIS File	
	n	%[a]	n	%[a]	n	%[a]
Not easily available/accessible	21	30.0	8	21.6	38	30.9
Not relevant	16	22.9	4	10.8	47	38.2
Skill in using computer hardware/software	4	5.7	2	5.4	3	2.4
Skill in using a database	6	8.6	1	2.7	6	4.9
Not timely or current	0	0.0	1	2.7	4	3.3
Get same information more easily from another source	15	21.4	4	10.8	26	21.1
Difficult to physically obtain what is in there	1	1.4	1	2.7	4	3.3
System is not user friendly	0	0.0	1	2.7	0	0.0
Other	11	15.7	4	10.8	15	12.2

[a]Percentages do not total 100 because respondents could provide more than one response.

answered these questions, we interpret the data to mean that most aerospace engineers and scientists are simply not aware of the products. These data further indicate that the direct links are either not used or, in a small number of cases, the aerospace engineers and scientists we surveyed let others do the searching for them.

Factors That Affect the Use of Bibliographic Products

The AIAA members who used the print and electronic products were asked to indicate how their use of these products was affected by seven factors (Tables 7.21 and 7.22). We focus this analysis on *STAR* because it is the product used most often by U.S. aerospace engineers and scientists. The differences among the factors that affect the use of *STAR* are not substantial, especially considering the small number of users. Accessibility, ease of use, and familiarity or experience were the factors mentioned most often. Only expense seemed to have less impact on the decision. We interpret this to indicate that the cost of using *STAR* reduces its use by U.S. aerospace engineers and scientists.

Table 7.21. Factors Affecting Use of Selected (Print) Announcement, Current Awareness, and Bibliographic Products by U.S. Aerospace Engineers and Scientists

	Overall Mean[a] Influence of Factor on Use of Print Products							
	NASA STAR		*NASA SP-7037*[b]		*DoD CAB*[b]		*NTIS GRA&I*	
Factors	n	x̄	n	x̄	n	x̄	n	x̄
Accessibility	213	3.8	60	3.8	17	3.3	33	3.5
Ease of Use	212	3.6	58	3.7	17	3.3	33	3.4
Expense	209	2.7	57	3.0	17	2.6	32	2.9
Familiarity or experience	211	3.6	58	3.3	17	3.2	33	3.3
Technical quality or reliability	211	3.5	59	3.8	18	3.6	31	3.7
Comprehensiveness	210	3.5	59	3.6	17	3.4	32	3.7
Relevance	211	3.5	59	3.4	17	3.6	32	3.6

[a]A 5-point scale was used to measure influence, where 5 indicates the greatest influence.
[b]Ceased publication.

Table 7.22. Factors Affecting Use of Selected (Electronic) Announcement, Current Awareness, and Bibliographic Products

| | Overall Mean[a] Influence of Factor on Use of Electronic Products | | | | | |
| | RECON | | DROLS | | NTIS File | |
Factors	n	x̄	n	x̄	n	x̄
Accessibility	103	4.1	30	3.8	153	3.8
Ease of Use	100	3.5	29	3.5	149	3.4
Expense	99	2.7	28	3.9	144	2.6
Familiarity or experience	101	3.3	29	3.2	148	3.3
Technical quality or reliability	102	3.6	29	3.5	150	3.5
Comprehensiveness	104	3.7	29	3.6	149	3.6
Relevance	103	3.7	29	3.6	148	3.5

[a]A 5-point scale was used to measure influence, where 5 indicates the greatest influence.

NTIS File is the electronic source used most often by U.S. aerospace engineers and scientists perhaps because it is more well-known and is easier to use than the other electronic products. The two factors that most influence its use are accessibility and comprehensiveness. The relatively small number of aerospace engineers and scientists who actually use these products makes interpretations tentative at best, but there is evidence across both formats that the bibliographic and current awareness products must be accessible if they are going to be used by end users.

From our study, we are not able to determine if these products are inaccessible to U.S. aerospace engineers and scientists because they or their organizations do not have them or access to them, or if they are available somewhere in the organization, but the participants in this study do not know about them. In either case, the findings indicate problems with the federal dissemination system. No aerospace engineer or scientist can be expected to know personally all the researchers involved in research relative to his or her expertise. Current awareness publications should help keep one aware of the relevant literature, especially technical report literature that has very limited end user distribution. The lack of use and even awareness of these products means that only if intermediaries are very active will aerospace engineers and scientists be kept abreast of the latest research in their areas.

CONCLUSIONS

We investigated the communication practices and information-related activities of U.S. aerospace engineers and scientists who represent the spectrum of aerospace from research through development and production to technical services. In doing so, we explored such topics as the number of hours spent communicating information, the kinds and amounts of information produced and used, the use of libraries and computer networks, the strategies employed in seeking information, the reasons for using specific information products, and the ratings of specific information products. We also collected data about aerospace engineers' and scientists' use of the bibliographic products designed to provide awareness of and linkages to the results of federally funded R&D.

Our investigation of U.S. aerospace engineers and scientists confirms Mailloux's (1989) findings concerning the time spent by engineers communicating information. Discipline and position in the organizational hierarchy notwithstanding, survey respondents consider the ability to communicate effectively to be important and indicate that as they have advanced professionally, the amount of time they devote to communicating technical information has increased. Furthermore, most of the respondents spend more time communicating information to others *orally* than they spend writing. Our findings support Allen's (1977) claims about the fundamental role of oral communication in engineering work. They also confirm the findings of Czepiel (1974), Rogers (1982), and Utterback (1971) about the importance of oral communication in technological innovation and the R&D process. These findings further underscore the vital role of oral communication in the exchange of *tacit* or "sticky" knowledge, which lies at the heart of most technology (Teece, 1981; von Hippel, 1994). Much engineering work is collaborative in nature (Adams, 1991; Florman, 1987) and, as Sabbagh (1996) found, the RD&P of LCA is also inherently collaborative. Findings from our study support both claims and show that the written communication process is collaborative as well.

The U.S. aerospace engineers and scientists who participated in this phase of our research use and produce a variety of formal and informal information products. Survey respondents reported using more information products than they produce, and they use more informal than formal ones. Of the information products they produce, considerably more are informal than formal products. Our findings support Allen's (1977) claims that most engineers are not expected to make original contributions to the published literature of their disciplines. The importance of oral communication is reinforced by the overall use of technical talks and presentations reported by the study participants across the spectrum of aerospace from research through development and production to technical services. We assume that most of the information products (e.g., in-house technical reports) are produced for internal distribution. The bulk of the "open literature" information products were produced by respondents who identified themselves as scientists and engineers engaged in research. Use by scientists and research engineers of the open literature and to some extent in-house and NASA technical reports and the considerable use of in-house technical reports and various forms of informal technical information like drawings and specifications by other engineers support our assertion in Chapter 5 of the existence of knowledge subcommunities within engineering. Regardless of education, professional duties performed, and technical discipline, survey respondents rely on the same three factors—relevance to task, technical quality, and comprehensiveness of data and information—for determining which information products they use.

Computer and information technology holds the promise of increasing interorganizational communication and improving personal and organizational productivity (Markus, 1994). According to Daft and Lengel (1986), they may also increase "information richness" by matching information processing requirements with the appropriate communication channel or source, thereby improving organizational effectiveness. The U.S. aerospace engineers and scientists we surveyed have access to computer networks. Based on the number of hours they are used each week, it appears that computer networks are vital to aerospace work. The heavy use of email and the Internet and the low use of computer networks for such functions as searching bibliographic databases, accessing library catalogs, and ordering documents from a library may indicate that U.S. aerospace engineers and scientists view and use computer networks as an extension of the informal, collegial approach to the search for information. Nevertheless, we interpret the data to mean that computers, computer networks, and related information technology should play an increased role in diffusing the results of federally funded aerospace R&D.

The importance of libraries to engineering and organizations involved in technological innovation and R&D is well established (Matarazzo, 1990, 1987). One of several critical functions performed by libraries is "boundary-spanning," a technique by which corporate intelligence and related technical and business information external to the organization is gathered and distributed internally. Most of the U.S. aerospace engineers and scientists in our study work in organizations that have a library. Most respondents reported having used a library in the six months prior to completing the survey. They indicated that a library is important in performing their professional duties and that workplace proximity to a library affects its use. Interaction with libraries and librarians appears to take place only after an *informal* and collegial search for information is complete. The most frequent reason given by those respondents who had not used a library in the past six months was that their information needs were more easily met some other way. We interpret these data to indicate that the "some other way" most likely means obtaining needed information from coworkers inside and colleagues outside their organizations.

The results of federally funded aerospace R&D are used by U.S. aerospace engineers and scientists in performing their present professional duties. Respondents in all disciplines used coworkers inside and colleagues outside their organizations and DoD and NASA technical reports to learn about or to obtain the results of federally funded aerospace R&D. Furthermore, these respondents reported that the time and effort it took to locate the results and the time and effort it took to obtain them were the primary difficulties they had with accessing the results of federally funded aerospace R&D. Most of the "producer initiated" activities and bibliographic products appear least effective in creating awareness of, linkages to, and promoting the use of the results of federally funded aerospace R&D. From the above, we conclude that (a) the same informal and collegial process that characterizes the information-seeking behaviors of U.S. aerospace engineers and scientists is also utilized to obtain the results of federally funded aerospace R&D; (b) the role played by libraries and librarians as information intermediaries in aerospace knowledge diffusion needs further examination; and (c) the problems associated with the lack of widespread use and awareness of federally funded aerospace R&D are specific to the design and implementation of the existing federal dissemination system.

Although the research reported in this chapter is consistent with the results of other research focusing on the communication practices and information-related activities of engineering and scientists working in R&D, it is neither exhaustive nor conclusive. There is simply far more to be learned about the production and use of information across the spectrum of U.S. aerospace than can be determined from a single study. There is much to be learned about the diffusion of knowledge at the (intra) organizational level and between small, medium, and large firms, both domestic and foreign, involved in the RD&P of LCA. In general, aeronautical R&T has become so sophisticated, broad, and expensive that even the largest manufacturer of LCA simply cannot afford to do it all. Consequently, LCA manufacturers, to remain competitive in a global marketplace, must seek to identify and obtain aeronautical R&T from a variety of external sources. As Leonard-Barton (1995) points out:

> The ability of a firm to recognize the value of new, external information, assimilate it, and apply it to commercial ends is critical to its innovative capabilities. (p. 136)

Given that the ability to identify and absorb information that is external to the firm has become more important as the world's economic borders dissolve and the market for knowledge expands, an understanding of the process by which the results of federally funded aerospace R&D is diffused within U.S. aerospace becomes an economic, as well as, a national imperative.

IMPLICATIONS FOR DIFFUSING THE RESULTS OF FEDERALLY FUNDED AEROSPACE R&D

This book is aimed at helping policymakers and information professionals determine how best to manage federally funded aerospace knowledge. We recognize that an incredible amount of aerospace knowledge exists. However, users of this knowledge tell us that locating and obtaining this knowledge is a major impediment to its utilization. Even firms with the most sophisticated capabilities to absorb knowledge would find it difficult to assimilate and apply the results of federally funded aerospace R&D. Based on what is known, the existing federal system should be modified extensively or replaced with a closed-loop, user-focused system that is based on a diffusion-oriented model.

CHAPTER REFERENCES

Adams, J.L. (1991). *Flying Buttresses, Entropy, and O-rings: The World of an Engineer*. Cambridge, MA: Harvard University Press.

Allen, T.J. (1977). *Managing the Flow of Technology: Technology Transfer and the Dissemination of Technological Information Within the R&D Organization*. Cambridge, MA: MIT Press.

Buckland, M.K. (1983). *Library Services in Theory and Context*. New York, NY: Pergamon Press.

Czepiel, J.A. (1974). "Word-of-Mouth Processes in the Diffusion of a Major Technological Innovation." *Journal of Marketing Research* (May) 11: 172-180.

Daft, R.L. and R.H. Lengel. (1986). "Organizational Information Requirements, Media Richness, and Structural Design." *Managerial Science* 32(5): 554-571.

David, P.A. (1986). "Technology Diffusion, Public Policy, and Industrial Competitiveness." In *The Positive Sum Strategy: Harnessing Technology for Economic Growth*. R. Landau and N. Rosenberg, eds. Washington, DC: National Academy Press, 373-391.

Dervin, B. and M.S. Nilan. (1986). "Information Needs and Uses." Chapter 1 in *Annual Review of Information Science and Technology. 21*. M.E. Williams, ed. New York, NY: John Wiley, 3-33.

Dougherty, R.M. (1990). "Needed: User-Responsive Research Libraries." *Library Journal* (January) 116(1): 59-62.

Florman, S.C. (1987). *The Civilized Engineer*. New York, NY: St. Martin's Press.

Gerstberger, P.G. and T.J. Allen. (1968). "Criteria Used By Research and Development Engineers in the Selection of an Information Source." *Journal of Applied Psychology* (August) 52(4): 272-279.

Glassman, N.A. and T.E. Pinelli. (1992). *Source Selection and Information Use by U.S. Aerospace Engineers and Scientists: Results of a Telephone Survey*. Washington, DC: National Aeronautics and Space Administration. NASA TM-107658. (Available NTIS: 92N33299.)

Hagstrom, W.O. (1965). *The Scientific Community*. New York, NY: Basic Books.

Harris, I.W. (1966). *The Influence of Accessibility on Academic Library Use*. Ph.D. Diss., Rutgers University. Available UMI 67-5262.

Kaufman, H.G. (1983). *Factors Related to the Use of Technical Information in Engineering Problem Solving*. Brooklyn, NY: Polytechnic Institute of New York.

Kemper, J.D. (1990). *Engineers and Their Profession*. 4th ed. Philadelphia, PA: W.B. Saunders College Publishing.

Kuhlthau, C.C. (1993). *Information Meaning: A Process Approach to Library and Information Services*. Norwood, NJ: Ablex Publishing.

Lancaster, F.W. (1991). "Has Technology Failed Us?" *Information Technology and Library Management: Festschrift in Honour of Margaret Bechman.* 13th International Essen Symposium, 22-25 October 1989. A.H. Helal and J.W. Weiss, eds. Essen, Germany: Essen University Library.

Leonard-Barton, D. (1995). *Wellsprings of Knowledge: Building and Sustaining the Sources of Innovations.* Boston, MA: Harvard Business School Press.

Lunsford, A. and L. Ede. (1990). *Singular Texts / Plural Authors.* Carbondale, IL: Southern Illinois University Press.

Mailloux, E.N. (1989). "Engineering Information Systems." Chapter 6 in *Annual Review of Information Science and Technology.* 25 M.E. Williams, ed. Amsterdam, The Netherlands: Elsevier Science Publishers, 239-266.

Markus, M.L. (1994). "Electronic Mail as the Medium of Managerial Choice." *Organizational Science* (November) 5(4): 502-527.

Matarazzo, J.M. (1990). *Valuing Corporate Libraries: A Survey of Senior Managers.* Washington, DC: Special Libraries Association.

Matarazzo, J.M. (1987). *President's Task Force on the Value of the Information Professional.* Washington, DC: Special Libraries Association.

Menzel, H. (1966). "Information Needs and Uses in Science and Technology." Chapter 3 in *Annual Review of Information Science and Technology.* 1 C.A. Cuadra, ed. New York, NY: John Wiley, 41-69.

Menzel, H. (1964). "The Information Needs of Current Scientific Research." *Library Quarterly* 34(1): 4-19.

Mick, C.K.; G.N. Lindsey; D. Callahan; and F. Spielberg. (1979). *Towards Usable User Studies: Assessing the Information Behavior of Scientists and Engineers.* Washington, DC: National Science Foundation. (Available NTIS; PB80-177165.)

Mowery, D.C. (1983). "Economic Theory and Government Technology Policy." *Policy Sciences* 16: 27-43.

Orr, R.H. (1970). "The Scientist as an Information Processor: A Conceptual Model Illustrated With Data on Variables Related to Library Utilization." In *Communication Among Scientists and Engineers.* C.E. Nelson and D.K. Pollack, eds. Lexington, MA: D.C. Heath, 143-189.

Paisley, W.J. (1968). "Information Needs and Uses." Chapter 1 in *Annual Review of Information Science and Technology.* 3 C.A. Cuadra, ed. Chicago, IL: Encyclopedia Britannica Inc, 1-30.

Parsaye, K. (1989). *Intelligent Databases: Object-Oriented, Deductive, Hypermedia Technologies.* New York, NY: John Wiley.

Pinelli, T.E. (1991). "The Information-Seeking Habits and Practices of Engineers." *Science & Technology Libraries* 11(3): 5-25.

Pinelli, T.E.; R.O. Barclay; M.L. Keene; J.M. Kennedy; and L.F. Hecht. (1995). "From Student to Entry-Level Professional: Examining the Role of Language and Written Communications in the Reacculturation of Aerospace Engineering Students." *Technical Communication* (Third Quarter) 42(3): 492-503.

Pinelli, T.E.; R.O. Barclay; and J.M. Kennedy. (1996a). *The Technical Communication Practices of U.S. Aerospace Engineers and Scientists: Results of Phase 1 Mail Survey: Aircraft Design Perspective.* Washington, DC: National Aeronautics and Space Administration. NASA TM-110235. (Available NTIS; 97N10132.)

Pinelli, T.E.; R.O. Barclay; and J.M. Kennedy. (1996b). *The Technical Communication Practices of U.S. Aerospace Engineers and Scientists: Results of Phase 1 Mail Survey: Avionics and Electrical Components and Subsystems Perspective.* Washington, DC: National Aeronautics and Space Administration. NASA TM-110232. (Available NTIS; 97N10134.)

Pinelli, T.E.; R.O. Barclay; and J.M. Kennedy. (1996c). *The Technical Communication Practices of U.S. Aerospace Engineers and Scientists: Results of Phase 1 Mail Survey: Flight Test Engineers Perspective.* Washington, DC: National Aeronautics and Space Administration. NASA TM-110231. (Available NTIS; 97N10135.)

Pinelli, T.E.; R.O. Barclay; and J.M. Kennedy. (1996d). *The Technical Communication Practices of U.S. Aerospace Engineers and Scientists: Results of Phase 1 Mail Survey: Human Factors and Crew Integration Perspective.* Washington, DC: National Aeronautics and Space Administration. NASA TM-110233. (Available NTIS; 97N10133.)

Pinelli, T.E.; R.O. Barclay; and J.M. Kennedy. (1996e). *The Technical Communication Practices of U.S. Aerospace Engineers and Scientists: Results of Phase 1 Mail Survey: Manufacturing and Production Perspective.* Washington, DC: National Aeronautics and Space Administration. NASA TM-110236. (Available NTIS: 97N10138.)

Pinelli, T.E.; R.O. Barclay; and J.M. Kennedy. (1996f). *The Technical Communication Practices of U.S. Aerospace Engineers and Scientists: Results of Phase 1 Mail Survey: Propulsion and Aircraft Engine Perspective.* Washington, DC: National Aeronautics and Space Administration. NASA TM-110234. (Available NTIS; 97N10130.)

Pinelli, T.E.; R.O. Barclay; and J.M. Kennedy. (1996g). *The Technical Communication Practices of U.S. Aerospace Engineers and Scientists: Results of Phase 1 Mail Survey: Service/Maintenance and Marketing/Sales Perspective.* Washington, DC: National Aeronautics and Space Administration. NASA TM-110268. (Available NTIS; 97N10145.)

Pinelli, T.E.; R.O. Barclay; and J.M. Kennedy. (1996h). *The Technical Communication Practices of U.S. Aerospace Engineers and Scientists: Results of Phase 1 Mail Survey: Structures and Materials Perspective.* Washington, DC: National Aeronautics and Space Administration. NASA TM-110237. (Available NTIS; 97N10131.)

Pinelli, T.E.; R.O. Barclay; and J.M. Kennedy. (1995). *The Technical Communications Practices of U.S. Aerospace Engineers and Scientists: Results of Phase 1 NASA Langley Research Center Mail Survey.* Washington, DC: National Aeronautics and Space Administration. NASA TM-110208. (Available NTIS; 96N18066.)

Pinelli, T.E.; R.O. Barclay; and J.M. Kennedy. (1994). *The Use of Selected Information Products and Services by U.S. Aerospace Engineers and Scientists: Results of Two Surveys.* Washington, DC: National Aeronautics and Space Administration. NASA TM-109022. (Available NTIS; 94N24649.)

Pinelli, T.E.; A.P. Bishop; R.O. Barclay; and J.M. Kennedy. (1993). "The Information-Seeking Behavior of Engineers." In *Encyclopedia of Library and Information Science.* A. Kent and C.M. Hall, eds., 52:15 New York, NY: Marcel Dekker, 167-201.

Pinelli, T.E.; J.M. Kennedy; and T.F. White. (1992). *Engineering Work and Information Use in Aerospace: Results of a Telephone Survey.* Washington, DC: National Aeronautics and Space Administration. NASA TM-107673. (Available NTIS; 92N34233.)

Rogers, E.M. (1982). "Information Exchange and Technological Innovation." Chapter 5 in *The Transfer and Utilization of Technical Knowledge.* D. Sahal, ed. Lexington, MA: D.C. Heath, 105-123.

Rosenberg, V. (1967). "Factors Affecting the Preferences of Industrial Personnel for Information Gathering Methods." *Information Storage and Retrieval* (July) 3: 119-127.

Rubbert, P.E. (1994). "CFD and the Changing World of Airplane Design." Paper presented at the *19th Congress of the International Council of the Aeronautical Sciences,* 18-23 September, held in Anaheim, CA. In ICAS Proceedings of 1994, Volume I, LVII-LXXXIII.

Sabbagh, K. (1996). *Twenty-First Century Jet: The Making and Marketing of the Boeing 777.* New York, NY: Scribner.

Sayer, J. (1965). "Do Present Information Systems Serve the Engineer?" *Data Processing* 7(2): 24-25, 64-65.

Storer, N.W. (1966). *The Social System of Science.* New York, NY: Holt, Rinehart and Winston.

Teece, D.J. (1981). "The Market for Know-How and the Efficient International Transfer of Technology." *The Annals of the American Academy of Political and Social Science* (November) 458: 81-96.

Tornatzky, L.G. and M. Fleischer. (1990). *The Process of Technological Innovation.* Lexington, MA: D.C. Heath.

Utterback, J.M. (1971). "The Process of Technological Innovation Within the Firm." *Academy of Management Journal* 14(1): 75-88.

Vincenti, W.G. (1990). *What Engineers Know and How They Know It: Analytical Studies From Aeronautical History.* Baltimore, MD: Johns Hopkins University Press.

Voight, M.J. (1961). *Scientists' Approaches to Information.* ACRL Monograph No. 24. Chicago, IL: American Library Association.

von Hippel, E. (1994). "'Sticky Information' and the Locus of Problem Solving: Implications for Innovation." *Management Science* (April) 40(4): 429-439.

Chapter 8

The Relationship between Technical Uncertainty and Information Use by Industry-Affiliated U.S. Aerospace Engineers and Scientists

Claire J. Anderson
Myron Glassman
Thomas E. Pinelli

SUMMARY

Chapter 8 presents selected results of three studies that investigated the relationship between technical uncertainty and information use by U.S. industry-affiliated engineers and scientists involved in aerospace research, development, and production. Specifically, the studies assessed the impact of technical uncertainty on information-seeking behavior in general and the use of federally funded aerospace research and development (R&D) in particular and determined the conditions under which the results of federally funded aerospace R&D are sought and used. Survey respondents included U.S. aerospace engineers and scientists: 341 worked in research and design, 872 in design and development, and 419 in manufacturing and production. Results of the studies revealed that the degree of technical uncertainty affected information-seeking behavior, increasing the frequency of use of information *internal* to the organization and extending the search for information *external* to the organization through formal channels. Use of federally funded aerospace R&D expanded under conditions of increasing technical uncertainty. As technical uncertainty increased, the "value" factors associated with information use—technical quality, comprehensiveness, and relevance—became more important in channel and source selection than the cost or ease of access of information. Conclusions drawn from an analysis of the data and implications for diffusing the results of (U.S.) federally funded aerospace R&D are presented.

INTRODUCTION

Organizations that are involved in technological innovation and research and development (R&D) can be viewed as open systems that must deal with complexity and sources of work-related uncertainty (Katz and Kahn, 1966). This proposition traces its origins to, among others, Galbraith (1977, 1973) and Duncan (1972), who have conceptualized organizations as open systems that collect, process, and use information to reduce uncertainty. Miller (1971) states that organizations, especially those involved in technical innovation and R&D, use *internally* and *externally* derived information to reduce uncertainty. Tyson (1992) and Mowery (1985) note that the aerospace industry, in particular the large commercial aircraft sector, is characterized by the high degree of systemic complexity embodied in the research, design, and production (RD&P) of its products. The large commercial aircraft sector must contend with both technical uncertainty, which is largely *internally* centered, and marketplace uncertainty, which is largely *external* to the organization.

Utterback (1974, 1971) states that to survive and grow, organizations must cope successfully with uncertainty. Information that is largely external to the organization is used to moderate and reduce uncertainty (March and Simon, 1958). However, because of a need for stability and control and to protect their intellectual property and competitive advantage, there is a tendency for organizations involved in technological innovation and R&D to isolate themselves from their external environment. These organizations frequently erect barriers that prohibit or limit access to information that resides in the *external* environment (Gerstenfeld and Burger, 1980). This is especially true for organizations involved in work that may be classified for reasons of national security. As Fischer (1980) points out, there is a real danger of organizations involved in technological innovation and R&D becoming isolated from their *external* environment and from information that resides *external* to the organization.

In this chapter, we report the results of our research that investigated the relationship between technical uncertainty and information use by U.S. industry-affiliated engineers and scientists involved in aerospace RD&P. Specifically, we looked at technical uncertainty and (a) the use of *internal* and *external* information, (b) the use of informal and formal information channels or sources, and (c) the influence of two sets of selected "value" factors—accessibility and quality—on information use.

BACKGROUND

Research related to uncertainty, information channel or source selection, and the impact of accessibility and quality on information use is reviewed in this section. The review was performed to establish a context for the investigation; it is comprehensive but not exhaustive.

Uncertainty

The study of uncertainty as a central concept in organizational research reached its peak in the early 1980s. As a variable, uncertainty has been investigated by researchers concerned with developing theories of organizational management and for explaining how organizations interact with their environments. The assumption that organizations are open systems that collect, process, and use information to reduce work-related uncertainty was advanced by Katz and Kahn, 1966; Thompson, 1967; and Weick, 1969. Zaltman, Duncan, and Holbek (1973) proposed that organizations must deal with sources of *internal* (i.e., organizational or work-related) and *external* (i.e., environmentally-based) uncertainty. Downey and Slocum (1975) defined uncertainty as the difference between information possessed and information required to complete a task. Adding to their work, Tushman and Nadler (1978) proposed three sources of work-related uncertainty: subunit task characteristics, subunit task environment, and inter-unit task interdependence. Stating that organizations must develop information processing mechanisms capable of dealing with external and internal sources of uncertainty, Tushman and Nadler (1980) proposed an information processing model as a means of designing and structuring organizations involved in technological innovation and R&D. Gifford, Bobbitt, and Slocum (1979) report that the greater the degree of uncertainty, the greater the information processing requirements and the greater the use of information *external* to the organization. Rogers (1983, 1982) reports that as an information processor, the engineer-scientist involved in technological innovation and R&D uses information to moderate (i.e., reduce) uncertainty.

Uncertainty is also used to explain the nature of the relationship between organizations and their environments (Jain, 1984). Huber, O'Connell, and Cummings (1975) treated perceived environmental uncertainty as a construct intervening between environmental characteristics (i.e., amount and specificity of information) and information search by organizational decision makers. Schmidt and Cummings (1976), stated that uncertainty involves a perceived inability to control or accurately predict outcomes of the interaction between an organization and its environment, and it implies a lack of information about future events, so that alternatives and their outcomes are unpredictable. Duncan (1972) identified five sources of uncertainty for a firm's *external* environment: customers, suppliers, competitors, sociopolitical climate, and technological milieu.

Brown and Utterback (1985) concluded that individuals who saw the *external* environment as uncertain sought greater contact with sources of information outside their organizations than did individuals who did not see the *external* environment as uncertain. In other words, the higher the degree of perceived environmental uncertainty, the more likely an organization is to collect, process, and use *external* information. Organizations use a variety of techniques or "boundary-spanning activities" to maintain contact with the *external* environment and to acquire information that is *external* to the organization (Boyd, 1989; Duncan, 1972; Hambrick; 1979; Auster and Choo, 1993). Individuals in organizations who perform "boundary-spanning activities" have been referred to as "technological gatekeepers" (Allen and Cohen, 1969; Keller and Holland, 1975; Tushman and Katz, 1980).

Information Channel and Source Selection

An information channel is any medium by which a message may be transmitted from a source to a receiver (Shannon and Weaver, 1949). Chakrabarti, Feineman, and Fuentevilla (1983) define an information channel as a means by which information is moved from one point to another. Three categories of channels are identified in the literature: interpersonal (i.e., oral), written (i.e., paper or text-based), and electronic (i.e., electronic networks). Nilakanta and Scamell (1990) define an information source as a medium in which data, information, or knowledge is stored. Sources include people (e.g., coworkers), documents (e.g., technical reports), and electronic media (e.g., databases). According to Swanson (1987), the use of an information channel or source is discretionary. A body of literature exists concerning information channel and source use and organizations involved in technological innovation and R&D (Holland, Stead, and Leibrock, 1976). Much of this literature is concerned with the use of channels and sources for information *acquisition.*

Researchers concerned with the diffusion of innovation have investigated the use of channels and sources for information *dissemination.* Zmud (1983) reported that different information channels have different levels of influence on the adopters of innovations. Ebadi and Utterback (1984) demonstrated that the diffusion of innovation can be affected by both information channel and source selection. Hauptman (1986); Chakrabarti, Feineman, and Fuentevilla (1983); and Czepiel (1975) reported that the characteristics of information channels and sources can influence the communication of information within an organization and the diffusion of innovation outside the organization.

Concerning the use of channels and sources for information *acquisition*, the literature posits two general models: the "cost-benefit" model and the "law of least effort" model (Hardy, 1982). The "cost-benefit" model has its roots in normative economics. This model proposes that information seeking is rational and that information seekers make an assessment of the expected benefits (i.e., amount or quality) and costs (i.e., social and economic) of using a particular information channel or source on that basis. The "least-effort model" proposes that channel or source selection lies in the accessibility or ease of using a particular information channel or source. The selection process has little to do with or little regard for the benefits (i.e., quality or usefulness or the amount of information obtained) derived or expected from a particular information channel or source. Several studies, most of which were conducted prior to the widespread availability of computerized databases, have found that the accessibility or the ease of using an information channel or source support the "least-effort model" (Allen, 1977; Allen, 1966; Gerstberger and Allen, 1968; Rosenberg, 1967).

Accessibility Versus Quality

The channel and source selection literature has been interpreted to mean that the value (i.e., quality) of the information plays no explanatory role in channel or source selection. In other words, the accessibility of an information channel or source is the dominant or sole criterion upon which information use is based. In what may appear to be an apparent paradox, an engineer or scientist would use an accessible channel or source even if it produced information of poor quality. Would someone seeking information knowingly use information that was not reliable or that was technically inaccurate? Argote (1982) and O'Reilly (1982) argued that due to the ambiguity inherent in available information and pressures to produce results, accessibility of information may dominate quality as a factor determining use. In his study of county welfare workers, O'Reilly suggested that accessibility had a greater influence on information use than did the technical quality or reliability of information.

Culnan (1983) argues that although channel and source selection appear to be influenced largely by accessibility, information "quality" must be viewed separately from channel and source selection. Orr (1970) agrees and states that for scientists the quality and reliability of information are critical. In his study of information use by engineers in problem solving, Kaufman (1983) reported that participants in his research identified technical quality or reliability followed by relevance as the criteria for selecting the most useful information source. However, Kaufman also reported that accessibility appears to be the most frequently used criterion in selecting an information channel or source, even if that channel or source produced information that was least useful in solving the problem.

RESEARCH DESIGN AND METHODOLOGY

The objective of this research was to examine the relationship between technical (task) uncertainty and information use by industry-affiliated U.S. aerospace engineers and scientists (Pinelli, et. al., 1993). Because the study investigated the relationship between task uncertainty and information use, a perceptual measure of uncertainty was used. (Uncertainty was measured using a 5-point scale with 5 being very uncertain.) Three research questions were addressed. The first research question asked whether U.S. aerospace engineers and scientists were able to make accurate assessments of the uncertainty they face in a project. (This needed to be addressed prior to any analysis of uncertainty as a contingency in information-seeking behavior.) The second research question looked at the relationship of perceived uncertainty to personal or task characteristics. The last research question flowed from the first two and addressed the impact of uncertainty on information-seeking behavior in general and the use of federally funded aerospace R&D in particular.

Data Collection

Data were collected from three mail (self-reported) surveys of U.S. aerospace engineers and scientists. Participation in the study was limited to U.S. industry-affiliated aerospace engineers and scientists. The survey instruments, which had only minor differences, had been pretested on two groups of aerospace engineers and scientists across the United States. A toll-free telephone number was provided to survey recipients who were instructed to contact the researchers if the survey was not relevant to them. Postcard reminders were sent to survey recipients who had not responded by completing and returning the survey by the established cut-off date.

Survey 1. A random sample of 750 members of the American Institute of Aeronautics and Astronautics (AIAA) served as the population for Survey 1 (this will be referred to in the text as the AIAA sample). By June 21, 1995, 341 usable surveys were returned. Eighty-nine questionnaires were returned as unusable because the recipient was unemployed, the recipient was not working in aerospace, the recipient had retired, the survey was not applicable, or the recipient was not employed at that company. The sample for Survey 1 consisted of 341 surveys, for an adjusted response rate of 53%. This segment of the study consisted of industry-affiliated engineers and scientists performing professional duties primarily in research, design and development, and management. Survey 1 was conducted between April and June 1995 (Pinelli, Barclay, and Kennedy, 1995).

Survey 2. A random sample of 2000 members of the Aerospace Division of the Society of Automotive Engineers (SAE) served as the population for Survey 2 (this will be referred to in the text as the SAE sample). After unusable questionnaires and public-sector respondents were eliminated, the sample for Survey 2 consisted of 946 surveys, for an adjusted response rate of 67%. This segment of the study consisted of engineers and scientists performing professional duties primarily in design and development, manufacturing and production, and management. Survey 2 was conducted between August and November 1993 (Pinelli, Barclay, and Kennedy, 1994a).

Survey 3. Survey 3 used a random sample of U.S. aerospace engineers and scientists whose names appeared on the Society of Manufacturing Engineers (SME) mailing list as subscribers to *Manufacturing Engineering* (not necessarily members of SME) and whose Standard Industry Classification (SIC) code indicated that they were employed in an aerospace organization (this will be referred to in the text as the SME sample). The group consisted of engineers and scientists performing professional duties in design and development, manufacturing and production, and management. After unusable questionnaires were discarded, the final sample for Survey 3 consisted of 419 surveys, for an adjusted response rate of 41%. Survey 3 was conducted between May and August 1994 (Pinelli, Barclay, and Kennedy, 1994b).

A variation of Flanagan's (1954) critical incident technique was used to guide the data collection. Each survey instrument asked participants to focus on "the most important job-related project, task, or problem you have worked on in the past six months." The rationale of this "event-oriented" technique is that it is easier for people to recall accurately what they did on a specific occasion than to remember what they do in general (Lancaster, 1978). Throughout this chapter, "project" will be used as a collective term for the project, task, or problem described by the respondents. Data were collected on the perceived technical uncertainty and complexity of the project and the nature of the project. Data on information-seeking behaviors and communication practices were also elicited.

Analysis

The statistical analyses adopted a probability level of $p \leq .05$ for significant findings. In the following discussion, the terms "statistically significant" and "significant" are used interchangeably. Two group comparisons used t-tests with a one-tailed analysis. To examine more than two groups, one-way analyses of variance were employed. F values are reported in the tables. Significant differences within groups were identified using *a priori* contrast tests. These tests systematically compare all possible pairs of group means. The least squares difference procedure was used for the contrast testing.

FINDINGS

Data are presented separately for the three surveys. Demographics appear first, then data about project characteristics. Data concerning technical uncertainty and information use follow.

Survey 1: AIAA–Research and Design Sector

Table 8.1 contains a description of the Survey 1 (AIAA) respondents. Of the 341 private sector respondents, about 91% and 84%

Table 8.1. Survey Demographics (N=341)

Characteristics	n	%
Education		
No degree	0	0.0
Bachelor's degree	102	29.9
Master's degree	178	52.2
Doctorate	57	16.7
Other	4	1.2
Years in Aerospace		
<1–5	18	5.3
6–10	47	13.9
11–15	63	18.6
16–20	37	10.9
21–30	82	24.2
31–40	80	23.6
≥ 41	12	3.5
Mean = 21.9 years		
Academic Preparation		
Engineer	310	90.9
Scientist	18	5.3
Other	13	3.8
Job Duties		
Engineer	285	83.6
Scientist	10	2.9
Other	46	13.5

were educated and worked as engineers. About 52% had masters degrees, about 17% held a doctorate, and about 1% had advanced their education to the post-graduate level. Their average work experience in aerospace was 21.9 years.

Table 8.2 contains data that characterize the project reported on by survey respondents. Over 90% of the respondents reported that they worked on the project in groups. The majority described their duties on the project as engineering (72.8%). Management accounted for the next highest share (22.3%); science accounted for only 2.1%. The categories of the project primarily involved design and development (56.1%), management (24.5%), and research (13.5%).

Table 8.2. Project Characteristics (N=341)

Characteristics	n	%
Worked on Project		
Alone	32	9.9
With others	294	90.2
Nature of Duties Performed on Project		
Science	7	2.1
Engineering	238	72.8
Management	73	22.3
Other	9	2.8
Category of the Technical Project		
Research	42	13.5
Design/development	174	56.1
Manufacturing/production	3	1.0
Computer applications	10	3.2
Management	76	24.5
Quality control/assurance	5	1.6

Research Question 1: Is Technical Uncertainty an Issue?

The first research question asked whether uncertainty is a valid construct for aerospace engineers and scientists working in the research and design sector. A perceptual definition (Downey and Slocum, 1975) of uncertainty was adopted, as the study focused on the linkage between individual behavior and uncertainty. Table 8.3 contains the respondents' perceptions of the technical uncertainty

Table 8.3. Technical Uncertainty and Complexity (*N*=341)

Think of the most important job-related project, task, or problem you have worked on in the past six months. On a scale of 1 (little uncertainty) to 5 (great uncertainty), how would you describe the overall technical uncertainty of this project?

	n	%
Great Uncertainty		
5	51	15.5
4	138	42.1
3	98	29.9
2	36	11.0
1	5	1.5
Little Uncertainty		

Overall mean uncertainty score = 3.59; s.d. = 0.93.

On a scale of 1 (very simple) to 5 (very complex), how would you describe the overall technical complexity of this project?

	n	%
Very Complex		
5	95	29.0
4	149	45.4
3	75	22.9
2	8	2.4
1	1	0.3
Very Simple		

Overall mean complexity score = 4.00; s.d. = 0.80.

of the project. The measure of uncertainty yielded a mean rating of 3.59 on a scale of 1 (little uncertainty) to 5 (great uncertainty). Answers to the uncertainty question were normally distributed but were skewed toward the "very uncertain" end of the scale, thus indicating that most respondents perceived that the project had a higher than average level of uncertainty. The distribution suggested that respondents were able to make a determination about the levels of uncertainty that they face. It also implied that answers were not corrupted by desirability or similarity biases. If these biases were present, one might expect less variance and higher levels of uncertainty.

The construct of "complexity" is sometimes referred to in the literature as one source of uncertainty (Tushman, 1979). Complexity involves the degree to which a task requires coordination and joint problem solving with other departments within the organization. While an in-depth analysis of both uncertainty and complexity is beyond the scope of this chapter, it was worthwhile to look at the distribution of the responses to the question dealing with complexity and the relationship between the two variables. Complexity, rated similarly to uncertainty on a scale of 1 (very simple) to 5 (very complex), yielded a mean of 4.00. Responses to the complexity question were normally distributed but were skewed toward the "very complex" end of the scale. Given that the respondents were aerospace engineers and scientists working in research and design, one would expect them to work on projects of considerable complexity.

Pearson's correlation between technical uncertainty and complexity was .48 ($p \leq .01$). Approximately 23% of the variance in one variable was accounted for by variance of the other. This suggests that although there was a statistically significant relationship, the two variables do not tap the same dimension and, hence, are not substitute measures of the same construct.

Research Question 2: Are Personal and Task Characteristics Related to Perceptions of Technical Uncertainty?

Research has established links between organizational, task, and personal variables and information-seeking patterns (Orr, 1970; Rosenberg, 1967; Zenger and Lawrence, 1989). Similarly, perceptions of uncertainty have also been postulated as correlates of personal variables (Downey and Slocum, 1975).

Three personal variables were tested: the respondents' education, years of work experience, and academic background. The amount of uncertainty did not vary as a function of the years of work experience or academic preparation for the present job. The only personal variable associated with uncertainty was education (Table 8.4). Those whose educational level extended beyond the baccalaureate degree experienced significantly more uncertainty than those whose educational level did not. This finding was not surprising, for those with a more advanced education are likely to be given more difficult projects.

Table 8.4. Technical Uncertainty and Education (N=341)

Education	Uncertainty Score[a]	
No degree	--	
Bachelor's degree	3.43	
Master's degree	3.54	
Doctorate	3.96	$F = 6.32$**

[a]Higher numbers indicate greater uncertainty.
**$p \leq .01$.

Academic preparation as either an engineer or a scientist was not significant in relation to uncertainty. Given this, the fact that few scientists were represented in the sample, and prior work on the lack of differences in information-seeking behavior of U.S. aerospace engineers and scientists (Pinelli, 1991), AIAA engineers and scientists were not treated differently in the analysis.

Three task variables were examined: category of the project, individual working alone or in groups on a project, and the type of duties involved in the project. A significant relationship was found to exist between category of the project and uncertainty (Table 8.5). Tushman (1978) noted that in R&D settings tasks can differ along several dimensions, such as Rosenbloom and Wolek's (1970) factors of time span of feedback, specific versus general problem orientation, and generation of new knowledge versus use of existing knowledge to classify tasks. Tushman developed four task categories from nonroutine to routine: basic research, applied research, development, and technical services. In this study we modified Tushman's typology to provide a more detailed categorization.

Uncertainty was found to be associated with the type of project on which an individual worked. The highest levels of uncertainty were found for research projects; the lowest were found for quality control or assurance projects (Table 8.5). Survey respondents having unspecified duties (the "other" category) were not included in the analysis. Respondents solving management problems experienced the second highest level of uncertainty, followed closely by those working on design or development problems and computer applications. Those working on manufacturing and production and quality control or assurance problems reported the lowest levels of uncertainty. This finding is consistent with Connolly's (1975) investiga-

Table 8.5. Technical Uncertainty and Project Characteristics (N=341)

Which category best describes your project?

Category	Uncertainty Score[a]	
Research	4.1	
Design/development	3.5	
Computer applications	3.1	
Management	3.7	
Quality control/assurance	2.6	
Manufacturing/production	2.7	$F = 5.219$**

Which of the following best describes your duties on the project?

Duty	Uncertainty Score	
Engineer	3.6	
Scientist	4.1	
Management	3.7	
Other	3.2	$F = 0.101$

On this project, did you work alone or with others?

Method	Uncertainty Score	
Alone	3.6	
With others	3.6	$t = 0.05$

[a]Higher numbers indicate greater uncertainty.
**$p \leq .01$.

tion of R&D laboratories, which found that increasingly applied work was associated with lower levels of uncertainty.

It is interesting to note that respondents working on managerial projects reported experiencing relatively high levels of uncertainty. This finding could reflect their lack of managerial training, given that the majority (90.9%) of the respondents had been trained as engineers and a small number had been trained as scientists (5.3%) and in the "other" category (3.8%). This finding may also reflect the high degree of uncertainty that is associated with managing individuals involved in performing R&D projects. Respondents who performed professional duties as scientists on the project reported experiencing more uncertainty than any of the others. Those who worked alone and with others reported experiencing the same levels (i.e., same uncertainty scores) of uncertainty.

Research Question 3a.   What Is the Impact of Technical Uncertainty on Information-Seeking Behavior in General?

Before investigating the impact of uncertainty on information-seeking behavior, the information-seeking patterns of the entire sample needed examination. Respondents were asked to rank order the steps that they had followed in obtaining the information they needed to complete the project (Table 8.6).

As expected, engineers and scientists seeking information go about searching in some orderly fashion. Specifically, one would expect "easy-to-obtain" sources to be consulted first with progressively more "difficult-to-obtain" sources consulted until the needed information is found. Respondents indicated that they first used their personal stores of technical information, then spoke with persons within the organization. If this approach did not provide adequate information, they next sought information from colleagues outside the organization. The information search then extended to literature such as conference papers, professional journals, and technical reports, and finally to a librarian. These findings are con-

Table 8.6.   Rank Order of Information Sources Used to Complete the Project (N=341)

Rank		Median Rank	% Using First
1	Used my personal store of technical information, including sources I keep in my office	1	64.8
2	Spoke with coworkers or people inside my organization	2	28.4
3	Used literature resources (e.g., conference papers, journals, technical reports) found in my organization's library	4	3.1
4	Spoke with colleagues outside my organization	3	2.3
5	Searched (or had someone search for me) an electronic (bibliographic) database	4	2.1
6	Spoke with a librarian or technical information specialist	4	0.0

sistent with Pinelli's (1991) research among U.S. aerospace engineers and scientists, Rosenberg's (1967) work with research and non-research personnel in government and industry, and Dewhirst's (1971) study of engineers and scientists at a U.S. government aeronautical research center.

To investigate whether uncertainty affected information channel or source selection in general, the sample was divided into users and nonusers of five information channels or sources to determine if varying levels of project uncertainty affected the use of each information channel or source (Table 8.7). The rationale rests in the assumption that as uncertainty increases, the search for information expands beyond internal and informal channels or sources to external and formal channels or sources.

As expected, almost all the respondents used their personal stores of information to help them complete the project. It appeared that the majority of respondents were unable to complete the project by relying solely on their own stores of information, for almost all spoke with coworkers within their organization. No statistically significant differences existed between levels of uncertainty among respondents who opted to use their own stores of technical information and those who did not, those who spoke with coworkers within the organization and those who did not, and those respondents who spoke with colleagues outside the organization and those who did not. Those who did not consult with colleagues outside the organization, consult the literature, search a database, or speak with a librarian reported less uncertainty than those who did.

The data suggest that many respondents were able to complete the project after consulting with coworkers. Still, 88% spoke to people outside the organization, 64% consulted the literature, and 38% consulted a librarian or an information specialist. As task uncertainty increased, the information search broadened beyond internal and informal sources to external and formal sources of information. In only one instance did respondents who used external and formal channels or sources report significantly more uncertainty than those respondents who did not use the channels or sources.

Another issue involved the relationship between technical uncertainty and the time spent communicating in writing or orally (Table 8.8). No statistically significant difference was found to exist between technical uncertainty and the number of hours that survey

Table 8.7. Technical Uncertainty and Information Source Use (N=341)

Where did you get the information you needed for the project?

Source	Uncertainty[a] Nonusers		Users	
Used my personal store of technical information	4.00	(8)	3.60	(302)
Spoke with my coworkers or people inside my organization	4.09	(11)	3.59	(299)
Spoke with colleagues outside my organization	3.41	(34)	3.64	(266)
Used literature resources (e.g., journals, conference papers, technical reports) found in my organization's library	3.53	(102)	3.69	(185)
Searched (or had someone search for me) an electronic (bibliographic) database	3.52	(145)	3.74	(140)*
Spoke with a librarian or technical information specialist	3.57	(181)	3.70	(107)

[a]Higher numbers indicate greater uncertainty.
*$p \leq .05$.

respondents spent sending or receiving technical information orally. Further, no statistically significant difference was found between technical uncertainty and the number of hours respondents spent sending or receiving written technical communications.

While one might expect to see a linear relationship between technical uncertainty and the number of hours that survey respondents spent sending and receiving information orally, this is not the case. The hours spent sending and receiving technical communications orally increased as technical uncertainty increased except in Response Category 4, the next step beyond the median rating. Respondents with moderate (Response Category 3) and high (Response Category 5) levels of technical uncertainty spent significantly more time communicating orally than did respondents in the other three categories. The same finding held for the number of hours that respondents spent receiving technical communications orally. The lack of linear relationships has two potential but not totally satisfactory explanations. Although the results are not totally com-

Table 8.8.   Technical Uncertainty and Communications Frequency (N=341)

Hours Spent Each Week Communicating Technical Information

	Low Uncertainty			High Uncertainty		
	1	2	3	4	5	
Written	7.5	13.2	11.4	11.3	11.9	F = .435
Oral	11.3	10.9	11.5	10.6	12.0	F = .314

Hours Spent Each Week Receiving Technical Information

Written	9.3	11.1	10.3	9.6	9.3	F = .281
Oral	5.7	9.2	8.6	7.7	9.2	F = .540

parable, Tushman (1979), in a study of a large R&D laboratory, found a weak inverse relationship between environmental variability and communication to areas outside the project. He suggested that this occurred because project members closed off their extra-project communication upon encountering change in the environment. Tushman also suggested that the decrease was a short-run response that might change over time. His alternative explanation was that as uncertainty sets in, individuals may seek more formal (nonverbal) information from outside the organization, thus reducing oral communication, but as the project accumulates a written (formal) engineering knowledge base, oral communication about that knowledge base increases.

The next concern centered on the relationship between technical uncertainty and the use of specific written (formal) information products: in-house technical reports, journal articles, conference papers, DoD technical reports, and NASA technical reports (Table 8.9). To examine the relationship between technical uncertainty and the use of these five information products, respondents were divided into low uncertainty and high uncertainty groups. Respondents reporting a middle level of technical uncertainty were excluded. Users of the formal information products reported significantly higher levels of technical uncertainty than nonusers reported.

Finally, we investigated the relationship between uncertainty and the frequency of use of 15 information products (Table 8.10).

Table 8.9. Technical Uncertainty and Information Product Use (*N*=341)

Product	Uncertainty[a]	
	Nonuse	Use
In-house technical reports	3.1	3.6*
Journal articles	3.2	3.7**
Conference papers	3.2	3.7**
DoD technical reports	3.4	3.7**
NASA technical reports	3.4	3.7**

[a]Higher numbers indicate greater uncertainty.
*$p \leq .05$. **$p \leq .01$.

These products were classified as internal or external sources of information. The results demonstrate a positive relationship between increased technical uncertainty and the frequency of use of both internal and external information products. Of the internal products, four of nine—letters, audiovisual materials, in-house technical reports, and technical proposals—were used more frequently in cases of high technical uncertainty; however, the relationships were not statistically significant except for audiovisual materials and technical manuals. For all of the external information products, use increased as technical uncertainty increased, but the increase was not statistically significant. Increased usage was noted for abstracts, journal articles, conference and meeting papers, technical talks or presentations, DoD technical reports, and NASA technical reports. These findings are consistent with reported information-seeking patterns (see Table 8.7) wherein the search extended beyond internal and informal to external and formal information sources as uncertainty increased.

Research Question 3b. What Is the Impact of Uncertainty on the Use of Federally Funded Aerospace R&D?

About 73% of the survey respondents reported using the results of federally funded aerospace R&D in their work. Some 60% of the AIAA survey respondents reported using the results of federally funded aerospace R&D in completing their projects. About 64% of

Table 8.10. Technical Uncertainty and the Use of Internal and External Information Products (N=341)

How many times in the past six months did you use the following?

| | Uncertainty | | | | |
| | Low | | | | High |
Product	1	2	3	4	5
Internal Sources					
Drawings/specifications	81.0	23.0	16.8	20.4	10.7*
Memoranda	70.6	22.3	26.4	26.0	56.0
Letters	20.8	7.9	14.4	16.3	29.3
Audiovisual materials	0.6	6.3	14.4	6.7	21.5*
Technical manuals	70.0	8.1	4.3	5.1	2.4**
In-house technical reports	4.0	4.6	6.5	6.9	8.2
Trade/promotional literature	20.0	3.1	9.7	6.8	5.3
Computer programs/ documentation	20.0	3.8	7.3	6.5	5.3
Technical proposals	0.0	1.6	2.6	3.7	5.4
External Sources					
Journal articles	0.2	2.1	2.8	6.9	12.2
Conference/meeting papers	0.4	1.7	4.3	5.5	6.7
Technical talks/presentations	1.0	4.8	7.1	6.7	11.5
Abstracts	0.0	0.7	1.5	4.8	6.7
DoD technical reports	0.0	0.6	1.3	1.6	2.8
NASA technical reports	0.0	1.5	1.1	2.1	4.1

$*p \leq .05.$ $**p \leq .01.$

the respondents who used this information to complete their projects found it in published DoD or NASA technical reports.

To examine the relationship between technical uncertainty and nine factors thought to influence the use of DoD and NASA technical reports, respondents were divided into low and high uncertainty groups, and the median rating was excluded (Table 8.11). For DoD technical report use, respondents in the high technical uncertainty group reported numerically higher influence scores for four of the nine factors than respondents in the low uncertainty group reported. One score was statistically significant, however, "important in

Table 8.11. Technical Uncertainty and the Factors Influencing Use of DoD and NASA Technical Reports (N=341)

Report	Importance Scores[a]	
	Low Uncertainty	High Uncertainty
DoD Technical Reports		
Easy to physically obtain	4.1	3.9
Easy to use or read	3.8	3.9
Expense	3.4	3.2
Technical quality	4.3	4.4
Comprehensive data and information	4.4	4.3
Relevance	4.7	4.5
Obtainable at nearby location/source	3.4	3.4
Good prior experience using them	3.1	3.2
Important in your work	2.5	3.0*
NASA Technical Reports		
Easy to physically obtain	4.0	4.0
Easy to use or read	3.8	4.0
Expense	3.3	3.2
Technical quality	4.5	4.4
Comprehensive data and information	4.3	4.3
Relevance	4.8	4.5
Obtainable at nearby location/source	3.5	3.4
Good prior experience using them	3.4	3.3
Important in your work	2.7	3.2*

[a]Influence was measured on a 5-point scale where 1 = not at all important and 5 = very important. The higher the number the greater the influence of the factor on report use.

* $p \leq .05$.

your work." Under conditions of low and high uncertainty, numerically higher influence scores were reported for "important in your work" for DoD technical reports.

For NASA technical report use, respondents in the high technical uncertainty group reported numerically higher influence scores for two of the nine factors (scores for two factors tied) than respondents in the low uncertainty group reported. Two scores were statistically significant—factors of "comprehensive data and information" and "important in your work." Similar to the data on DoD technical reports for both high and low uncertainty groups, the numerically highest influence scores were reported for technical quality, comprehensiveness of the data, and the relevance of NASA technical reports to the work.

Although not statistically significant, the data demonstrate that "value" factors such as comprehensiveness, relevance, and importance influence information use under conditions of uncertainty. Neither ease of obtaining the data, prior experience, nor expense was a factor influencing use under conditions of increased uncertainty. This finding failed to uphold both the "ease of use" thesis and the "cost-benefit" explanation of information-seeking behavior.

Survey 2: SAE–Design and Development Sector

Table 8.12 contains a description of the Survey 2 (SAE) respondents. Slightly over 90% of the 872 private-sector respondents were educated and worked as engineers. Somewhat more than 50% had undergraduate degrees, 33% of the respondents held master's degrees, and 5% had advanced their education to the post-graduate level. Their average work experience in aerospace was 18.7 years.

Table 8.13 contains data that characterize the project reported on by survey respondents. Over 80% of the respondents reported that they worked on the project in groups. The majority described their duties in the project as engineering (71.9%). Management accounted for the next highest share (24.4%); science accounted for only 2.3%. The categories of the project primarily involved design and development (56.4%), manufacturing and production (11.5%), and management (14.3%).

Research Question 1: Is Technical Uncertainty an Issue?

The first research question asked whether uncertainty is a valid construct for aerospace engineers and scientists working in the design and development sector. A perceptual definition (Downey and Slocum, 1975) of uncertainty was adopted, as the study focused on the linkage between individual behavior and uncertainty. Table 8.14 contains the respondents' perceptions of the technical uncertainty of the project. The measure of uncertainty yielded a mean

Table 8.12. Survey Demographics (*N*=872)

Characteristics	*n*	%
Education		
No degree	50	5.7
Technical/vocational degree	22	2.5
Bachelor's degree	458	52.5
MBA	63	7.2
Master's degree	232	26.6
Post graduate	47	5.4
Years in Aerospace		
<1–5	85	9.7
6–10	206	23.6
11–15	134	15.4
16–20	81	9.3
21–30	187	21.4
31–40	145	16.6
≥ 41	34	3.9
Mean = 18.7 years		
Academic Preparation		
Engineer	791	90.7
Scientist	64	7.3
Other	17	1.9
Job Duties		
Engineer	792	90.8
Scientist	18	2.1
Other	62	7.1

rating of 3.19 on a scale of 1 (little uncertainty) to 5 (great uncertainty). Answers to the uncertainty question were normally distributed, indicating that most respondents perceived the project had a moderate level of technical uncertainty. The distribution suggested that respondents were able to make a determination about the levels of uncertainty that they face. It also implied that answers were not corrupted by desirability or similarity biases. If these biases were present, one might expect less variance and higher levels of uncertainty.

Table 8.13. Project Characteristics (N=872)

Characteristics	n	%
Worked on Project		
Alone	151	17.3
With others	721	82.7
Nature of Duties Performed on Project		
Science	20	2.3
Engineering	627	71.9
Management	213	24.4
Other	12	1.4
Category of the Technical Project		
Research	78	8.9
Design/development	492	56.4
Manufacturing/production	100	11.5
Computer applications	37	4.2
Management	125	14.3
Education	13	1.5
Other	27	3.2

The construct of "complexity" is sometimes referred to in the literature as one source of uncertainty (Tushman, 1979). Complexity involves the degree to which a task requires coordination and joint problem solving with other departments within the organization. While an in-depth analysis of both uncertainty and complexity is beyond the scope of this chapter, it was worthwhile to look at the distribution of the responses to the question dealing with complexity and the relationship between the two variables. Complexity, rated similarly to uncertainty on a scale of 1 (very simple) to 5 (very complex), yielded a mean of 3.70. Complexity scores were normally distributed but were skewed toward the "very complex" end of the scale. Given that the respondents were aerospace engineers and scientists working in design and development, one would expect them to work on projects of considerable complexity.

Pearson's correlation between technical uncertainty and complexity was .47 ($p \leq .01$). Approximately 22% of the variance in one variable was accounted for by variance of the other. This suggests that although there was a statistically significant relationship, the two variables do not tap the same dimension and, hence, are not substitute measures of the same construct.

Table 8.14. Technical Uncertainty and Complexity *(N=872)*

Think of the most important job-related project, task, or problem you have worked on in the past six months. On a scale of 1 (little uncertainty) to 5 (great uncertainty), how would you describe the overall technical uncertainty of this project?

	n	%
Great Uncertainty		
5	94	10.9
4	248	28.4
3	299	34.3
2	187	21.4
1	44	5.0
Little Uncertainty		

Overall mean uncertainty score = 3.19; s.d. = 1.05.

On a scale of 1 (very simple) to 5 (very complex), how would you describe the overall technical complexity of this project?

	n	%
Very Complex		
5	154	17.7
4	381	43.7
3	266	30.5
2	64	7.3
1	7	.8
Very Simple		

Overall mean complexity score = 3.70; s.d. = .87.

Research Question 2: Are Personal and Task Characteristics Related to Perceptions of Technical Uncertainty?

Research has established links between organizational, task, and personal variables and information-seeking patterns (Orr, 1970; Rosenberg, 1967; Zenger and Lawrence, 1989). Similarly, perceptions of uncertainty have also been postulated as correlates of personal variables (Downey and Slocum, 1975).

Three personal variables were tested: the respondents' education, years of work experience, and academic background. The amount of uncertainty did not vary as a function of the years of work experience or academic preparation for the present job. The only personal variable associated with uncertainty was education (Table 8.15). Those whose educational level extended beyond the baccalaureate degree experienced significantly more uncertainty than those whose educational level did not. This finding was not surprising, for those with a more advanced education are likely to be given more difficult projects.

Academic preparation as either an engineer or a scientist was not significant in relation to uncertainty. Given this, the fact that few scientists were represented in the sample, and prior work on the lack of differences in information-seeking behavior of U.S. aerospace engineers and scientists (Pinelli, 1991), SAE engineers and scientists were not treated differently in the analysis.

Three task variables were examined: category of the project, individual working alone or in groups on a project, and the type of duties involved in the project. The analyses found significant relationships between all variables tested and uncertainty (Table 8.16). Tushman (1978) noted that in R&D settings, tasks can differ along several dimensions such as Rosenbloom and Wolek's (1970) factors of time span of feedback, specific versus general problem orientation, and generation of new knowledge versus use of existing knowledge to classify tasks. Tushman developed four task categories from nonroutine to routine: basic research, applied research,

Table 8.15. Technical Uncertainty and Education ($N=872$)

Education	Uncertainty Score[a]	
No degree	2.86	
Technical/vocational degree	2.86	
Bachelor's degree	3.06	
MBA	3.47	
Master's degree	3.41	
Doctorate	3.42	
Other	2.80	$F = 8.13$**

[a]Higher numbers indicate greater uncertainty.
**$p \leq .01$.

development, and technical services. In this study we modified Tushman's typology to provide a more detailed categorization.

Uncertainty was found to be associated with the type of project on which an individual was working. The highest levels of uncertainty were found among researchers; the lowest levels were found among educators (Table 8.16). Survey respondents whose duties were not specified (the "other" category) were not included in the analysis. Respondents involved in R&D problems experienced the highest level of uncertainty, followed closely by those working on computer application and management problems. Those working

Table 8.16. Technical Uncertainty and Project Characteristics (N=872)

Which category best describes your project?

Category	Uncertainty Score[a]	
Research	3.5	
Development	3.4	
Computer applications	3.2	
Management	3.2	
Quality control/assurance	3.1	
Design	3.0	
Manufacturing	3.0	
Production	2.8	
Education	2.7	$F = 4.46$**

Which of the following best describes your duties on the project?

Duty	Uncertainty Score	
Engineer	3.2	
Scientist	3.9	
Management	3.2	
Other	2.6	$F = 4.22$**

On this project, did you work alone or with others?

Method	Uncertainty Score	
Alone	2.8	
With others	3.3	$t = 5.28$**

[a]Higher numbers indicate greater uncertainty.
**$p \le .01$.

on manufacturing, production, and education problems reported the lowest levels of uncertainty. This finding is consistent with Connolly's (1975) investigation of R&D laboratories, which found that increasingly applied work was associated with lower uncertainty.

It is interesting that respondents working on managerial projects experienced relatively high levels of uncertainty. This finding could reflect their lack of managerial training, given that the majority (90.7%) of the respondents were trained as engineers and a small number as scientists (7.3%) and in the "other" category (1.9%). This finding may also reflect the high degree of uncertainty associated with managing those involved in performing R&D projects. Respondents performing professional duties as scientists on the project reported significantly more uncertainty than any of the others. Those who worked with others reported significantly higher levels of uncertainty than those who worked alone.

Research Question 3a.   What Is the Impact of Technical Uncertainty on Information-Seeking Behavior in General?

Before investigating the impact of uncertainty on information-seeking behavior, the information-seeking patterns of the entire sample needed examination. Respondents were asked to rank order the steps that they had followed in obtaining the information they needed to complete the project (Table 8.17).

As expected, engineers and scientists seeking information go about searching in some orderly fashion. Specifically, one would expect "easy-to-obtain" sources to be consulted first with progressively more "difficult-to-obtain" sources consulted until the needed information is found. Respondents indicated that they first used their personal stores of technical information, then spoke with persons within the organization. If this approach did not provide adequate information, they next sought information from colleagues outside the organization. The information search then extended to literature such as conference papers, professional journals, and technical reports, and finally to a librarian. These findings were consistent with Pinelli's (1991) research among U.S. aerospace engineers and scientists, Rosenberg's (1967) work with research and non-research personnel in government and industry, and Dewhirst's (1971) study of engineers and scientists in a U.S. government aeronautical research center.

Table 8.17. Rank Order of Information Sources Used to Complete the Project (N=872)

Rank		Median Rank	% Using First
1	Used my personal stores of technical information, including sources I keep in my office	1	60.6
2	Spoke with coworkers or people inside my organization	2	26.9
3	Spoke with colleagues outside my organization	3	5.4
4	Used literature resources (e.g., conference papers, journals, technical reports) found in my organization's library	4	4.6
5	Spoke with a librarian or technical information specialist	4	3.1

To investigate whether uncertainty affected information channel or source selection in general, the sample was divided into users and nonusers of five information channels or sources to determine if varying levels of project uncertainty affected the use of each information channel or source (Table 8.18). The rationale rests in the assumption that as uncertainty increases, the search for information expands beyond internal and informal channels or sources to external and formal channels or sources.

As expected, almost all the respondents used their personal stores of information to help them complete the project. It appeared that the majority of respondents were unable to complete the project by relying solely on their own stores of information, for almost all spoke with coworkers within their organization. No statistically significant differences existed between levels of uncertainty among respondents who opted to use their own stores of technical information and those who did not. Almost all respondents spoke with coworkers within the organization; however, those who did not consult with others outside the organization reported significantly less uncertainty than those who did consult with others outside the organization.

Table 8.18. Technical Uncertainty and Information Source Use (N=872)

Where did you get the information you needed for the project?

Source	Uncertainty[a]			
	Nonusers		Users	
Used my personal store of technical information	3.20	(76)	3.18	(796)
Spoke with my coworkers or people inside my organization	2.97	(88)	3.21	(784)*
Spoke with colleagues outside my organization	2.89	(243)	3.30	(629)**
Used literature resources (e.g., journals, conference papers, technical reports) found in my organization's library	2.97	(323)	3.31	(549)**
Spoke with a librarian or technical information specialist	3.08	(366)	3.33	(506)**

[a]Higher numbers indicate greater uncertainty.
*$p \leq .05$.
**$p \leq .01$.

The data suggest that many respondents were able to complete the project after consulting with coworkers. Still, 72% spoke to people outside the organization, 63% consulted the literature, and 58% consulted a librarian or an information specialist. As task uncertainty increased, the information search broadened beyond internal and informal sources to external and formal sources of information. In all instances, respondents who used external and formal channels or sources reported significantly more uncertainty than those respondents who did not use the channels or sources.

Another issue involved the relationship between technical uncertainty and the time spent communicating in writing or orally (Table 8.19). A statistically significant difference was found to exist between technical uncertainty and the number of hours that survey respondents spent sending or receiving technical information orally. However, no statistically significant difference was found between technical uncertainty and the number of hours respondents spent sending or receiving written technical communications.

Table 8.19.   Technical Uncertainty and Communications Frequency ($N=872$)

Hours Spent Each Week Communicating Technical Information

	Low Uncertainty			High Uncertainty		
	1	2	3	4	5	
Written	7.1	8.1	8.5	9.0	9.6	$F = 1.53$
Oral	8.0	10.2	10.9	9.5	11.7	$F = 2.96*$

Hours Spent Each Week Receiving Technical Information

Written	6.6	6.5	8.0	7.8	9.0	$F = 1.97$
Oral	5.0	6.0	7.4	6.4	8.0	$F = 3.60**$

$*p \leq .05.$ $**p \leq .01.$

While one might expect to see a linear relationship between technical uncertainty and the number of hours that survey respondents spent sending and receiving information orally, this is not the case. The hours spent sending and receiving technical communications orally increased as technical uncertainty increased except in Response Category 4, the next step beyond the median rating. Respondents with moderate (Response Category 3) and high (Response Category 5) levels of technical uncertainty spent significantly more time communicating orally than did respondents in the other three categories. The same finding held for the number of hours that respondents spent receiving technical communications orally. The lack of linear relationships has two potential but not totally satisfactory explanations. Although the results are not totally comparable, Tushman (1979), in a study of a large R&D laboratory, found a weak inverse relationship between environmental variability and communication to areas outside the project. He suggested that this occurred because project members closed off their extra-project communication upon encountering change in the environment. Tushman also suggested that the decrease was a short-run response that might change over time. His alternative explanation was that as uncertainty sets in, individuals may seek more formal (nonverbal) information from outside the organization, thus reducing oral communication, but as the project accumulates a written (formal) engineering knowledge base, oral communication about that knowledge base increases.

The next concern centered on the relationship between technical uncertainty and the use of specific written (formal) information products: in-house technical reports, journal articles, conference papers, AGARD technical reports, DoD technical reports, and NASA technical reports (Table 8.20). To examine the relationship between technical uncertainty and the use of these six information products, respondents were divided into low uncertainty and high uncertainty groups. Respondents reporting a middle level of technical uncertainty were excluded. Users of the formal information products reported significantly higher levels of technical uncertainty than nonusers reported, except for DoD technical reports.

Finally, we investigated the relationship between technical uncertainty and the frequency of use of 15 information products (Table 8.21). These products were classified as internal or external sources of information. The results demonstrate a positive relationship between increased technical uncertainty and the frequency of use of both internal and external information products. Of the internal products, five of nine—memoranda, letters, in-house technical reports, trade and promotional literature, and computer documentation—were used more frequently in cases of high technical uncertainty; however, the relationships were not statistically significant except for in-house technical reports. For all but one of the external information products, AGARD technical reports, use increased significantly as technical uncertainty increased. Usage of

Table 8.20. Technical Uncertainty and Information Product Use (N=872)

Product	Uncertainty[a]	
	Nonuse	Use
In-house technical reports	2.88	3.27**
Journal articles	3.01	3.28**
Conference papers	2.93	3.37**
AGARD[b] technical reports	3.16	3.39*
DoD technical reports	3.15	3.23
NASA technical reports	3.05	3.36**

[a]Higher numbers indicate greater uncertainty.
[b]Advisory Group for Aerospace Research and Development
*$p \leq .05$. **$p \leq .01$.

Table 8.21. Technical Uncertainty and the Use of Internal and External Information Products (N=872)

How many times in the past six months did you use the following?

Product	Uncertainty Low 1	2	3	4	High 5
Internal Sources					
Drawings/specifications	31.7	36.1	32.9	31.5	28.3
Memoranda	18.3	16.5	24.3	28.8	21.0
Letters	11.5	10.8	16.3	16.3	20.6
Audiovisual materials	5.1	3.7	4.8	4.3	6.6
Technical manuals	7.2	7.3	7.0	6.5	5.8
In-house technical reports	6.6	7.7	8.6	10.0	15.8*
Trade/promotional literature	4.6	7.1	6.1	9.5	8.9
Computer programs/ documentation	3.8	2.8	4.8	5.1	6.2
Technical proposals	3.1	2.3	3.2	2.4	3.3
External Sources					
Journal articles	4.1	4.7	5.8	9.1	10.7*
Government technical reports	4.5	2.8	4.5	4.6	6.2*
Conference/meeting papers	3.5	2.5	4.1	3.2	6.3*
Technical talks/presentations	2.8	3.5	4.8	6.7	6.6*
Abstracts	2.5	1.2	1.7	5.7	5.2*
AGARD technical reports	.1	.2	.2	.3	.3

*$p \leq .05$.

abstracts, journal articles, conference papers, U.S. government technical reports, and technical talks or presentations increased. These findings are consistent with reported information-seeking patterns (See Table 8.17) wherein the search extended beyond internal and informal to external and formal information sources as uncertainty increased.

Research Question 3b. What Is the Impact of Uncertainty on the Use of Federally Funded Aerospace R&D?

About 42% of the survey respondents reported using the results of federally funded aerospace R&D in their work. Some 25% of the

SAE survey respondents reported using the results of federally funded aerospace R&D in completing their projects. A majority of the respondents who used this information to complete their projects found it in published DoD or NASA technical reports.

To examine the relationship between technical uncertainty and nine factors thought to influence the use of DoD and NASA reports, respondents were divided into low and high uncertainty groups, and the median rating was excluded (Table 8.22). For DoD technical report use, respondents in the high technical uncertainty group reported numerically higher influence scores for eight of the nine factors than respondents in the low uncertainty group reported. Only three scores were statistically significant however—factors of quality, relevance, and importance. Under conditions of low and high uncertainty, numerically higher influence scores were reported for technical quality, comprehensiveness of the data, and the relevance of the DoD technical reports to the work.

For NASA technical report use, respondents in the high technical uncertainty group reported numerically higher influence scores for eight of the nine factors than respondents in the low uncertainty group reported. Three were statistically significant— factors of technical quality, comprehensiveness, and importance. Similar to the data on DoD technical reports for both high and low uncertainty groups, the numerically highest influence scores were reported for technical quality, comprehensiveness of the data, and the relevance of the NASA technical reports to the work.

Statistically the data demonstrate that "value" factors such as comprehensiveness, relevance, and importance influence information use under conditions of uncertainty. Neither ease of obtaining the data, prior experience, nor expense was a factor influencing use under conditions of increased uncertainty. This finding failed to uphold both the "ease of use" thesis and the "cost-benefit" explanation of information-seeking behavior.

Survey 3: SME–Manufacturing and Production Sector

Survey 3 used a random sample of U.S. aerospace engineers and scientists whose names appeared on the membership list of the Society of Manufacturing Engineers (SME) and whose SIC code indicated that they were employed in an aerospace organization. The

Table 8.22. Technical Uncertainty and the Factors Influencing Use of DoD and NASA Technical Reports (N=872)

Report	Importance Scores[a]	
	Low Uncertainty	High Uncertainty
DoD Technical Reports		
Easy to physically obtain	3.57	3.59
Easy to use or read	3.68	3.73
Expense	3.00	3.10
Technical quality	3.93	4.16*
Comprehensive data and information	3.92	4.12*
Relevance	3.96	4.18*
Obtainable at nearby location/source	3.35	3.36
Good prior experience using them	3.21	3.17
Important in your work	2.42	2.81**
NASA Technical Reports		
Easy to physically obtain	3.60	3.60
Easy to use or read	3.76	3.78
Expense	3.03	3.09
Technical quality	4.03	4.24*
Comprehensive data and information	3.99	4.20*
Relevance	4.05	4.19
Obtainable at nearby location/source	3.37	3.32
Good prior experience using them	3.23	3.18
Important in your work	2.31	2.74**

[a]Influence was measured on a 5-point scale where 1 = very unimportant and 5 = very important. The higher the number the greater the influence of the factor on report use.

*$p \leq .05$. **$p \leq .01$.

group consisted of engineers and scientists performing professional duties in design and development, manufacturing and production, and management. After discarding the responses from public sector respondents and other nonusable questionnaires, the final sample for Survey 3 consisted of 419 surveys, for an adjusted response rate of 41% (Pinelli, Barclay, and Kennedy, 1994b).

Table 8.23 is a demographic description of the Survey 3 respondents. Average work experience in aerospace was 16.7 years. Most

Table 8.23. Survey Demographics *(N=419)*

Characteristics	*n*	%
Education		
No degree	95	22.7
Bachelor's degree	196	46.8
Master's degree	84	20.0
Post graduate	6	1.4
Other	38	9.1
Years in Aerospace		
<1–5	4	.9
6–10	56	13.4
11–15	96	22.9
16–20	131	31.3
21–30	84	20.0
31–40	44	10.5
$\geq$ 41	4	.9
Mean = 16.7 years		
Academic Preparation		
Engineer	308	73.5
Scientist	10	2.4
Other	101	24.1
Job Duties		
Engineer	302	72.1
Scientist	5	1.2
Other	112	26.7

(46.8%) held undergraduate degrees; 20% held master's degrees, and 1.4% had advanced their education to the post-graduate level. Job duties were mostly in engineering (72.1%) and only a few respondents (1.2%) worked as scientists. These duties closely paralleled the respondents' academic preparation: 73.5% were educated as engineers and 2.4% as scientists.

Table 8.24 contains data that characterize the project reported on by survey respondents. A majority (74.7%) of respondents reported that they worked on the project in groups. Most described their duties in the project as engineering (62.3%). Management accounted for the next highest share (25.3%); science accounted for only 2.4%. The project primarily involved manufacturing and production (49.4%), design and development (22%), and management (10.5%).

Research Question 1: Is Technical Uncertainty an Issue?

The first research question asked whether uncertainty is a valid construct for aerospace engineers and scientists working in the manufacturing or production sector. Table 8.25 contains the respondents' perceptions of the technical uncertainty of the project. The measure of uncertainty yielded a mean rating of 3.22 on a scale of 1 (little uncertainty) to 5 (great uncertainty). Answers to the uncertainty question were fairly normally distributed although less so than the answers of the design and development group in Survey 2 (SAE). The distribution suggested that the SME respondents, like the SAE respondents, were able to make a determination about the levels of uncertainty that they face. It also implied that answers were not contaminated by desirability or similarity biases.

The distribution of the responses to the question dealing with complexity showed that complexity, rated on a scale of 1 (very simple) to 5 (very complex), yielded a mean of 3.83. Responses to the complexity question were skewed toward the "very complex" end of

Table 8.24. Project Characteristics ($N=419$)

Characteristics	n	%
Worked on Project		
Alone	106	25.3
With others	313	74.7
Nature of Duties Performed on Project		
Engineering	261	62.3
Science	10	2.4
Management	106	25.3
Other	42	10.0
Category of the Technical Project		
Research	20	4.8
Design/development	92	22.0
Manufacturing/production	207	49.4
Computer applications	18	4.3
Quality control/assurance	38	9.1
Management	44	10.5

Table 8.25. Technical Uncertainty and Complexity (N=419)

Think of the most important job-related project, task, or problem you have worked on in the past six months. On a scale of 1 (little uncertainty) to 5 (great uncertainty), how would you describe the overall technical uncertainty of this project?

Great		n	%
Uncertainty	5	38	9.1
	4	139	33.2
	3	145	34.6
Little	2	74	17.7
Uncertainty	1	23	5.5

Overall mean uncertainty score = 3.22, s.d. = 1.02.

On a scale of 1 (very simple) to 5 (very complex), how would you describe the overall technical complexity of this project?

		n	%
Very Complex	5	81	19.3
	4	197	47.1
	3	132	31.8
	2	7	1.7
Very Simple	1	2	.2

Overall mean complexity score = 3.83, s.d. = .77.

the scale. Given that the respondents were aerospace engineers and scientists involved in manufacturing and production and design and development, we expected that survey respondents would be involved in complex projects. An examination of the relationship between complexity and uncertainty yields a Pearson's correlation between technical uncertainty and complexity of .27 ($p \leq .05$). In other words, approximately 7% of the variance in one variable was accounted for by the variance of the other. This measure suggests that, although there was a weak statistically significant relationship, the two variables do not tap the same dimension and, hence, are not substitute measures of the same construct.

Research Question 2: Are Personal and Task Characteristics Related to Perceptions of Technical Uncertainty?

Three personal variables were used to test the second research question: the respondent's education, years of work experience, and academic background. The amount of uncertainty did not vary as a function of the years of work experience or academic preparation for the present job. The only personal variable associated with uncertainty was education (Table 8.26). Similar to the findings from the AIAA and SAE studies, as the respondents' education extended beyond the baccalaureate degree, they reported significantly more uncertainty than those whose education level was a baccalaureate degree or less. Also similar to the AIAA and the SAE findings, no linear or curvilinear relationship could be found between work experience and uncertainty. Again, the explanation may rest in the nature of the aerospace industry that results in education, rather than experience, being a correlate of uncertainty. This confirms Downey and Slocum's (1975) suggestion that duration of the experience may not be as important as the variety of experience.

Academic preparation as either an engineer or scientist was not significant in relation to uncertainty for the SME survey respondents. Given this, the fact that few scientists were represented in the sample, and prior work on the lack of differences in information-seeking behavior between U.S. aerospace engineers and scientists (Pinelli, 1991), SME engineers and scientists were not treated differently in the analysis.

Three task variables were examined: category of the project, those working alone or in groups on a project, and the type of duties involved in the project. Uncertainty differed significantly according to the category of the project (Table 8.27). Those working on proj-

Table 8.26. Technical Uncertainty and Education (N=419)

Education	Uncertainty Score[a]	
No degree	2.9	
Bachelor's degree	3.4	
Master's degree	3.3	
Doctorate	4.0	$F = 7.32$

[a]Higher numbers indicate greater uncertainty.
**$p \leq .01$.

Table 8.27. Technical Uncertainty and Project Characteristics (N=419)

Which category best describes your project?

Category	Uncertainty Score[a]	
Research	3.2	
Design	2.9	
Development	3.5	
Manufacturing	3.1	
Production	3.3	
Computer applications	3.5	
Quality control/assurance	3.6	
Management	3.3	$F = 2.64**$

Which of the following best describes your duties on the project?

Duty	Uncertainty Score	
Engineer	3.2	
Scientist	3.1	
Management	3.3	
Other	3.2	$F = .34$

On this project, did you work alone or with others?

Method	Uncertainty Score	
Alone	3.1	
With others	3.3	$t = 1.33$

[a]Higher numbers indicate greater uncertainty.
$**p \leq .01$.

ects involving quality control or assurance, computer applications, and development had the highest levels of uncertainty (3.6, 3.5, and 3.5, respectively), and those projects involving design projects had the lowest (2.9). The manufacturing and production focus of the SME sample might account for the lower uncertainty rating for design because once manufacturing begins, the design becomes "frozen" and will remain so unless the original design criteria are changed. Those who reported their project as "other" were not included in the analysis.

Like the AIAA study, no significant differences were found among uncertainty levels and the category of the project; the performance of duties on the project as either an engineer, scientist, or manager; and whether the respondent worked alone or with others (in a group) on the project. The former finding may be explained by the small number of scientists in the AIAA sample, the nature of uncertainty in aerospace production and manufacturing (which differs from uncertainty in aerospace research and design), and the fact that about 90% of the AIAA survey respondents worked with others in completing the project.

Research Question 3a. What Is the Impact of Technical Uncertainty on Information-Seeking Behavior in General?

To investigate the impact of uncertainty on information-seeking behavior, the information-seeking patterns of the SME sample were examined (Table 8.28). Respondents were asked to rank order the steps they had followed in obtaining the information they needed to complete the project. They first used their personal stores of technical information; then they spoke with coworkers within the orga-

Table 8.28. Rank Order of Information Sources Used to Complete the Project (N=419)

Rank	Median Rank	% Using[a] First
1 Used my personal stores of technical information, including sources I keep in my office	1	54.3
2 Spoke with coworkers or people inside my organization	2	34.8
3 Spoke with colleagues outside my organization	3	5.7
4 Searched (or had someone search for me) an electronic (bibliographic) database	4	4.9
5 Used literature resources (e.g., conference papers, journals, technical reports) found in my organization's library	4	4.0
6 Spoke with a librarian or technical information specialist	4	1.3

[a]Percentages of those reporting thus eliminating cases with missing data and unusable responses such as don't know and unable to estimate.

nization. If this approach did not provide adequate information, they next sought information from colleagues outside the organization. The information search then extended to formal (written) channels and sources as conference and meeting papers, professional journals, and technical reports, and finally to a librarian.

To determine if technical uncertainty affects information channel or source selection, the sample was divided into users and non-users of each channel or source (Table 8.29). The rationale for doing this is based on the assumption that as technical uncertainty increases, the search extends from the use of internal and informal to external and formal information channels and sources.

As expected, almost all the respondents used their personal stores of information to complete the project. The findings indicate

Table 8.29. Technical Uncertainty and Information Source Use (N=419)

Where did you get the information you needed for the project?

Source	Uncertainty[a]			
	Nonusers		Users	
Used my personal store of technical information	3.57	(19)	3.22	(373)
Spoke with my coworkers or people inside my organization	3.21	(14)	3.23	(379)
Spoke with colleagues outside my organization	3.08	(79)	3.28	(307)
Used literature resources (e.g., journals, conference papers, technical reports) found in my organization's library	3.15	(242)	3.38	(121)*
Searched (or had someone search for me) an electronic (bibliographic) database	3.15	(220)	3.37	(134)*
Spoke with a librarian or technical information specialist	3.17	(148)	3.27	(220)

[a]Higher numbers indicate greater uncertainty.
[b]Differences in numbers reported are due to missing data and unusable responses such as don't know and unable to estimate.
*$p \leq .05$.

that the majority of survey respondents were unable to complete the problem based solely on the use of their personal stores of technical information, for almost all of the respondents spoke with others within and outside their organization. Yet, no significant differences were noted between users and nonusers of these informal channels and sources as far as uncertainty was concerned.

The data suggest that many respondents were able to solve the problem after consulting with coworkers in their organization. Still, 79% of the respondents spoke with colleagues outside of their organization, 33% consulted the literature, 37% used an electronic (bibliographic) database, and 59% consulted an information specialist. As task uncertainty increased, the information search broadened beyond the use of internal and informal channels and sources. Those respondents who used the formal literature or searched (or had someone search) an electronic (bibliographic) database also reported working on projects having significantly higher technical uncertainty than those who did not use these information channels and sources. The single exception was among those respondents who spoke with a librarian or a technical information specialist.

Another issue involved the relationship between technical uncertainty and the hours spent each week communicating technical information in writing or orally (Table 8.30). Like the AIAA survey (Survey 1) but unlike the SAE survey (Survey 2), no statistically significant differences were found in the SME survey between the number of hours spent communicating technical information orally

Table 8.30. Technical Uncertainty and Communications Frequency (N=419)

Hours Spent Each Week Communicating Technical Information

	Low Uncertainty			High Uncertainty		
	1	2	3	4	5	
Written	15.4	11.7	10.6	9.7	10.3	$F = 1.93$
Oral	11.3	13.7	12.2	14.9	12.5	$F = 1.29$

Hours Spent Each Week Receiving Technical Information

| Written | 15.6 | 10.6 | 10.5 | 10.0 | 8.6 | $F = 1.76$ |
| Oral | 6.0 | 8.5 | 7.7 | 9.4 | 9.5 | $F = 1.09$ |

and in writing and technical uncertainty scores. No meaningful trends in the data could be interpreted.

Next, we examined the relationship between technical uncertainty and the use of specific information products: journal articles, conference papers, in-house technical reports, DoD technical reports, and NASA technical reports. To examine the relationship between technical uncertainty and the use of these five information products, respondents were divided into low uncertainty and high uncertainty groups (Table 8.31). Respondents who reported a median level of uncertainty were excluded. Under conditions of higher uncertainty, the use of journal articles and NASA technical reports increased significantly.

Significant findings concerning the increased use of journal articles and NASA technical reports were similar to those of the SAE survey. However, the SAE sample also found significantly greater use of in-house technical reports and conference papers under increased conditions of uncertainty; the SME sample did not.

Finally, we investigated the relationship between technical uncertainty and the frequency of use of 15 information products (Table 8.32). The products were classified as internal and external sources of information. The results demonstrate a positive but lim-

Table 8.31. Technical Uncertainty and Information Product Use (N=419)

Product	Uncertainty[a]	
	Nonuse	Use
In-house technical reports	3.1	3.3
Journal articles	3.0	3.4**
Conference papers	3.2	3.3
DoD technical reports	3.3	3.3
NASA technical reports	3.2	3.5*

[a]Higher numbers indicate greater uncertainty.

$*p \leq .05.$ $**p \leq .01.$

Table 8.32. Technical Uncertainty and the Use of Internal and External Information Products (N=419)

How many times in the past six months did you use the following?

| | Uncertainty | | | | |
| | Low | | | | High |
Product	1	2	3	4	5
Internal Sources					
Drawings/specifications	116.5	119.6	89.6	41.4	68.3*
Memoranda	16.8	29.8	38.6	30.3	35.5
Letters	11.6	32.2	20.7	18.0	18.8
Audiovisual materials	143.7	8.6	8.6	13.0	23.1**
Technical manuals	24.2	26.0	21.3	18.2	8.7
In-house technical reports	18.4	12.7	14.6	9.2	7.4
Trade/promotional literature	8.4	34.8	10.7	20.6	13.1
Computer programs/ documentation	26.8	32.8	20.6	18.4	12.9
Technical proposals	9.9	10.6	7.4	7.1	4.9
External Sources					
Journal articles	5.7	10.9	6.5	7.4	5.4
Conference/meeting papers	13.6	10.6	5.4	8.9	5.3
Technical talks/presentations	3.2	9.2	5.9	6.2	5.9
Abstracts	7.9	4.6	2.9	4.7	1.4
DoD technical reports	5.2	18.6	1.9	2.0	4.6
NASA technical reports	0.0	6.8	1.1	1.4	2.2

*$p \leq .05$. **$p \leq .01$.

ited relationship between technical uncertainty and the frequency of use of some internal and external information products.

Of the internal products, two of nine—drawings and specifications and audiovisual materials—were used significantly more frequently under conditions of increased technical uncertainty. Of the external products, none was found to demonstrate statistically significant differences in usage that could be related to conditions of technical uncertainty. Unlike the results in the SAE survey, increases in project technical uncertainty did not result in a corresponding increase in the frequency of use of external information products on the part of the SME survey respondents.

Research Question 3b. **What Is the Impact of Technical Uncertainty on the Use of Federally Funded Aerospace R&D?**

To examine the relationship between technical uncertainty and the nine factors thought to influence the use of DoD and NASA technical reports, respondents were divided into groups of low and high uncertainty (Table 8.33). The median rating was eliminated.

Table 8.33. Factors Influencing Use of DoD and NASA Technical Reports and Technical Uncertainty (*N*=419)

	Importance Scores[a]	
Report	Low Uncertainty	High Uncertainty
DoD Technical Reports		
Easy to physically obtain	3.97	3.89
Easy to use or read	4.04	4.03
Expense	3.62	3.54
Technical quality	4.28	4.20
Comprehensive data and information	4.32	4.20
Relevance	4.43	4.20
Obtainable at nearby location/source	3.76	3.54
Good prior experience using them	3.55	3.25*
Important in your work	2.31	2.41
NASA Technical Reports		
Easy to physically obtain	3.79	3.79
Easy to use or read	3.92	3.98
Expense	3.49	3.59
Technical quality	4.17	4.19
Comprehensive data and information	4.18	4.16
Relevance	4.30	4.20
Obtainable at nearby location/source	3.66	3.49
Good prior experience using them	3.44	3.22
Important in your work	1.89	2.29**

[a]Influence was measured on a 5-point scale where 1 = very unimportant and 5 = very important. The higher the number the greater the influence of the factor on report use.
*$p \leq .05$. **$p \leq .01$.

For DoD technical report use, only one factor—good prior experience—demonstrated a statistically significant difference between high and low uncertainty scores and DoD technical report users. Among users and nonusers, the numerically highest influence scores were reported for technical quality, comprehensiveness, and relevance.

For NASA technical report use, the highest numerical influence scores reported by both users and nonusers were in the categories of technical quality, comprehensiveness of the data, and the relevance of the NASA technical reports to their work. In comparing users to nonusers, only the dimension of "importance to my work" demonstrated a statistically significant difference in levels of technical uncertainty.

Overall, about 31% of the SME survey respondents reported having used the results of federally funded aerospace R&D in their work. About 19% of the SME survey respondents reported having used the results of federally funded aerospace R&D in completing their projects. A majority of the respondents who used this information to complete their projects indicated that they had found it in published DoD or NASA technical reports.

CONCLUSIONS

Three separate surveys were conducted that investigated the relationship between technical uncertainty and information seeking and use by engineers and scientists working in the research and design, the design and development, and the manufacturing and production sectors of the U.S. aerospace industry. The three industry surveys yielded similar findings although the composition of the three samples varied. A greater percentage of the AIAA sample (research and design) than the SAE sample (design and development) and the SME sample (manufacturing and production) held advanced degrees. The AIAA sample also had more aerospace work experience than the other two samples. About equal percentages of the AIAA and SAE samples had been educated as engineers or scientists; a considerably smaller percentage of the SME sample had been educated as engineers. A majority of AIAA and SAE respondents categorized the nature of their technical project (i.e., the "critical" event) as design or development. The greatest percentage of the SME sample categorized the nature of their technical project as manufacturing and production. Also, higher percentages of AIAA and SAE survey respondents than SME survey respondents worked in groups on the project. Members of the three samples reported performing similar duties on the technical project.

Survey respondents in the three samples reported similar patterns of information-seeking. As part of a general search for information, members of the three samples used internal and informal information channels and sources first. If the search proved inadequate to that point, they next consulted external and formal information channels and sources. The analysis was then extended to determine linkages between project uncertainty and information-seeking behavior. The mean complexity scores for the technical project were higher than the mean uncertainty scores among all three samples. A greater mean uncertainty score was reported by the AIAA sample members than by the SAE and SME sample members. Analysis of the relationships between technical uncertainty and information-seeking behavior showed marked similarities, with only minor differences being observed among the three samples.

Overall technical (task) uncertainty played a role in information channel and source selection. However, technical uncertainty was not a strong predictor of the intensity (frequency) of use of any information channel or source. Findings were mixed on the impact of technical uncertainty on the number of hours spent communicating technical information in writing or orally.

Statistically, high technical (task) uncertainty resulted in a greater use of federally funded aerospace R&D contained in DoD and NASA technical reports. For members of the SAE and SME samples, use was statistically significant for NASA technical reports only. Analysis of the findings hints at some of the factors contributing to the use of DoD or NASA technical reports. Although the three samples did not yield identical findings, the data do show that such "value" factors as technical quality, comprehensiveness, and relevance influence use. This finding suggests that neither the "ease of use" nor the "cost-benefit" model fully explains the use of DoD and NASA technical reports by U.S. aerospace engineers and scientists working in research and design, design and development, or manufacturing and production. The findings support Culnan's (1983) argument that information "quality" must be viewed separately from channel and source selection, although accessibility does influence channel and source selection. They affirm Orr's

(1970) claim that the quality and reliability of information are critical factors in its use.

IMPLICATIONS FOR DIFFUSING THE RESULTS OF FEDERALLY FUNDED AEROSPACE R&D

Although the findings of these three studies are preliminary, the data show that technical uncertainty affects the information-seeking behaviors of U.S. aerospace engineers and scientists. The general information-seeking behaviors of the U.S. aerospace engineers and scientists reported in the three studies supports the findings from the literature that accessibility (i.e., ease of use) is an important factor in information use. Survey respondents opted first for internal and informal information channels and sources, then expanded the search using external and formal information channels and sources. However, access to information may not be the primary determinant for use. Quality, reliability, and relevance of information found in one's personal stores and through coworkers may actually be the key factors that determine use.

As technical uncertainty increased, respondents made significantly more use of external and formal information channels and sources. Under conditions of high technical (task) uncertainty, respondents were more likely to use the results of federally funded aerospace R&D found in DoD and NASA technical reports than under conditions of low technical uncertainty. Finally, those respondents who used federally funded aerospace R&D did so because of the perceived technical quality, comprehensiveness, and relevance of the data to their work rather than the ease of use, accessibility, or the cost of the DoD or NASA technical reports.

The findings point to a relationship between technical uncertainty and the use of DoD and NASA technical reports, and suggest that neither the "ease of use" nor the "cost-benefit" model should be used exclusively in making decisions about diffusing the results of federally funded aerospace R&D. Obviously, the results of federally funded aerospace R&D will not have maximal use if users find the results difficult to obtain, costly to obtain in terms of effort and monetary expense, and not easy to use once they are obtained. Conversely, simply increasing accessibility and reducing cost by making DoD and NASA technical reports available electronically will not guarantee their use by U.S. aerospace engineers and scientists who elect to use them based on the technical quality, comprehensiveness, and relevance of the data contained therein.

Finally, although the findings can be used to help develop a strategic plan for diffusing the results of federally funded R&D to the U.S. aerospace industry, they may be specific to the aerospace industry only and, in particular, to the commercial aviation sector, which has a high degree of complexity and technical uncertainty embodied in its products.

CHAPTER REFERENCES

Allen, T.J. (1977). *Managing the Flow of Technology: Technology Transfer and the Dissemination of Technological Information Within the R&D Organization*. Cambridge, MA: MIT Press.

Allen, T.J. (1966). "Performance of Information Channels in the Transfer of Technology." *Industrial Management Review* 8: 87-98.

Allen, T.J. and S.I. Cohen. (1969). "Information Flow in R&D Laboratories." *Administrative Science Quarterly* 14: 12-19.

Argote, L. (1982). "Input Uncertainty and Organizational Coordination in Hospital Emergency Rooms." *Administrative Science Quarterly* 27: 420-434.

Auster, E. and C.W. Choo. (1993). "Environmental Scanning by CEOs in Two Canadian Industries." *Journal of the American Society for Information Science* (May) 44(4): 194-203.

Boyd, B.K. (1989). *Perceived Uncertainty and Environmental Scanning: A Structural Model*. Ph.D. Diss., Los Angeles, CA: University of Southern California (USC). (Available exclusively from Micrographics Dept., Doheny Library, USC.)

Brown, J.W. and J.M. Utterback. (1985). "Uncertainty and Technical Communication Patterns." *Management Science* 31: 301-311.

Chakrabarti, A.K.; K.S. Feineman; and W. Fuentevilla. (1983). "Characteristics of Sources, Channels, and Contents for Scientific and Technical Information Systems in Industrial R and D." *IEEE Transactions on Engineering Management* (May) EM-22(2): 83-88.

Connolly, T. (1975). "Communication Nets and Uncertainty in R&D Planning." *IEEE Transactions on Engineering Management* (May) EM-22(2): 50-54.

Culnan, M.J. (1983). "Environmental Scanning: The Effects of Task Complexity and Source Accessibility on Information Gathering." *Decision Sciences* 14(2): 194-206.

Czepiel, J.A. (1975). "Patterns of Interorganizational Communications and the Diffusion of a Major Technological Innovation in a Competitive Environment." *Academy of Management Journal* 18: 6-24.

Dewhirst, H.D (1971). "Influence of Perceived Information-Sharing Norms on Communication Channel Utilization." *Academy of Management Journal* 13: 305-315.

Downey, H.K. and J.W. Slocum, Jr. (1975). "Uncertainty: Measures, Research and Sources of Variation." *Academy of Management Journal* 18: 562-578.

Duncan, R.B. (1972). "Characteristics of Organizational Environments and Perceived Uncertainty." *Administrative Science Quarterly* 17: 313-327.

Ebadi, Y.M. and J.M. Utterback. (1984). "The Effects of Communication on Technical Innovation." *Management Science* (May) 30(5): 572-585.

Fischer, W.A. (1980). "Scientific and Technical Information and the Performance of R&D Groups." In *Management of Research and Innovation*. B.V. Dean and J.L. Goldhar, eds. New York, NY: North-Holland, 67-89.

Flanagan, J.C. (1954). "The Critical Incident Technique." *Psychological Bulletin* (July) 51(4): 327-358.

Galbraith, J.K. (1977). *Organization Design*. Reading, MA: Addison-Wesley.

Galbraith, J.K. (1973). *Designing Complex Organizations*. Reading, MA: Addison-Wesley.

Gerstberger, P.G. and T.J. Allen. (1968). "Criteria Used by Research and Development Engineers in the Selection of an Information Source." *Journal of Applied Psychology* (August) 52(4): 272-279.

Gerstenfeld, A. and P. Berger. (1980). "An Analysis of Utilization Differences for Scientific and Technical Information." *Management Sciences* 26(2): 165-179.

Gifford, W.; H. Bobbitt; and S. Slocum. (1979). "Message Characteristics and Perceptions of Uncertainty by Organizational Decision Makers." *Academy of Management Journal* 22(1): 458-481.

Hambrick, D.C. (1979). *Environmental Scanning, Organizational Strategy and Executive Roles: A Study in Three Industries.* Ph.D. Diss., Pennsylvania State University. UMI 79-22294.

Hardy, A.P. (1982). "The Selection of Channels When Seeking Information: Cost-Benefit vs Least-Effort." *Information Processing & Management* 18(6): 289-293.

Hauptman, O. (1986). "Influence of Task Type on the Relationship Between Communication and Performance: The Case for Software Development." *R & D Magazine* 16(2):127-139.

Holland, W.E.; B.A. Stead; and R.C. Leibrock. (1976). "Information Channel/Source Selection as a Correlate of Technical Uncertainty in Research and Development Organizations." *IEEE Transactions on Engineering Management* (November) EM-23(4): 163-167.

Huber, G.P.; M.J. O'Connell, and L.L. Cummings. (1975). "Perceived Environmental Uncertainty: Effects of Information and Structure." *Academy of Management Journal* (December) 18(3): 725-740.

Jain, S.C. (1984). "Environmental Scanning in U.S. Corporations." *Long Range Planning* 17(2): 117-128.

Katz, D. and R.L. Kahn. (1966). *The Social Psychology of Organizations*. New York, NY: John Wiley.

Kaufman, H.G. (1983). *Factors Related to the Use of Technical Information in Engineering Problem Solving*. Brooklyn, NY: Polytechnic Institute of New York.

Keller, R.T. and W.E. Holland. (1975). "Boundary-Spanning Roles in a Research and Development Organization." *Academy of Management Journal* 18: 388-393.

Lancaster, F. W. (1978). *Critical Incident Technique*. Urbana, IL: University of Illinois Graduate School of Library and Information Science.

March, J. and H. Simon. (1958). *Organizations*. New York, NY: John Wiley.

Miller, R.E. (1971). *Innovation, Organization, and Environment: A Study of Sixteen American and West European Steel Firms*. Sherbrooke, PQ, Canada: University of Sherbrooke Press.

Mowery, D.C. (1985). "Federal Funding of R&D in Transportation: The Case of Aviation." Paper commissioned for a workshop on *The Federal Role in Research and Development* sponsored by the National Academy of Sciences, National Academy of Engineering, and Institute of Medicine in Washington, DC, 21-22 November.

Nilakanta, S. and R.W. Scamell. (1990). "The Effect of Information Sources and Communication Channels on the Diffusion of Innovation in a Data Base Development Environment." *Management Science* (January) 36(1): 24-40.

O'Reilly, C.A. (1982). "Variations in Decision Makers' Use of Information Sources: The Impact of Quality and Accessibility on Information." *Academy of Management Journal* 25(4): 756-771.

Orr, R.H. (1970). "The Scientist as an Information Processor: A Conceptual Model Illustrated with Data on Variables Related to Library Utilization." In *Communication Among Scientists and Engineers*. C.E. Nelson and D.K. Pollock, eds. Lexington, MA: D.C. Heath, 143-189.

Pinelli, T.E. (1991). *The Relationship Between the Use of U.S. Government Technical Reports by U.S. Aerospace Engineers and Scientists and Selected Institutional and Sociometric Variables*. Washington, DC: National Aeronautics and Space Administration. NASA TM-102774. (Available NTIS; 91N18898.)

Pinelli, T.E.; R.O. Barclay; and J.M. Kennedy. (1995). *The Technical Communications Practices of U.S. Aerospace Engineers and Scientists: Results of the Phase 1 AIAA Mail Survey*. Washington, DC: National Aeronautics and Space Administration. NASA TM-110180. (Available NTIS; 95N34217.)

Pinelli, T.E.; R.O. Barclay; and J.M. Kennedy. (1994a). *The Technical Communications Practices of U.S. Aerospace Engineers and Scientists: Results of the Phase 1 SAE Mail Survey*. Washington, DC: National Aeronautics and Space Administration. NASA TM-109088. (Available NTIS; 94N32837.)

Pinelli, T.E.; R.O. Barclay; and J.M. Kennedy. (1994b). *The Technical Communications Practices of U.S. Aerospace Engineers and Scientists: Results of the Phase 1 SME Mail Survey.* Washington, DC: National Aeronautics and Space Administration. NASA TM-109169. (Available NTIS; 95N20163.)

Pinelli, T.E.; N.A. Glassman; L. O. Affelder; L.M. Hecht; J.M. Kennedy; and R.O. Barclay. (1993). *Technical Uncertainty and Project Complexity as Correlates of Information Use by U.S. Industry-Affiliated Aerospace Engineers and Scientists: Results of an Exploratory Investigation.* Washington, DC: National Aeronautics and Space Administration. NASA TM-1075693. (Available NTIS; 94N17291.)

Rogers, E.M. (1983). *Diffusion of Innovations.* 3rd ed. New York, NY: The Free Press.

Rogers, E.M. (1982). "Information Exchange and Technological Innovation." Chapter 5 in *The Transfer and Utilization of Technical Knowledge.* D. Sahal, ed. Lexington, MA: D.C. Heath, 105-123.

Rosenberg, V. (1967). "Factors Affecting the Preferences of Industrial Personnel for Information Gathering Methods." *Information Storage and Retrieval* (July) 3: 119-127.

Rosenbloom, R.S. and F.W. Wolek. (1970). *Technology and Information Transfer: A Survey of Practice in Industrial Organizations.* Boston, MA: Harvard University Press.

Schmidt, S.M. and L.L. Cummings. (1976). "Organizational Environment, Differentiation and Perceived Environmental Uncertainty." *Decision Sciences* 7: 447-467.

Shannon, C.E. and W. Weaver. (1949). *The Mathematical Theory of Communication.* Urbana, IL: University of Illinois Press.

Swanson, E.B. (1987). "Information Channel Disposition and Use." *Decision Sciences* 18: 131-145.

Thompson, J.D. (1967). *Organizations in Action.* New York, NY: McGraw-Hill.

Tushman, M.L. (1979). "Impacts of Perceived Environmental Variability on Patterns of Work-Related Communication." *Academy of Management Journal* (September) 22(3): 482-500.

Tushman, M.L. (1978). "Technical Communication in R&D Laboratories: The Impact of Project Work Characteristics." *Academy of Management Journal* (December) 21(4): 624-645.

Tushman, M.L. and R. Katz. (1980). "External Communications and Project Performance: An Investigation into the Role of Gatekeepers." *Management Science* 26: 1071-1085.

Tushman, M.L. and D.A. Nadler. (1980). "Communication and Technical Roles in R&D Laboratories: An Information Processing Model." In *Management of Research and Innovation.* B.V. Dean and J.L. Goldhar, eds. New York, NY: North-Holland, 91-112.

Tushman, M.L. and D.A. Nadler. (1978). "Information Processing as an Integrating Concept in Organizational Design." *Academy of Management Review* (January) 3(1): 613-624.

Tyson, L.D. (1992). "Industry Policy and Trade Management and the Commercial Aircraft Industry." Chapter 5 in *Who's Bashing Whom? Trade Conflict in High Technology Industries*. L.D. Tyson, ed. Washington, DC: Institute for International Economics, 155-216.

Utterback, J.M. (1974). "Innovation in Industry and the Diffusion of Technology." *Science* 183: 620-626.

Utterback, J.M. (1971). "The Process of Innovation: A Study of the Origination and Development of Ideas for New Scientific Instruments." *IEEE Transactions on Engineering Management* EM18: 124-131.

Weick, K.E. (1969). *The Social Psychology of Organizing*. Reading, MA: Addison-Wesley.

Zaltman, G.; R. Duncan; and J. Holbek. (1973). *Innovation and Organizations*. New York, NY: John Wiley.

Zenger, T.R. and B.S. Lawrence. (1989). "Organizational Demography: The Differential Effects of Age and Tenure Distributions on Technical Communication." *Academy of Management Journal* 32(1): 353-378.

Zmud, R.W. (1983). "The Effectiveness of External Information Channels in Facilitating Innovation Within Software Development Groups." *MIS Quarterly* (June) 7(2): 43-58.

Chapter 9

The Differential Effects of Workplace Experience on Information Production and Use—A Comparison of New and Established U.S. Aerospace Engineers and Scientists

C. J. Anderson
M. Glassman
E. W. Morrison

SUMMARY

Chapter 9 reports selected results from an investigation that focused on the communications practices and information-related activities of early career-stage (i.e., new) U.S. aerospace engineers and scientists in the workplace. Undertaken as a Phase 1 activity of the *NASA/DoD Aerospace Knowledge Diffusion Research Project*, the investigation used mail (self-reported) surveys to collect data from 264 members of the American Institute of Aeronautics and Astronautics (AIAA) who had converted their AIAA memberships from student to professional status and who had an average of 2.4 years of aerospace work experience. To learn more about the concomitant communications norms, we compared the results of this investigation with data we collected from 1673 student members of the AIAA and with data we collected from 341 established members of the AIAA who had an average of 21.9 years of aerospace work experience. Selected results from these studies are presented about career goals; the importance, receipt, and helpfulness of communications and information-use skills instruction; collaborative writing; information products produced and used; factors impeding the production of written technical communication; library use; the use of electronic (computer) networks; and the use of federally funded aerospace research and development (R&D) in completing or solving their most recent project, task or problem. The chapter concludes with implications for diffusing the results of (U.S.) federally funded aerospace R&D.

INTRODUCTION

Engineers in the world of work report that the communication (i.e., production, transfer, and use) of information takes up as much as 80% of their time, that the communication of information is an essential element of successful engineering practice, and that the ability to communicate information effectively is critical to professional success and advancement (Mailloux, 1989). Feedback from professional engineers and from engineers' supervisors concerning engineering competencies ranks communications and information-use skills—the ability to write effectively, to make oral presentations, and to search out and acquire information—*high* in terms of importance to engineering practice. This same feedback, however, ranks the communications and information-seeking skills of entry-level engineers *low* (Bakos, 1986; Chisman, 1987; Katz, 1993; Kimel and Monsees, 1979).

Although communications and information-seeking skills are important to successful technical performance, these same skills are also extremely important to the socialization of "newcomer" engineers (Gundry, 1993). Engineers entering the "world of work" need these skills to obtain information about workplace norms: what they believe they are expected to do to fit into an organization and to work effectively with the people employed by these organizations. Newly hired engineers use communications and information-seeking skills to obtain information about role expectations and organizational values: how other members of the organization think and what they value, how new members are expected to behave, and what is expected of members to fit in and meet organizational expectations, values, and goals.

Whereas communications and information-seeking skills are important for all newly hired professionals, there is reason to believe that these skills are especially important for entry-level engineers: the job of an engineer requires the continuous production, transfer, and use of information to deal with a constant state of work-related uncertainty. A few studies have focused on engineering manpower and the role of information in career selection (American Association of Engineering Societies, 1986) and the use of and uses for information within the engineering profession (Allen, 1977; Rosenbloom and Wolek, 1970). Our review of the literature produced little information, however, about engineers as organizational newcomers and the role of communications and information-seeking in the socialization process of entry-level engineers.

Current discussions of communications and information-use and -seeking skills and competencies for engineers lack four important elements: (a) clear agreement from the professional engineering community about what constitutes "acceptable and desirable communications and information-seeking norms" within that community; (b) ample and generalizable data from engineering students about the communications and information-use and -seeking skills instruction and training they receive; (c) generalizable data from entry-level engineers about the adequacy and usefulness of the instruction and training they received as students and what skills they require as organizational newcomers; and (d) a higher-level theoretical framework, a comprehensive understanding of the nature of knowledge and learning, within which the interpretation of such data can take on consistent and fuller meaning.

This chapter focuses on information production and use by early career-stage U.S. aerospace engineers and scientists. This chapter provides the fourth missing element, a comprehensive theoretical framework, by adopting Bruffee's (1993) description of the socially constructed nature of human knowledge. Specific elements in Bruffee's model of particular interest to our work are the concept of "reacculturation" (Bruffee's word for the process through which we switch membership from one culture to another), and the important role of communication and information in the reacculturation process. As Bruffee defines it,

> Reacculturation involves giving up, modifying, or renegotiating the language, values, knowledge, and mores that are constructed, established, and maintained by the community one is coming from, and becoming fluent instead in the language and so on of another community. (p. 225)

For Bruffee,

> Members of a knowledge community *construct knowledge* in the language that constitutes that community by justifying beliefs that they mutually hold *[emphasis added]*. But they do not justify those beliefs by testing them against a 'foundation'—either a presumed mental structure or a presumed reality. They justify them socially in conversation with one another. (pp. 221-222)

Bruffee's model is particularly appropriate to our work both because the model assumes a relationship between communication and information-seeking and career success in much the same way that studies relate communication and information-seeking competency to the professional success of engineers and because Bruffee looks directly at the process by which people move from one knowledge community to another—the process which, applied to entry-level engineers and their transition from engineering students to engineering professionals.

This chapter also supplies the third missing element, adequate and generalizable data from entry-level engineers. We surveyed 264 members of the American Institute of Aeronautics and Astronautics (AIAA) who had converted their AIAA memberships from student to professional status and who had an average of 2.4 years of aerospace work experience. To learn more about the concomitant communications norms, we compared the results of this investigation with data we collected from 1673 student members of the AIAA and with data we collected from 341 professional (i.e., established) members of the AIAA who had an average of 21.9 years of aerospace work experience. Selected results from these studies are presented about career goals; the importance, receipt, and helpfulness of communications and information-use skills instruction; collaborative writing; information products produced and used; factors impeding the production of written technical communication; library use; the use of electronic (computer) networks; and the use of federally funded aerospace research and development (R&D) in completing or solving their most recent project, task or problem.

BACKGROUND

In keeping with Bruffee's concept of reacculturation as a process that enables recent graduates to join professional communities whose understanding may differ from that of the community they recently left, we reviewed literature that focused on the socialization of newcomers and the factors that motivate and impede information-seeking by early career-stage professionals. We did not review the literature pertaining to the communications practices and information-seeking behaviors of engineering students and professionals. Chapters 5, 7, 10, and 11 contain reviews of this literature.

Environment

In recent years, there has been a growing recognition that organizational newcomers actively seek information as a strategy for assimilating into new work environments (Ashford and Black, 1992; Miller and Jablin, 1991; Morrison, 1995, 1993a, 1993b; and Ostroff and Kozlowski, 1992). Within the fields of psychology and communications, information seeking is seen as a coping strategy essential for dealing with uncertain environments (Berger, 1979; Folkman and Lazarus, 1980; White, 1974). Because the period of adjustment into a new organization is typically characterized by uncertainty, information seeking is an activity that newcomers can use to facilitate their own socialization and adaptation (Miller and Jablin, 1991; Morrison, 1993a). Information seeking has been found to increase newcomers' sense of knowledge and job mastery and their satisfaction, performance, organizational commitment, and intentions to remain in the organization (Morrison, 1993b; Ostroff and Kozlowski, 1992). Unlike cases in which the information is provided to newcomers without their seeking it, active information seeking gives newcomers control over the type, amount and timing of the information they receive (Morrison, 1993a). It also enables them to compensate in areas where information is not forthcoming. This may be important, as newcomers often feel that they receive less information than they need (Jablin, 1984).

Hanser and Muchinsky (1978) suggested that work situations can be conceptualized as information environments, full of information and cues that individuals can use to improve their work performance and to achieve valued work goals. The information environment contains different types of work-related information, which differ in their availability and usefulness, as well as various information sources, which differ along such attributes as expertise and accessibility. Newcomers' perceptions of these various types and sources of information may well have an important impact on their information seeking, which in turn has an effect on socialization outcomes as adjustment and commitment. Whereas past research on newcomer information-seeking has investigated different ways in which newcomers seek information and outcomes associated with this activity (Morrison, 1995; 1993a; 1993b; Ostroff and Kozlowski, 1992), scant attention has been devoted to understanding the factors that motivate and impede entry-level engineers as newcomers seeking information.

Information Types

Ostroff and Kozlowski (1992) and Morrison (1993a; 1993b) assessed information seeking with respect to five types of work-related information: task, role, social, organizational, and feedback information. Task information is defined as information about how to perform specific job tasks and assignments; role information is defined as information about the expectations and responsibilities associated with the job; social information is defined as information about coworkers and about how to behave within the workgroup; organizational information is defined as information about the organization, including policies, procedures, structures, and objectives; and feedback is defined as information about how well one is performing.

Newcomers rely heavily on feedback from supervisors and coworkers to successfully assimilate into their organization. Research on feedback-seeking behavior (Ashford, 1986; Morrison and Cummings, 1992; Morrison and Weldon, 1990) has highlighted the importance of this information for enabling employees to assess how well they are performing and to modify their behavior as necessary. Feedback is particularly important for newcomers who are more uncertain about performance-related issues than their more experienced colleagues (Louis, 1980; Miller and Jablin, 1991). The literature on socialization has identified other types of information that newcomers need to obtain as well. Ostroff and Kozlowski (1992) proposed that there are four primary learning domains within socialization related to the newcomer's job, role, workgroup, and organization, respectively. These four domains correspond to the information types that Morrison (1993a; 1993b) assessed in her research on newcomer information seeking. Morrison proposed that in addition to feedback, newcomers need to obtain information on the following: task duties and procedures; expectations and responsibilities; group norms and interaction patterns; and organizational norms, values, processes, and structures. Her research indicated that newcomers seek each of these information types, with the amount of seeking related to such outcomes as job mastery and role clarity.

Information Sources and Modes

Within a newcomer's information environment are several potential sources of information. Supervisors, coworkers, and mentors have been shown to play important roles in helping newcomers to learn about and adapt to their new environment (Louis, Posner, and Powell, 1983; Morrison, 1993a; Ostroff and Kozlowski, 1992), and newcomers may seek each of the five types of information from these sources through what Ashford and Cummings (1983) referred to as inquiry (Morrison, 1993; Ostroff and Kozlowski, 1992). Inquiry entails directly asking another person for information. A second mode of information seeking is monitoring or observing which entails observing the environment for informational cues. An advantage of monitoring relative to inquiry is that it is far less obtrusive, and newcomers do not have to worry about others making references about their competency or interpersonal skills, or what Ashford and Cummings (1983) referred to as the "social costs" related to inquiry. A third way in which newcomers can actively obtain information is by consulting written reports, documents, and handbooks (Morrison, 1993b; Ostroff and Kozlowski, 1992). Although certain types of information do not exist in documented form, this strategy, like monitoring, enables newcomers to avoid the social costs that may be incurred with inquiry.

The decision to seek information is typically depicted as a conscious and rational one. Ashford and Cummings' (1983) model of feedback seeking conceptualizes information as a resource that people use to accomplish various ends. In deciding whether to seek that resource actively, people consider both anticipated costs and anticipated benefits. Vancouver and Morrison's (1995) research on feedback seeking indicates that people also consider costs and benefits when deciding whether to obtain information from a particular source. The next section discusses several factors that affect the perceived costs and benefits of information seeking and, hence, the degree to which newcomers engage in this behavior.

Costs and Benefits of Information Seeking

As with any resource, individuals ascribe value or importance to information based on the perceived utility of that information for achieving valued objectives. Although newcomers tend to see some types of information as more valuable than others for adjusting to a new organization (Morrison, 1995), there also will be variance in how valuable newcomers see work-related information in general. Some newcomers will see information as highly critical to their performance and assimilation, whereas others will see it as less so. These perceptions depend on such factors as the nature of the job and work environment, the employee's level of experience and mastery, and his or her tolerance for uncertainty. To the extent that newcomers see work-related information as important or useful, it is predicted that they will be more willing to exert more effort to obtain information, and will thus engage in more information seeking.

A fundamental motive behind information seeking is the desire to reduce uncertainty (Berger, 1979; Miller and Jablin, 1991). Uncertainty is described as a state of having insufficient or inconsistent information (Ashford, 1986). For most people, uncertainty is unpleasant and anxiety provoking (Epstein, 1972; Kahn, Quinn, Wolfe, Snoek, and Rosenthal, 1964; McGrath, 1976; Nelson, 1987). Several researchers, therefore, have proposed that one of the most common ways by which individuals try to reduce uncertainty is seeking information (Berger, 1979; White, 1974; Zemore and Shepel, 1987). Further, uncertainty has been found to motivate feedback-seeking behavior (Ashford, 1986). Building on the existing literature, we expect that newcomers will seek information to the extent that they view their work environment as having a high degree of uncertainty.

Although information seeking is beneficial for reducing newcomers' uncertainty and helping them to achieve valued work goals, it can also be costly (Ashford, 1986). Information seeking requires time and effort (Ashford and Cummings, 1983). Furthermore, information is difficult to obtain in some cases either because the information is informal and tacit or because the locus of the information is unclear (Morrison, 1995). There is evidence that when trying to obtain technical information, people do so in a way that requires the least amount of effort possible (Gerstberger and Allen, 1968; O'Reilly, 1982). Thus, newcomers predictably will be less willing to seek information that they perceive to be difficult to obtain.

Newcomers have several potential sources of information at their disposal, including supervisors, peers, mentors, and printed documents. Hence, newcomers must determine not only whether it is beneficial to seek information in general, but also whether it is useful to seek information from a particular source (Morrison and Bies, 1991; Vancouver and Morrison, 1995). Several studies have demonstrated that employees seek different types and amounts of information from supervisors, peers, mentors, and documents. Because these sources differ from one another along several dimensions, however, it is not clear which source attributes actually determine whether a newcomer will select a given source.

Source expertise is defined as the extent to which a given source possesses accurate and useful knowledge about a particular domain. It is predicted that a newcomer's perception of a source's expertise will have a motivating impact on information seeking. Newcomers seek information in order to obtain knowledge that will be of value to them (Ashford, 1986). Expertise will play an important role in this process because it affects the quality of information that the source is able to provide. It is expected that newcomers will seek more information from a given source to the extent that they see that source as possessing relevant expertise. There is some indirect evidence to support this prediction. Vancouver and Morrison (1995) found that people have a strong preference for sources with high expertise when they are trying to obtain performance feedback, and Pinelli, Bishop, Barclay, and Kennedy (1993) reported that engineers select sources of technical information based, in part, on considerations of quality, relevance, and comprehensiveness.

A second source attribute that will affect information seeking is perceived accessibility. Accessibility refers to the ease with which one is able to locate and utilize a particular information source. Research on how engineers and scientists obtain scientific and technical information indicates that accessibility is the most important determinant of information use (Allen, 1977; Gerstberger and Allen, 1968; Young and Harriot, 1979). Source accessibility has also been found to have an effect, albeit, a small one, on feedback seeking (Vancouver and Morrison, 1995). It can be expected that accessibility is an important determinant of newcomer information seeking, and that it plays a role across all of the types of information that newcomers try to obtain. Information seeking is largely a process of minimizing costs (Pinelli, Bishop, Barclay, and Kennedy, 1993). To the extent that a given source is easily accessible, the costs of the information seeking are less. Hence, information seeking should be greater when accessibility is perceived to be high.

It has been hypothesized that newcomers with less tenure (i.e., fewer years of employment) with an organization would engage in more information seeking than newcomers with longer tenure. There are two primary reasons for this prediction. *First*, the socialization literature suggests that uncertainty is at its highest during the period immediately following entry (Louis, 1980; Miller and Jablin, 1991). As they gain tenure and experience, newcomers acquire increasing amounts of information and uncertainty decreases. The result is that newcomers will come to regard information seeking as an increasingly less valuable coping strategy. *Second*, as newcomers gain tenure, their colleagues and supervisors may expect them to be knowledgeable and competent, and may thus be increasingly less tolerant of information seeking. Hence, the social costs associated with information seeking may increase with tenure, at least in cases where information must be sought from other persons (Ashford, 1986).

An important assumption underlying research on newcomer information seeking is that this activity helps newcomers to obtain information that will facilitate their assimilation. If successful, information seeking should therefore enhance newcomers' knowledge and sense of competence. Ostroff and Kozlowski (1992) did not make a clear distinction between information seeking and passive modes of information acquisition, yet they found that the more information that newcomers acquired, (whether actively *or* passively), the greater their knowledge of and sense of adjustment to their work situation. It has been hypothesized that the overall amount of information seeking in which newcomers engage will increase their level of knowledge in the four socialization domains (task, role, workgroup, and organization), and their sense of adjustment to their job. It has also been hypothesized that by reducing newcomers' uncertainty and facilitating their assimilation, information seeking will enhance both their job satisfaction and organizational commitment.

Entry and socialization to an organization is a time of learning through confusion and uncertainty and of coming to understand expectations and surviving in an unfamiliar setting (Gundry, 1993). Organizational newcomers are more apt to seek information, and thereby to facilitate their assimilation process, to the extent that they perceive their information environment as uncertain, information as important, and available sources as accessible and knowledgeable. Organizations desiring that newcomers emerge from the socialization process knowledgeable, well assimilated, and committed to their organization may wish to stress the importance of information, encourage information seeking by newcomers, and ensure that supervisors, colleagues, and other sources of accurate and reliable information are accessible to organizational newcomers.

METHODS

Self-administered (self-reported) surveys were mailed to 700 U.S. aerospace engineers and scientists who had changed their AIAA membership from student to professional status in the past five years. The study was undertaken as a Phase 1 activity of the *NASA/DoD Aerospace Knowledge Diffusion Research Project* (Pinelli, Kennedy, and Barclay, 1991). The survey was conducted during April-July 1995. By July 27, 1995, 264 usable questionnaires had been received; we heard from 163 AIAA members who did or could not participate. The adjusted completion rate for the survey was 49% (Pinelli, Barclay, and Kennedy, 1995a).

As a Phase 3 activity of the *NASA/DoD Aerospace Knowledge Diffusion Research Project*, we sent self-administered (self-reported) questionnaires to 4300 aerospace engineering and science students who were (student) members of the AIAA. The survey was conducted during March-September 1993. By September 15, 1993, we received 1673 completed surveys. Because of the summer break, only one mailing was possible. After reducing the sample size for incomplete addresses and other mailing problems, the response rate for the survey was 42% (Pinelli, Hecht, Barclay, and Kennedy, 1994). Self-administered (self-reported) questionnaires were sent to a sample of 700 industry-affiliated U.S. aerospace engineers and scientists who had been members of the AIAA for more than five years. The survey was conducted between April-June, 1995 as Phase 1 activity of the *NASA/DoD Aerospace Knowledge Diffusion Research Project*. We received 341 usable questionnaires by the established cutoff date of June 21, 1995. Eighty-nine surveys were returned as unusable for various reasons. The adjusted response rate was 53% (Pinelli, Barclay, and Kennedy, 1995b).

Student t-tests with a one-tailed analysis were used to estimate if observed differences between two groups were statistically different. A significant test result ($p \leq .05$) indicates that there is only a 5% probability that the observed differences between the distribution of responses for the two groups can be attributed to chance. A significant result is therefore interpreted that participant's responses on the factors or variables in question are influenced by (vary systematically with) a participant's status in one of two group comparisons: engineering students and new engineers or new engineers and established engineers. In the discussion that follows, the terms *statistically significant* and *significant* are used interchangeably.

SAMPLE DEMOGRAPHICS

Table 9.1 shows demographic information for the new and the established U.S. aerospace engineers and scientists. Of the 264 new U.S. aerospace engineers and scientists, 52% (136 respondents) worked in industry, 32% (83 respondents) worked in government, 9% (24 respondents) worked in academia, and 8% (20 respondents) had some other affiliation. The following "composite" participant profile was developed for the new U.S. aerospace engineers and scientists: works in industry (51.7%), has a bachelor's degree (51.9%), has an average of 2.4 years of work experience in aerospace, was educated as and works as an engineer (98.1%, 85.2%), works in design and development (49.4%), and is male (89%).

Of the 341 established U.S. aerospace engineers and scientists, 100% (341 respondents) worked in industry. The following "composite" participant profile was developed for the established U.S. aerospace engineers and scientists: works in industry (100.0%), has a master's degree (52.2%), has an average of 21.9 years of work experience in aerospace, was educated as and works as an engineer (90.9%, 83.6%), works in design and development (53.4%), and is male (89.5%).

Of the 1673 aerospace engineering and science majors who were student members of the AIAA, 948 were undergraduate students and 707 were graduate students (18 respondents did not report their status). The majority of respondents were male. About 82%

Table 9.1. Survey Demographics (*N*=264; 341)

Demographics	New Engineers		Established Engineers	
	n	%	*n*	%
Gender				
Female	29	11.0	5	1.5
Male	235	89.0	336	98.5
Degree				
Bachelor's or less	131	0.4	102	29.9
Master's	89	51.2	178	52.2
Doctorate	34	38.2	61	17.9
Educational preparation as				
Engineer	249	98.0	310	90.9
Scientist	2	0.8	18	5.3
Other	3	1.2	13	3.8
Primary duties				
Engineer	217	85.4	285	83.6
Scientist	11	4.3	10	2.9
Other	2	10.2	46	13.5
Work best classified as				
Teaching/academic	6	2.4	0	0.0
Research	51	20.2	44	12.9
Management/supervision	22	8.7	72	21.1
Design/development	123	48.6	182	53.4
Manufacturing/production	10	4.0	3	0.9
Service/maintenance	1	0.4	3	0.9
Sales/marketing	3	1.2	12	3.5
Quality assurance/control	11	4.3	6	1.8
Other	26	10.3	13	3.8
Work performed in				
Academia	24	9.1	0	0.0
Government	83	31.6	0	0.0
Industry	136	51.7	341	100.0
Other	20	7.6	0	0.0
Is English your first (native) language				
Yes	227	89.4	---	---
No	27	10.6	---	---
Current work federally funded				
Yes	163	67.6	248	74.7
No	78	32.4	84	25.3

of the undergraduates and 87% of the graduate students were male. Most respondents reported that they were studying to become engineers. Among undergraduates, about 95% were preparing to become engineers; about 2% reported that they were preparing to become scientists. About 90% of the graduate students were preparing for careers as engineers; a slightly higher percentage of graduate students, about 7%, reported that they were preparing to become scientists. Most AIAA student members were U.S. citizens; about 92% of the undergraduates and 81% of the graduate students indicated that they were U.S. citizens. English is the first (native) language for most of the student participants. About 87% of the undergraduate students indicated that English is their first (native) language and about 77% of the graduate students indicated that English is their first (native) language. The U.S. was the native country of most AIAA student respondents. About 84% of the undergraduate students and about 73% of the graduate students indicated that the U.S. was their native country.

PRESENTATION OF THE DATA

To learn more about the concomitant communication norms among the three groups, we compared selected results from the student and professional (i.e., established) engineering studies with the results obtained from the survey of early career-stage (i.e., new) U.S. aerospace engineers and scientists. Data are presented for the following topics: career goals; the importance, receipt, and helpfulness of communications and information-use skills instruction; collaborative writing; information products produced and used; factors impeding the production of written technical communication; library use; the use of electronic (computer) networks; and the use of federally funded aerospace R&D in completing or solving their most recent project, task or problem.

Career Goals

Students and new engineers rated the importance of 15 goals to a successful career. The list includes aspirations that are classified as engineering, science, or management goals. Table 9.2 shows the mean importance ratings for each goal. Both students and new engineers gave high ratings to the engineering-related goals and aspirations. The ordering of mean importance ratings for these factors, from highest to lowest, is similar for both engineering students

Table 9.2. Career Goals of U.S. Aerospace Engineering Students and New U.S. Aerospace Engineers

Goals	Engineering Students		New Engineers	
	n	$\bar{x}^a$	n	$\bar{x}^a$
Engineering				
Have the opportunity to explore new ideas about technology or systems	1628	6.3	260	6.1**
Advance to high level staff technical position	1607	5.4	259	5.2*
Have the opportunity to work on complex technical problems	1634	5.8	263	5.9*
Work on projects that utilize the latest theoretical results in your specialty	1628	5.6	261	5.0**
Work on projects that require learning new technical knowledge	1635	5.9	263	5.9
Science				
Establish a reputation outside your organization as an authority in your field	1620	5.3	259	5.3
Receive patents for your ideas	1632	4.3	257	3.7**
Publish articles in technical journals	1619	4.8	257	4.2**
Communicate your ideas to others in your profession by presenting papers at professional meetings	1630	5.0	256	4.3**
Be evaluated on the basis of your technical contributions	1615	5.4	263	5.5
Leadership (management)				
Become a manager or director	1603	5.0	258	4.7*
Plan and coordinate the work of others	1606	5.0	261	5.0
Advance to a policy-making position in management	1646	4.7	255	4.4*
Plan projects and make decisions affecting the organization	1614	5.3	259	5.2
Be the technical leader of a group of less experienced professionals	1612	5.2	260	5.3

[a]Participants used a 7-point scale to rate the importance of each goal, where 7 indicates the highest rating.

*$p \leq 0.05$. **$p \leq 0.01$.

and new engineers. The opportunity to explore new ideas about technology or systems ranked highest for both groups ($\bar{x}$ = 6.3 for students and $\bar{x}$ = 6.1 for new engineers), followed by working on projects that require learning new technical knowledge ($\bar{x}$ = 5.9 for both students and new engineers) and having the opportunity to work on complex technical problems ($\bar{x}$ = 5.8 for students and $\bar{x}$ = 5.9 for new engineers). Science goals—receiving patents for your ideas, publishing articles in technical journals, and communicating your ideas to others in your profession by presenting papers at professional meetings—were significantly more important to the engineering students than to the new engineers. Management goals—becoming a manager or director and advancing to a policy making position in management—were significantly more important to the engineering students than to the new engineers.

Importance of and Time Spent Communicating

One aspect of the surveys was designed to determine the significance of technical communication in the careers of new and established aerospace engineers and scientists. Table 9.3 shows each sample's mean response to survey questions about the importance of effective technical communication in the performance of the respondents' duties. Importance was measured on a 5-point scale, with 1 being not at all important and 5 being very important. The established engineers assigned a significantly higher rating to the importance of communicating technical information effectively than the new engineers did.

More so than their rating of its importance, the amount of time that engineers and scientists spend communicating technical information might be an effective indicator of its significance. Overall, the established engineers spent more hours each week communicating technical information to others and working with technical information received from others than did the new engineers. The established engineers spent significantly more hours each week communicating technical information in writing to others than did the new engineers. As shown in Table 9.3, about 72% of the new engineers and about 58% of the established engineers indicated that the amount of time they spent communicating technical information had increased in the last five years. About 66% of the new and about 61% of the established engineers reported that as they

Table 9.3. Technical Communication Practices of New and Established U.S. Aerospace Engineers and Scientists

Communication Factors	New Engineers		Established Engineers	
	n	$\bar{x}$	n	$\bar{x}$
Importance of communicating technical information effectively	264	4.6	337	4.7*
Hours spent each week communicating technical information in writing	257	9.7	328	11.5**
Hours spent each week communicating technical information orally	252	10.5	320	11.0
Hours spent each week working with written technical information received from others	253	9.9	332	9.8
Hours spent each week working with technical information received orally from others	249	8.0	322	8.3
Changes Over Time	n	%	n	%
Compared to 5 years ago, time spent communicating technical information has				
Increased	181	71.5	196	58.3
Stayed the same	57	22.8	97	28.9
Decreased	15	5.9	43	12.8
As you have advanced professionally, the amount of time spent working with technical information received from others has				
Increased	167	65.5	202	60.5
Stayed the same	74	29.0	86	25.7
Decreased	14	5.5	46	13.8

*$p \leq 0.05$. **$p \leq 0.01$.

have advanced professionally, the amount of time they spent on communicating technical information had actually increased.

Collaborative Writing

A series of questions was formulated to determine the role of collaborative writing in the professional lives of the new and established engineers responding to the survey. Table 9.4 shows that an average of about 72% of all respondents reported that they did at least some of their writing alone, and Tables 9.5 and 9.6 indicate that about 31% of the new engineers and about 29% of the established wrote only by themselves.

There were high percentages of each sample who also wrote with one other person (57.9% and 54.7%) and in groups of two to five people (44.4% and 48.3%). Thirteen percent of the new engineers and about 21% of established engineers reported doing some writing with a group of more than five people. The percentages of new engineers and established engineers who participated in various collaborative writing practices some of the time were similar.

About 44% of the new engineers and about 36% of the established engineers indicated that they routinely prepared written technical communications with the same group. Established engineers worked with different groups to prepare written technical communications more often than did the new engineers. Statistical differ-

Table 9.4. Collaborative Writing Practices of New and Established U.S. Aerospace Engineers and Scientists (N=261; 329)

Collaborative Practices	New Engineers $\bar{x}$%	%[a]	n	Established Engineers $\bar{x}$%	%[a]	n
I write alone	71.2	95.8	251	72.8	98.3	323
I write with one other person	14.2	57.9	151	12.5	54.7	179
I write with a group of two to five people	11.3	44.4	116	10.3	48.3	157
I write with a group of more than five people	3.3	13.0	34	4.2	21.4	70

[a]Percentages do not total 100.

Table 9.5. Influence of Group Participation on Writing Productivity of New and Established U.S. Aerospace Engineers and Scientists

Productivity	New Engineers		Established Engineers	
	n	%	*n*	%
A group is more productive than writing alone	99	38.5	96	29.4
A group is about as productive as writing alone	49	19.1	64	19.6
A group is less productive than writing alone	30	11.7	71	21.7
I only write alone	79	30.7	96	29.4

Table 9.6. Production of Written Technical Communications as a Function of Number of Groups and Group Size for New and Established U.S. Aerospace Engineers and Scientists

Number of Groups	New Engineers		Established Engineers	
	n	%	*n*	%
Worked with same group				
Yes	112	44.4	119	36.2
No	61	24.3	113	34.5
I only work alone	79	31.3	96	29.3
Size of Group	*n*	$\bar{x}$	*n*	$\bar{x}$
Number of people in group	112		117	
Mean		3.6		4.7**
Median		3.0		4.0
Number of groups	60		110	
Mean		3.0		4.5**
Median		3.0		3.0
Number of people in each group	59		110	
Mean		3.4		5.3**
Median		3.0		4.0

*$p \leq 0.05$. **$p \leq 0.01$.

ences were found in the number and size of work groups for new and established engineers (Table 9.6).

The differences were more noticeable in the responses of various samples to questions about the productivity of these collaborative writing practices (Table 9.5). A small majority of the new engineers and the established engineers indicated that writing in groups was more productive than writing alone. Nearly as many new engineers (19.1%) as established engineers (19.6%) thought group writing was merely on a par with solo writing. About 12% of the new engineers and about 22% of the established engineers thought group writing was less productive than individual efforts.

Use of Technical Information Products

The new and established engineers were polled on the numbers of each of 13 specific types of technical information products they had used in the six months prior to the survey. The list contained 10 informal and 3 formal (i.e., conference-meeting papers, journal articles, and technical reports) information products. Table 9.7 gives the mean (median) number of each of the 13 products used during the 6-month period by the two groups of respondents.

Most striking of the reported data is the high mean number of drawings and specifications ($\bar{x}$ = 45.7) used by the new engineers and the mean number of memoranda ($\bar{x}$ = 30.3) used by the established engineers in the past six months. New engineers also reported high use rates for technical manuals, memoranda, and computer programs and documentation. Established engineers reported high use rates for drawings and specifications, letters, and audiovisual materials. The reported uses of conference-meeting papers and journal articles in the past six months were comparable.

Production of Technical Information Products

The respondents were also polled about the number of informal and formal technical information products that they had produced in the six months prior to the survey. Table 9.8 shows the mean number of each type of technical information product produced by the new and established engineers who participated in the study. Some prominent mean numbers include the number of drawings and specifications, memoranda, and letters ($\bar{x}$ = 9.3, 8.9, 8.7) that

Table 9.7. Mean (Median) Number of Technical Information Products Used in the Past Six Months by New and Established U.S. Aerospace Engineers and Scientists

Product	New Engineers		Established Engineers	
	$\bar{x}$	Median	$\bar{x}$	Median
Abstracts	4.0	0.0	3.5	0.0
Journal articles	6.0	0.0	5.7	1.0
Conference/meeting papers	4.0	0.0	4.7	2.0
Trade/promotional literature	4.8	0.0	7.0	0.0
Drawings/specifications	45.7	6.0	18.4	3.0
Audiovisual material	5.0	0.0	11.2	1.0
Letters	10.5	2.0	16.5	4.0
Memoranda	14.6	5.0	30.3	10.0
Technical proposals	2.4	3.0	3.3	0.0
Technical manuals	19.4	3.0	5.6	0.0
Computer programs/documentation	12.8	2.0	6.4	0.0
In-house technical reports	---	---	6.6	3.0
Technical talks/presentations	6.2	1.0	7.3	2.0

new engineers produced alone. The numbers of these same products produced by new engineers in a group, as opposed to alone, were substantially lower. New engineers produced few technical information products in a group. Memoranda and letters ($\bar{x} = 15.2$ and 10.2) were the technical information products produced most often by established engineers. The numbers of these same products produced by established engineers in a group, as opposed to alone, were substantially lower. Considering the new engineers and the technical information products they produced in a group, trade and promotional literature, computer programs and documentation, and technical talks and presentations were prepared by the largest (average) number of people per group. Considering the established engineers and the technical information products they produced in a group, technical proposals, drawings and specifications, and computer programs and documentation were prepared by the largest (average) number of people per group.

Importance of Communications and Information–Use Skills

To learn about the importance of and the receipt of communications and information-use skills, we asked engineering students and new engineers three questions about those skills and their im-

Table 9.8. Mean (Median) Number of Technical Information Products Produced in the Past Six Months by New and Established U.S. Aerospace Engineers and Scientists

Product	Alone		In a Group		Average Number in Group	
	$\bar{x}$	Median	$\bar{x}$	Median	$\bar{x}$	Median
New Engineers						
Abstracts	.64	0.0	.28	0.0	2.77	3.0
Journal articles	.19	0.0	.17	0.0	2.65	2.0
Conference/meeting papers	1.21	0.0	.44	0.0	2.73	0.0
Trade/promotional literature	.21	0.0	.04	0.0	4.29	4.0
Drawings/specifications	9.32	0.0	2.67	0.0	3.01	3.0
Audiovisual material	2.24	0.0	.53	0.0	3.05	3.0
Letters	8.71	2.0	.49	0.0	2.42	2.0
Memoranda	8.93	3.0	.63	0.0	2.67	2.0
Technical proposals	1.01	0.0	1.05	0.0	3.70	3.0
Technical manuals	1.06	0.0	.37	0.0	4.15	3.0
Computer programs/ documentation	2.93	0.0	.32	0.0	3.36	3.0
In-house technical reports	---	---	---	---	---	---
Technical talks/ presentations	2.52	1.0	1.06	0.0	4.00	3.0
Established Engineers						
Abstracts	.34	0.0	.22	0.0	2.80	2.0
Journal articles	.07	0.0	.08	0.0	2.50	2.0
Conference/meeting papers	.43	0.0	.28	0.0	3.50	3.0
Trade/promotional literature	.20	0.0	.16	0.0	3.50	3.0
Drawings/specifications	2.52	0.0	1.90	0.0	5.30	3.0
Audiovisual material	4.88	0.0	2.13	0.0	4.70	3.0
Letters	10.15	3.0	1.54	0.0	2.90	3.0
Memoranda	15.16	6.0	1.99	0.0	3.30	0.0
Technical proposals	.75	0.0	1.03	0.0	10.40	5.0
Technical manuals	.22	0.0	.18	0.0	5.30	4.0
Computer programs/ documentation	.58	0.0	.15	0.0	2.90	0.0
In-house technical reports	1.44	0.0	1.04	0.0	4.00	3.0
Technical talks/ presentations	3.87	1.0	2.47	0.0	4.70	3.0

portance to professional success. Importance was measured using a 7-point scale with 7 indicating the highest rating. Students and new engineers were asked if they had received instruction in these skills; they were also asked to rate the helpfulness (usefulness) of the instruction. Table 9.9 shows the mean importance rating assigned to each of the six communications and information-use skill competencies by the engineering students and new engineers.

Both groups of respondents assigned high importance ratings to the six communications and information-use skills. Effectively using computer, communication and information technology was rated highest by both groups of respondents, followed by effectively communicating technical information in writing and orally. The ability to search an electronic (bibliographic) database was assigned the lowest mean importance rating by the engineering students and new engineers, alike. Overall, the engineering students attributed higher mean importance ratings to the six communications and information-use skills than did the new engineers. There are statisti-

Table 9.9. Importance of Selected Communications/Information-Use Skills to U.S. Aerospace Engineering Students and New U.S. Aerospace Engineers and Scientists

Importance of	Engineering Students		New Engineers	
	n	$\bar{x}^a$	*n*	$\bar{x}^a$
Effectively communicating technical information in writing	1628	6.3	262	6.2
Effectively communicating technical information orally	1628	6.3	263	6.2
Having a knowledge and understanding of engineering/science information resources and materials	1622	6.2	263	5.7**
Ability to search electronic (bibliographic) databases	1649	5.5	259	4.3**
Ability to use a library that contains engineering/science information resources and materials	1623	5.8	262	4.6**
Effectively using computer, communication and information technology	1629	6.5	262	6.3**

[a]Participants used a 7-point scale to rate the importance of each skill, where 7 indicates the highest rating.
*$p \leq 0.05$. **$p \leq 0.01$.

cal differences in the mean importance ratings assigned by the engineering students and the new engineers to four specific skills.

Receipt of Communications and Information-Use Skills Instruction

Higher percentages of the new engineers reported having received instruction and training in the six communications and information-use skills than did the engineering students (Table 9.10). No less than 50% of the engineering students and about 61% of the new engineers had received instruction or training in all of the six communications and information-use skills. About 84% of the new engineers reported having received instruction or training in technical writing or communication and in using computer, communication, and information technology; about 61% reported that they had received instruction or training in searching electronic (bibliographic) databases. Conversely, about 72% and 83% respectively of the engineering students reported having received instruction or training in technical writing or communication and in using computer, communication, and information technology; about 50% reported that they had received instruction or training in searching electronic (bibliographic) databases.

Table 9.10. Communications/Information-Use Skills Instruction Received by U.S. Aerospace Engineering Students and New U.S. Aerospace Engineers and Scientists (N=1673; 264)

Instruction	Engineering Students		New Engineers	
	n	%	*n*	%
Technical writing/communication	1250	72.2	224	84.4
Speech/oral communication	1076	62.2	199	75.4
Using a library that contains engineering/science information resources and materials	1037	59.9	198	75.3
Using engineering/science information resources and materials	1100	63.6	212	80.6
Searching electronic (bibliographic) databases	869	50.2	160	60.8
Using computer, communication, and information technology	1433	82.9	222	84.4

Helpfulness of Communications and Information-Use Skills Instruction

In terms of the helpfulness of the instruction or training received in the six communications and information-use skills, Table 9.11 shows some interesting differences in the mean helpfulness scores reported by the two groups. (Helpfulness of the instruction or training was measured using a 7-point scale where 7 is most helpful.) The mean helpfulness scores for the instruction or training received in technical writing or communication, speech and oral communication, and in using engineering or science information resources and materials were approximately equal. The two groups reported significantly different scores for the helpfulness of the instruction or training they had received in (a) using a library containing engineering or science information resources and materials; (b) searching electronic (bibliographic) databases; and (c) using computer, communication, and information technology.

Table 9.11. Helpfulness of Communications/Information-Use Skills Instruction Received by U.S. Aerospace Engineering Students and New U.S. Aerospace Engineers and Scientists

Instruction	Engineering Students		New Engineers	
	n	$\overline{x}^a$	n	$\overline{x}^a$
Technical writing/communication	1248	5.5	224	5.5
Speech/oral communication	1092	5.5	201	5.6
Using a library that contains engineering/science information resources and materials	1042	5.1	201	4.8**
Using engineering/science information resources and materials	1104	5.3	212	5.1
Searching electronic (bibliographic) databases	900	5.0	167	4.6**
Using computer, communication, and information technology	1416	5.9	225	6.1**

[a]Participants used a 7-point scale to rate the helpfulness of each competency, where 7 indicates the highest rating.
*$p \leq 0.05$. **$p \leq 0.01$.

Factors Impeding Production of Written Technical Communication

Engineering students and new engineers were polled on eight factors that might impede their ability to produce effective written technical communication (Table 9.12). (Respondents used a 7-point scale to measure the extent to which lack of knowledge of each principle impedes their ability to produce written technical communication, where 7 indicates greatly impedes.) These factors correspond to the eight principles of written communication in Table 9.14 that the new and established engineers were asked to consider for inclusion in an undergraduate course in technical communication for aerospace engineering and science students.

Table 9.12 shows the percentages of engineering students and new engineers who indicated that a lack of mastery of these princi-

Table 9.12. Factors Impeding the Ability of U.S. Aerospace Engineering Students and New U.S. Aerospace Engineers and Scientists to Produce Written Technical Communication

Principles	Engineering Students		New Engineers	
	n	$\bar{x}^a$	n	$\bar{x}^a$
Defining the purpose of the communication	1554	3.7	239	3.6
Assessing the needs of the reader	1585	3.9	238	3.9
Preparing/presenting information in an organized manner	1595	3.6	241	4.0**
Developing paragraphs (introductions, transitions, conclusions)	1600	3.4	243	3.4
Writing grammatically correct sentences	1604	3.1	243	3.3
Notetaking and quoting	1561	3.1	230	2.8**
Editing and revising	1554	3.3	237	3.5

[a]Participants used a 7-point scale to measure the extent to which each principle impedes their ability to produce written technical communications, where 7 indicates greatly impedes.
*$p \leq 0.05$. **$p \leq 0.01$.

ples impeded their ability to produce written technical communication. Considering that a 7-point rating scale was used, the mean impedance scores are moderate, indicating that neither the engineering students nor the new engineers hold the opinion that lack of mastery of the eight principles impedes their ability to produce written technical communication. For the engineering students and the new engineers, the highest mean scores were reported for assessing the needs of the reader, followed by defining the purpose of the communication and preparing or presenting information in an organized manner. Interestingly and statistically, the inability to prepare or present information in an organized manner appears to be a greater impediment for the new engineers than for the engineering students. Finally, the lack of mastery in notetaking and quoting appears to be a greater impediment to producing written technical communications for the engineering students than for the new engineers.

Professional Recommendations for an Undergraduate Technical Communication Course

The new and established engineers were asked to recommend principles, mechanics, and on-the-job products and skills to be taught in an undergraduate technical communication course for aerospace science and engineering students. Their recommendations are listed in Table 9.13 according to the percentage of respondents in each sample who endorsed a particular principle, mechanic, or skill. There was general agreement among the new and established engineers concerning three principles that should be included in an undergraduate technical communication course. Both groups of respondents recommended the inclusion of preparing and presenting information in an organized manner (97.7% and 96.5%), defining the purpose of the communication (92.6% and 90.7%), and assessing the needs of the reader (92.2% and 81.7%). Noticeable differences were reported for new and established engineers concerning the inclusion of developing paragraphs (65.2% and 86.2%), writing grammatically correct sentences (49.0% and 80.0%), and editing and revising (26.3% and 77.8%).

The new engineers recommended the inclusion of the following three mechanics—references, symbols, and punctuation—and the established engineers recommended the inclusion of the following

Table 9.13. **Principles, Mechanics, and On-the-Job Skills Recommended for Inclusion in an Undergraduate Technical Communications Course by New and Established U.S. Aerospace Engineers and Scientists**

Principles	New Engineers		Established Engineers	
	n	%	*n*	%
Defining the purpose of the communication	237	92.6	547	90.7
Assessing the needs of the reader	236	92.2	490	81.7
Preparing/presenting information in an organized manner	251	97.7	582	96.5
Developing paragraphs (introductions, transitions, conclusions)	167	65.2	520	86.2
Writing grammatically correct sentences	125	49.0	483	80.0
Notetaking and quoting	138	54.8	299	50.0
Editing and revising	221	26.3	469	77.8
Mechanics	*n*	%	*n*	%
Abbreviations	149	59.1	304	51.4
Acronyms	164	65.1	295	49.7
Capitalization	122	48.8	361	61.0
Numbers	148	59.4	286	48.7
Punctuation	166	65.6	450	75.9
References	234	92.9	455	76.7
Spelling	119	47.4	386	65.1
Symbols	194	79.5	339	57.3
On-the-Job Products/Skills	*n*	%	*n*	%
Abstracts	202	80.2	406	69.0
Letters	168	65.9	412	69.4
Memoranda	186	72.7	463	77.8
Technical instructions	213	84.2	340	57.6
Journal articles	121	47.8	275	46.4
Conference/meeting papers	143	56.7	243	43.8
Literature reviews	105	42.0	220	37.3
Technical manuals	194	77.0	287	48.3
Newsletter/newspaper articles	60	24.4	143	24.3
Oral (technical) presentations	249	96.9	567	95.3
Technical specifications	205	81.3	330	55.7
Technical reports	242	94.5	398	58.1
Use of information sources	216	86.7	468	79.1

three mechanics—references, punctuation, and spelling. The on-the-job products and skills recommended for inclusion by the two groups were similar. Oral (technical) presentations received the highest recommendation for inclusion from both groups (96.9% and 95.3%). The new engineers also recommended including technical reports (94.5%), use of information sources (86.7%), and technical specifications (81.3%). The established engineers also recommended including use of information sources (79.1%), memoranda (77.8%), and letters (69.4%).

Library Use

New and established engineers were asked to indicate the number of times that they had used a library during a 6-month period (Table 9.14). Despite the difference in their years of aerospace work experience, there was little difference in either their overall library use or in the average number of times either group of respondents had used a library in the past six months.

We also asked the new and established engineers who had not used a library during the past six months to indicate their reasons for nonuse (Table 9.15). About 87% of the new engineers and about 89% of the established engineers reported that "my information needs were more easily met some other way" as their reason for not

Table 9.14. Use of a Library in the Past Six Months by New and Established U.S. Aerospace Engineers and Scientists

Visits	New Engineers		Established Engineers	
	n	%	*n*	%
0	73	28.4	76	25.1
1–5	105	40.9	125	41.3
6–10	38	14.8	50	16.5
11–25	26	10.1	33	10.9
26–50	10	3.9	13	4.3
51 or more	3	1.9	6	2.0
Mean		7.3		7.6
Median		3.0		3.0

*$p \leq 0.05$.

Table 9.15. Reasons Why New and Established U.S. Aerospace Engineers and Scientists Did Not Use A Library During the Past Six Months

Reasons	New Engineers		Established Engineers	
	n	%	*n*	%
I had no information needs	34	50.7	28	47.5
My information needs were more easily met some other way	59	86.8	59	89.4
Tried the library once or twice before but I couldn't find the information I needed	9	14.8	10	20.0
The library is physically too far away	17	28.3	--	--
The library staff is not cooperative or helpful	2	3.5	0	0.0
The library staff does not understand my information needs	2	3.5	4	8.2
The library did not have the information I need	14	25.5	15	30.0
I have my own personal library and do not need another library	19	32.8	10	19.2
The library is too slow in getting the information I need	6	10.5	14	27.5
We have to pay to use the library	2	3.4	3	5.8
We are discouraged from using the library	2	3.4	2	3.8

using a library. About 51% of the new engineers and about 48% of the established engineers reported that they "had no information needs" as the second most frequently reported reason for not using a library. Other reasons for nonuse reported by the *new engineers* included having a personal library (32.8%), library is physically too far away (28.3%), and the library did not have needed information (25.5%). Other reasons for nonuse of a library reported by the *established engineers* included the library did not have needed information (30%), library too slow in getting needed information (27.5%), and tried the library once or twice but could not find needed information (20%).

Use of Selected Information Technologies

New and established engineers were asked about their use and nonuse of a wide range of information technologies (Table 9.16). Specifically, they were asked to indicate if they "already use it," "don't use it but may in the future," and "don't use it and doubt if I will." Both new and established engineers made considerable use of fax or telex (92.7% and 97.0%) and electronic mail (82.8% and 83.3%). Information technologies used least by new and established engineers included audiotapes and cassettes (6.2% and 19.4%), motion picture film (3.5% and 15.1%), and micrographics and microforms (33.6% and 21.7%). Established engineers made greater use of video conferencing (54.3%) than did the new engineers (20.2%). New engineers made greater use of desktop/electronic publishing (87.8%) than did the established engineers (51.1%).

Considering information technologies not used at present, about 68% of the new engineers indicated that they may use video conferencing in the future. Almost 47% and 45% of the new engineers indicated that they might use electronic bulletin boards and videotapes in the future. About 52% and 39% of the established engineers indicated that they might use electronic bulletin boards and desktop/electronic publishing in the future. In the "don't use it, and doubt if I will" category, about 76% and 70% of the new engineers identified audiotapes and cassettes and motion picture films. Established engineers also identified motion picture films (65.0%) and audiotapes and cassettes (58.8%).

Use of Electronic (Computer) Networks

About equal percentages of both new (89.2%) and established engineers (87%) reported using electronic (computer) networks in performing their present professional duties. Small percentages of both groups reported "lack of access" as a reason for not using electronic (computer) networks.

New and established engineers use electronic (computer) networks for a variety of purposes (Table 9.17). About equal percentages of the new engineers used electronic (computer) networks for electronic mail (94.8%) and electronic file transfer (94.4%), and to connect to geographically distant sites (78.6%). New engineers used electronic (computer) networks least to control equipment (19.1%), to order documents from the library (24.8%), and to prepare scien-

Table 9.16. Use, Nonuse, and Potential Use of Information Technologies by New and Established U.S. Aerospace Engineers and Scientists

Information Technologies	Already Use It		Don't Use It, But May In Future		Don't Use It, And Doubt If Will	
	n	%	n	%	n	%
New Engineers						
Audiotapes and cassettes	16	6.2	45	17.4	197	76.4
Motion picture films	9	3.5	69	27.0	178	69.5
Videotapes	93	36.2	115	44.7	49	19.1
Desktop/electronic publishing	230	87.8	26	9.9	6	2.3
Computer cassettes/cartridge tapes	90	35.2	77	30.1	89	34.8
Electronic mail	217	82.8	44	16.8	1	0.4
Electronic bulletin boards	122	47.3	122	47.3	14	5.4
FAX or TELEX	240	92.7	17	6.6	2	0.8
Electronic databases	185	71.7	69	26.7	4	1.6
Video conferencing	52	20.2	175	68.1	30	11.7
Micrographics and microforms	86	33.6	90	35.2	80	31.3
Established Engineers						
Audiotapes and cassettes	62	19.4	70	21.9	188	58.8
Motion picture films	48	15.1	63	19.9	206	65.0
Videotapes	189	57.3	84	25.5	57	17.3
Desktop/electronic publishing	167	51.1	127	38.8	33	10.1
Computer cassettes/cartridge tapes	133	43.2	96	31.2	79	25.6
Electronic mail	279	83.3	49	14.6	7	2.1
Electronic bulletin boards	115	36.4	165	52.2	36	11.4
FAX or TELEX	324	97.0	3	0.9	7	2.1
Electronic databases	215	66.6	95	29.4	13	4.0
Video conferencing	178	54.3	126	38.4	24	7.3
Micrographics and microforms	67	21.7	105	34.0	137	44.3

tific papers with colleagues at geographically distant sites (26.6%). The established engineers made the greatest use of electronic (computer) networks for electronic mail (92.4%), to connect to geographically distant sites (74.4%), to log onto computers (56.8%), and to access or search the library's catalog (43.3%). The established en-

Table 9.17.   Uses of Electronic Networks by New and Established U.S. Aerospace Engineers and Scientists

Purpose of Use	New Engineers		Established Engineers	
	n	%	*n*	%
Connect to geographically distant sites	180	78.6	206	74.4
Electronic mail	219	94.8	267	92.4
Electronic bulletin boards or conferences	128	55.7	110	41.2
Electronic file transfer	218	94.4	---	---
Log onto computers for computational analysis or to use design tools	155	67.4	150	56.8
Control equipment such as laboratory instruments or machine tools	44	19.1	12	4.7
Access/search the library's catalog	143	61.9	117	43.3
Order documents from the library	57	24.8	50	19.0
Search electronic (bibliographic) databases	---	---	91	34.4
Information search and data retrieval	161	69.7	153	56.0
Prepare scientific and technical papers with colleagues at geographically distant sites	61	26.6	77	28.9

gineers made the least use of electronic (computer) networks to control equipment such as laboratory instruments or machine tools (4.7%) and to order documents from the library (19%).

New and established engineers who used electronic (computer) networks were asked to identify the groups with whom they exchanged messages or files (Table 9.18). Both groups of respondents made the greatest use of electronic (computer) networks to exchange messages or files within their organizations: with members of their own work groups (85.8% and 89.6%), with people in their organization at the *same* geographic site who are not in their work group (82.3% and 84.5%), and with other people in their organization at a *different* geographic site who are not in their work group

Table 9.18. Use of Electronic Networks by New and Established U.S. Aerospace Engineers and Scientists to Exchange Messages or Files

Exchange With—	New Engineers		Established Engineers	
	n	%	*n*	%
Members of your work group	199	85.8	259	89.6
Other people in your organization at the *same* geographic site who are not in your work group	190	82.3	240	84.5
Other people in your organization at a *different* geographic site who are not in your work group	158	68.7	195	69.1
People outside of your work group and outside of your organization	193	83.5	226	80.1

(68.7% and 69.1%). However, high percentages of both groups of respondents reported exchanging messages or files with people outside of their work group and outside their organization (83.5% and 80.1%).

Using the Results of Federally Funded R&D

New and established engineers were asked a series of questions designed to learn about their use of information resulting from federally funded aerospace R&D in completing or solving the most important project, task, or problem they had worked on in the past six months. Survey respondents were asked to categorize their most important project, task, or problem (Table 9.19). Most of the respondents in both groups categorized their project, task, or problem as design and development (45.8% and 52.9%). About 22% of the new engineers categorized their project, task, or problem as research and about 23% of the established engineers categorized their project, task, or problem as management.

Survey respondents were asked to categorize the nature of the duties they performed in completing or solving the most important project, task, or problem they had worked on in the past six months. About 81% of the new engineers and about 73% of the established engineers categorized their duties as engineering. About

Table 9.19.   Most Important Project, Task, or Problem Completed by New and Established U.S. Aerospace Engineers and Scientists in the Past Six Months

Factors	New Engineers		Established Engineers	
	n	%	n	%
Categories of project, task, or problem:				
Research	58	22.1	42	12.8
Design/development	120	45.8	174	52.9
Manufacturing/production	15	5.7	3	0.9
Quality assurance/control	11	4.2	5	1.5
Computer applications	22	8.4	10	3.0
Management	14	5.3	76	23.1
Other	22	8.4	19	5.8
Nature of duties performed:				
Engineering	211	80.8	238	72.8
Science	14	5.4	7	2.1
Management	19	7.3	73	22.3
Other	17	6.5	9	2.8
Worked on project, task, or problem:				
Alone	53	20.3	32	9.8
With others	208	79.7	294	90.2
Number/Size of Groups	n	$\bar{x}$	n	$\bar{x}$
Worked on project, task, or problem:				
Mean number of groups	203	2.1	284	3.6**
Mean number of people/group	202	5.4	280	6.9**

**$p \leq 0.01$.

7% of the new engineers also categorized their duties as management. About 22% of the established engineers categorized their duties as management. Further, about 80% of the new engineers and about 90% of the established engineers indicated that they worked with others in completing or solving their most important project, task, or problem. New engineers reported working with an average of 2.1 groups, with each group containing an average of 5.4 people. The established engineers reported working with 3.6 groups, with each group containing an average of 6.9 people. The average number of groups and the number of people per group for both new and established engineers are statistically different.

Survey respondents rated the overall complexity of their most important project, task, or problem. (See Table 9.20.) Mean complexity scores of $\bar{x} = 3.8$ and $\bar{x} = 4.0$ (out of a possible 5.00) were reported by the new and established engineers. New and established engineers were also asked to rate the amount of technical uncertainty they faced when they started their most important project, task, or problem. The average (mean) technical uncertainty scores were $\bar{x} = 3.2$ and $\bar{x} = 3.6$ (out of a possible 5.00). The differences between the complexity and uncertainty scores reported by the new and established engineers were statistically significant. Correlation coefficients (Pearson's r) were calculated to compare overall level of project, task, or problem complexity and technical uncertainty and the level of project, task, or problem by category

Table 9.20. Correlation of Project Complexity and Technical Uncertainty by Categorization of Most Important Project, Task, or Problem Completed by New and Established U.S. Aerospace Engineers and Scientists in the Past Six Months

	New Engineers		Established Engineers	
Complexity–Uncertainty Correlations	*n*	*r*	*n*	*r*
Overall	260	.4	328	.5
Research	58	.3	42	.6*
Design/development	118	.4	174	.4
Manufacturing/production	15	.7	3	.8
Quality assurance/control	11	.5	5	.4
Computer applications	22	.4	10	.2
Management	14	.3	76	.5
Other	22	.5	18	.5
Overall Score	*n*	$\bar{x}$	*n*	$\bar{x}$
Complexity	262	3.8	328	4.0**
Uncertainty	260	3.2	328	3.6**

* r values are statistically significant at $p \leq 0.05$.
** $p \leq 0.01$.

and technical uncertainty. The correlation coefficients appear in Table 9.20. Positive and significant correlations exist for the projects, tasks, or problems categorized as research.

Survey respondents were given a list of the following information sources: (a) used personal stores of technical information, (b) spoke with coworkers inside the organization, (c) spoke with colleagues outside the organization, (d) spoke with a librarian or technical information specialist, (e) used literature resources in the organization's library, and (f) searched (or had someone search for me) an electronic (bibliographic) database. New and established engineers were asked to identify the steps they had followed to obtain the information used in completing or solving the most important project, task, or problem they had worked on in the past six months by sequencing these items (i.e., #1, #2, #3, #4, and #5) (Table 9.21). They were instructed to place an "X" beside any information source not used during the process of obtaining the information.

Some interesting differences appear for the two groups. The new engineers consulted coworkers as their 1st, 2nd, and 3rd steps. As their 4th step, they used literature resources found in a library. Searching an electronic (bibliographic) database was their 5th step, and speaking to a librarian was their 6th step. Overall, 73.9% of the new engineers did not speak with a librarian, and 62.3% did not search (or have searched for them) an electronic (bibliographic) database as part of their information search. The established engineers used their personal stores of technical information as their 1st step, spoke with coworkers as their 2nd step, and spoke with colleagues as their 3rd step. As their 4th and 5th steps, they used literature resources found in a library. Speaking to a librarian was their 6th step. Overall, 62.8% of the established engineers did not speak with a librarian, 50.5% did not search (or have searched for them) an electronic (bibliographic) database, and 35.5% did not use literature resources found in a library as part of their information search.

About 57% of the new engineers and about 73% of the established engineers used the results of federally funded aerospace R&D in their work. Both groups of respondents, who used the results of federally funded aerospace R&D in their work, were given a list of 12 sources. They were asked to indicate if they used these sources

Table 9.21. Information Sources Used by New and Established U.S. Aerospace Engineers and Scientists to Complete Most Important Project, Task, or Problem in the Past Six Months

Information Source	Used 1st %	Used 2nd %	Used 3rd %	Used 4th %	Used 5th %	Used 6th %	Used 7th %	Did Not Use %
New Engineers								
Used personal store of technical information	39.3	29.1	16.2	5.7	2.0	0.4	0.0	7.3
Spoke with coworker	48.8	36.2	9.3	2.0	1.6	0.0	0.0	2.0
Spoke with colleague	3.7	16.5	32.6	13.6	4.5	2.9	0.0	26.0
Spoke with a librarian	0.4	2.6	4.7	6.0	6.8	5.6	0.0	73.9
Used literature resources found in a library	5.4	12.5	19.6	19.6	5.8	2.1	0.0	35.0
Searched (or had someone search for me) an electronic (bibliographic) database in the library	3.0	3.8	8.1	11.0	8.9	3.0	0.0	62.3
Used none of the above steps	0.0	0.0	0.0	0.0	0.0	0.0	0.0	100.0
Established Engineers								
Used personal store of technical information	64.8	17.4	11.0	2.3	1.0	1.0	0.0	2.6
Spoke with coworker	28.4	49.7	11.9	4.5	1.3	0.6	0.0	3.5
Spoke with colleague	2.3	20.0	49.3	10.0	5.0	2.0	0.0	11.3
Spoke with a librarian	0.0	2.4	4.9	9.0	10.8	10.1	0.0	62.8
Used literature resources found in a library	3.1	7.3	11.8	22.6	13.6	5.9	0.0	35.5
Searched (or had someone search for me) an electronic (bibliographic) database in the library	2.1	4.9	8.7	17.8	11.5	3.8	0.7	50.5
Used none of the above steps	6.0	0.0	0.0	0.0	0.0	0.0	1.5	92.5

to learn about the results of federally funded aerospace R&D (Table 9.22). Personal contacts: coworkers inside my organization (95.3% and 90.2%) and colleagues outside my organization (62.2% and 72.9%) were the sources used most often to learn about the results of federally funded aerospace R&D. NASA and DoD technical reports and NASA and DoD contacts were the third most frequently

Table 9.22.   Sources Used to Learn about the Results of Federally Funded Aerospace R&D by New and Established U.S. Aerospace Engineers and Scientists to Complete Most Important Project, Task, or Problem in the Past Six Months

Source	New Engineers		Established Engineers	
	n	%[a]	n	%[a]
Coworkers inside my organization	101	95.3	166	90.2
Colleagues outside my organization	62	62.0	129	72.9
NASA and DoD contacts	54	56.3	124	67.4
Publications such as NASA *STAR*	6	6.8	21	12.3
NASA and DoD technical reports	65	65.7	128	72.3
Professional and society journals	47	48.5	121	66.9
Librarians inside my organization	25	27.8	59	33.5
Trade journals	16	17.4	72	40.7
Searches of computerized databases	39	41.5	82	47.1
Professional and society meetings	23	25.6	93	52.5
Visit to NASA and DoD facilities	31	34.8	89	49.7

[a]Includes combined "frequently" and "sometimes" responses.

used source for new engineers. NASA and DoD technical reports and NASA and DoD contacts were the third most frequently used source for new and established engineers. Publications such as NASA *STAR* were used least by new engineers (6.8%) and established engineers (12.3%) to find out about the results of federally funded aerospace R&D.

Those respondents who reported using the results of federally funded aerospace R&D were asked if they used these results in completing their most important job-related project, task, or problem they had worked on in the past six months. About 45% of the new engineers and about 60% of the established engineers reported that they had used the results. Those respondents who reported using these results were asked to rate the importance of the results in completing their most important job-related project, task, or problem. Importance was measured on a 5-point scale where 5 is very important. The mean importance ratings was 4.0 for both new and established engineers. About 64% of the new engineers and about 55% of the established engineers indicated that the results they used were published in either a NASA or DoD technical report.

New and established engineers who used the results of federally funded aerospace R&D in completing or solving the most important project, task, or problem they had worked on in the past six months were asked which problems, if any, they encountered in using these results (Table 9.23). From a list of six obstacles to obtaining the results of federally funded aerospace R&D, new and established engineers identified the time and effort it took to locate (46.1% and 52.3%) and obtain the results (42.6% and 52.8%) as the biggest obstacles. Problems concerning the legibility or readability of the results (13.9%); the accuracy, precision, and reliability of the results (16.5%); and the organization or format of the results (18.3%) were reported least often by the new engineers. Problems concerning the legibility or readability of the results (17.1%), the organization or format of the results (24.9%), and distribution limitations or security restrictions of results (29.0%) were reported least often by established engineers.

Table 9.23. Problems Related to Using the Results of Federally Funded Aerospace R&D by New and Established U.S. Aerospace Engineers and Scientists to Complete Most Important Project, Task, or Problem in the Past Six Months

Obstacles	New Engineers		Established Engineers	
	n	%	*n*	%
Time and effort to locate results	53	46.1	101	52.3
Time and effort to obtain results	49	42.6	102	52.8
Accuracy, precision, and reliability of results	19	16.5	58	30.1
Distribution limitations or security restrictions of results	28	24.3	56	29.0
Organization or format of results	21	18.3	48	24.9
Legibility or readability of results	16	13.9	33	17.1

CONCLUSIONS

This chapter focused on information production and use by early career-stage U.S. aerospace engineers and scientists. The professional engineering community ranks information-use skills and communications proficiencies high in terms of their importance to engineering practice. In addition to being critical for successful technical performance, these same skills and proficiencies also play an extremely important role in the socialization process of newly hired engineers.

To investigate reacculturative changes that occur when aerospace engineering students make the transition to professional status, we examined the professional goals and career aspirations of students and engineering professionals. They appear to share very similar professional aspirations and career goals. Factors specifically related to the engineering aspects of their careers are most important to students and new engineers alike. However, it appears that the importance of these factors moderates somewhat after students make the transition from an academic engineering community to the professional engineering community.

We observed a number of differences among new and established engineers in their attitudes toward the importance of communication and the time they devoted to communicating technical information. Not only did established engineers spend more time communicating technical information than new hires did, but they also assigned a higher importance rating to the ability to communicate technical information effectively than the new hires did. Both the new and established engineers reported that they do write collaboratively. Although the number of people per group, the number of groups, and the number of people in each collaborative writing group differed, both the new and established engineers found collaborative writing to be more productive than writing alone.

Both engineering students and new engineers were asked about the importance of communications and information-use skills to future professional success. They were also asked if they had received instruction in these skills and to rate the helpfulness of the instruction. Both groups assigned high importance ratings to six specific communications and information-use skills. However, the students assigned a significantly higher rating to four of the six skills than the new hires did. Both groups gave high importance ratings to writing and oral communication. At least 50% of both the engineering students and the new engineers had received instruction or training in the six communications and information-use skills. With the exception of the instruction or training received in writing and oral communication, engineering students rated the helpfulness of the instruction or training higher than the new engineers rated it.

The ability to search out and acquire information is essential for successful engineering practice; therefore, comparing the information-seeking practices of new and established engineers should provide some insight into the socialization process. Overall library use by new and established engineers was about equal, however. For both groups, the most frequently reported reasons for not using a library were "My information needs were more easily met some other way" and "I had no information needs." Further, in searching for information to complete the most important project or task or solve a key problem within the past six months, about 35% of both the new and established engineers had not used a library. Additionally, about 74% of the new engineers and about 63% of the established engineers had not consulted a librarian as part of the task-completion or problem-solving process.

The new and established engineers reported statistically significant differences in the complexity and technical uncertainty scores for their most important project, task, or problem in the past six months. In searching for information needed to complete the most important project or task or to solve a key problem, the new engineers made greater use of coworkers and colleagues as part of their search when searching than did the established engineers. The established engineers relied more on their personal stores of information. Both groups relied on coworkers and colleagues to learn about the results of federally funded aerospace R&D. Both new and established engineers reported the greater use of NASA and DoD technical reports than other sources of written information to learn about the results of federally funded aerospace R&D. The time and effort it takes to locate and to obtain the results of federally funded aerospace R&D were perceived as problems by both new and established engineers.

IMPLICATIONS FOR DIFFUSING THE RESULTS OF
FEDERALLY FUNDED AEROSPACE R&D

Pickett (1995) notes that the size of the aerospace workforce is declining and the employment outlook for aerospace engineers and scientists remains grim. He notes that over 200 000 highly skilled jobs in aerospace are being lost each year. Many of these highly educated and trained individuals are aerospace engineers and scientists; many are being forced into career paths that take them outside the aerospace industry. How can the skills and competencies of those engineers and scientists remaining in the workforce be maintained? What role, if any, can and should the U.S. government play in preserving and enhancing their skills and competencies? Can the results of federally funded aerospace R&D be better used to maintain these skills and competencies? Perhaps one way to put the results of federally funded aerospace R&D to better use in maintaining the skills and competencies of U.S. aerospace engineering professionals would be to decrease the time and effort it takes both new and established engineers and scientists to locate and to obtain those results.

CHAPTER REFERENCES

Allen, T.J. (1977). *Managing the Flow of Technology: Technology Transfer and the Dissemination of Technological Information Within the R&D Organization*. Cambridge, MA: MIT Press.

American Association of Engineering Societies, Engineering Manpower Commission. (1986). *Toward the More Effective Utilization of American Engineers*. Washington, DC: American Association of Engineering Societies.

Ashford, S.J. (1986). "Feedback-Seeking in an Individual Adaptation: A Resource Perspective." *Academy of Management Journal* 29(3): 465-487.

Ashford, S.J. and J.S. Black. (1992). "Self-Socialization: Individual Tactics to Facilitate Organizational Entry Paper." Presented at the Annual Meeting of the Academy of Management, Las Vegas, Nevada.

Ashford, S.J. and L.L. Cummings. (1983). "Feedback as an Individual Resource: Personal Strategies of Creating Information." *Organizational Behavior and Human Performance* 32: 370-398.

Bakos, J.D. (1986). "A Departmental Policy for Developing Communication Skills of Undergraduate Engineers." *Engineering Education* 77(2): 101-104.

Berger, C.R. (1979). "Beyond Initial Interaction: Uncertainty, Understanding, and the Development of Interpersonal Relationships." In *Language and Social Psychology.* H. Giles and R.N. St. Clair, eds. Oxford, UK: Basil Blackwell, 122-144.

Bruffee, K.A. (1993). *Collaborative Learning: Higher Education, Interdependence, and the Authority of Knowledge.* Baltimore, MD: Johns Hopkins University Press.

Chisman, J.A. (1987). "Helping Students to Speak and Write." *The Applied Journal of Engineering Education* 3(2): 187-188.

Epstein, S. (1972). "The Nature of Anxiety With Emphasis Upon Its Relationship to Expectancy." In Vol. 2 of *Anxiety: Current Trends in Theory and Research.* C.D. Spielberger, ed. New York, NY: Academic Press, 291-337.

Folkman, S. and R.S. Lazarus. (1980). "An Analysis of Coping in a Middle-Aged Community Sample." *Journal of Health and Social Behavior* 21: 219-239.

Gerstberger, P.G. and T.J. Allen. (1968). "Criteria Used by Research and Development Engineers in the Selection of an Information Source." *Journal of Applied Psychology* (August) 52(4): 272-279.

Gundry, L.K. (1993). "Fitting Into Technical Organizations: The Socialization of Newcomer Engineers." *IEEE Transactions on Engineering Management.* (November) 40(4): 335-345.

Hanser, L.M. and P.M. Muchinsky. (1978). "Work as an Information Environment." *Organizational Behavior and Human Performance* 21: 47-60.

Jablin, F.M. (1984). "Assimilating New Members into Organizations." In *Communication Yearbook 8.* R.N. Bostrom, ed. Beverly Hills, CA: Sage, 594-626.

Kahn, R.L.; R.P. Quinn; D.M. Wolfe; J.D. Snoek; and R.A. Rosenthal. (1964). *Organizational Stress: Studies in Role Conflict and Ambiguity.* New York, NY: John Wiley.

Katz, S.M. (1993). "The Entry-Level Engineer: Problems in Transition From Student to Professional." *Journal of Engineering Education* 82(3): 171-174.

Kimel, W.R. and M.E. Monsees (1979). "Engineering Graduates: How Good Are They?" *Engineering Education* (November) 70(2): 210-212.

Louis, M.R. (1980). "Surprise and Sense-Making: What Newcomers Experience in Entering Unfamiliar Organizational Settings." *Administrative Science Quarterly* (June) 25: 226-251.

Louis, M.R.; B.Z. Posner; and G.N. Powell. (1983). "The Availability and Helpfulness of Socialization Practices." *Personnel Psychology* 36: 857-866.

Mailloux, E.N. (1989). "Engineering Information Systems." Chapter 6 in *Annual Review of Information Science and Technology.* 25 M.E. Williams, ed. Amsterdam, The Netherlands: Elsevier Science Publishers, 239-266.

McGrath, J.E. (1976). "Stress and Behavior in Organizations." In *Handbook of Industrial and Organizational Psychology*. M.D. Dunnette, ed. Chicago, IL: Rand McNally, 1351-1396.

Miller, V.D. and F.M. Jablin. (1991). "Information Seeking During Organizational Entry: Influences, Tactics, and a Model of the Process." *Academy of Management Review* 16(1): 92-120.

Morrison, E.W. (1995). "Information Usefulness and Acquisition During Organizational Encounter." *Management Communication Quarterly* 9: 131-155.

Morrison, E.W. (1993a). "A Longitudinal Study of the Effects of Information Seeking on Newcomer Socialization." *Journal of Applied Psychology* (April) 78(2): 173-183.

Morrison, E.W. (1993b). "A Longitudinal Study of Newcomer Information Seeking: Exploring Types, Modes, Sources, and Outcomes." *Academy of Management Journal* 36(3): 557-589.

Morrison, E.W. and R.J. Bies. (1991). "Impression Management in the Feedback-Seeking Process: A Literature Review and Research Agenda." *Academy of Management Review* 16: 522-541.

Morrison, E.W. and L.L. Cummings. (1992). "The Impact of Diagnosticity and Performance Expectations on Feedback Seeking Behavior." *Human Performance* 5(4): 251-264.

Morrison, E.W. and E. Weldon. (1990). "The Impact of an Assigned Performance Goal on Feedback Seeking Behavior." *Human Performance* 3: 37-50.

Nelson, D.L. (1987). "Organizational Socialization: A Stress Perspective." *Journal of Organizational Behavior* 8: 311-324.

O'Reilly, C.A. (1982). "Variations in Decision Makers' Use of Information Sources: The Impact of Quality and Accessibility of Information." *Academy of Management Journal* 25(4): 756-771.

Ostroff, C. and W.J. Kozlowski. (1992). "Organizational Socialization as a Learning Process: The Role of Information Acquisition." *Personnel Psychology* (Winter) 45(4): 849-847.

Pickett, G.E. (1995). "Aerospace Skills: Can We Afford to Lose Them?" *Aerospace America* (February) 33(2): B24.

Pinelli, T.E.; R.O. Barclay; and J.M. Kennedy. (1995a). *How Early Career-Stage U.S. Aerospace Engineers and Scientists Produce and Use Information*. Washington, DC: National Aeronautics and Space Administration. NASA TM-110181. (Available NTIS: 96N10999.)

Pinelli, T.E.; R.O. Barclay; and J.M. Kennedy. (1995b). *The Technical Communications Practices of U.S. Aerospace Engineers and Scientists: Results of the Phase 1 AIAA Mail Survey*. Washington, DC: National Aeronautics and Space Administration. NASA TM-110180. (Available NTIS: 95N34217.)

Pinelli, T.E.; A.P. Bishop; R.O. Barclay; and J.M. Kennedy. (1993). "The Information Seeking Behavior of Engineers." *Encyclopedia of Library and Information Science*. A. Kent and C.M. Hall, eds. Vol. 52, Supplement 15. New York, NY: Marcel Dekker, 167-201.

Pinelli, T.E.; L.M. Hecht; R.O. Barclay; and J.M. Kennedy. (1994). *The Technical Communication Practices of Aerospace Engineering Students: Results of the Phase 3 AIAA National Student Survey*. Washington, DC: National Aeronautics and Space Administration. NASA TM-109121. (Available NTIS; 95N18950.)

Pinelli, T.E.; J.M. Kennedy; and R.O. Barclay. (1991). "The NASA/DoD Aerospace Knowledge Diffusion Research Project." *Government Information Quarterly* 8(2): 219-233.

Rosenbloom, R.S. and F.W. Wolek. (1970). *Technology and Information Transfer: A Survey of Practice in Industrial Organizations*. Boston, MA: Harvard University Press.

Vancouver, J.B. and E.W. Morrison. (1995). "Feedback Inquiry: The Effect of Source Attributes and Individual Differences." *Organizational Behavior and Human Decision Processes* 62: 276-285.

White, R.W. (1974). "Strategies for Adaptation: An Attempt at Systematic Description." In *Coping and Adaptation*. G.V. Coelho; D.A. Hamburg; and J.E. Adams, eds. New York, NY: Basic Books, 47-68.

Young, J.F. and L.C. Harriot. (1979). "The Changing Technical Life of Engineers." *Mechanical Engineering* (January) 101(1): 20-24.

Zemore, R. and L.F. Shepel. (1987). "Information Seeking and Adjustment to Cancer." *Psychological Reports* (June) 60(3, Part 1): 874.